T0234046

Introductory
Mathematical Analysis
for Quantitative Finance

CHAPMAN & HALL/CRC
Financial Mathematics Series

Aims and scope:
The field of financial mathematics forms an ever-expanding slice of the financial sector. This series aims to capture new developments and summarize what is known over the whole spectrum of this field. It will include a broad range of textbooks, reference works and handbooks that are meant to appeal to both academics and practitioners. The inclusion of numerical code and concrete real-world examples is highly encouraged.

Series Editors

M.A.H. Dempster
Centre for Financial Research
Department of Pure Mathematics and Statistics
University of Cambridge

Dilip B. Madan
Robert H. Smith School of Business
University of Maryland

Rama Cont
Department of Mathematics
Imperial College

An Introduction to Computational Risk Management of Equity-Linked Insurance

Runhuan Feng

Derivative Pricing

A Problem-Based Primer

Ambrose Lo

Portfolio Rebalancing

Edward E. Qian

Interest Rate Modeling

Theory and Practice, 2nd Edition

Lixin Wu

Metamodeling for Variable Annuities

Guojun Gan and Emiliano A. Valdez

Modeling Fixed Income Securities and Interest Rate Options
Robert A. Jarrow

Financial Modelling in Commodity Markets
Viviana Fanelli

Introductory Mathematical Analysis for Quantitative Finance
Daniele Ritelli, Giulia Spaletta

For more information about this series please visit: *https://www.crcpress.com/Chapman-and-Hall CRC-Financial-Mathematics-Series/book-series/CHFINANCMTH*

Introductory Mathematical Analysis for Quantitative Finance

Daniele Ritelli
Giulia Spaletta

CRC Press
Taylor & Francis Group
Boca Raton London New York

CRC Press is an imprint of the
Taylor & Francis Group, an **informa** business
A CHAPMAN & HALL BOOK

CRC Press
Taylor & Francis Group
6000 Broken Sound Parkway NW, Suite 300
Boca Raton, FL 33487-2742

First issued in paperback 2022

© 2020 by Taylor & Francis Group, LLC
CRC Press is an imprint of Taylor & Francis Group, an Informa business

No claim to original U.S. Government works

ISBN-13: 978-0-815-37254-7 (hbk)
ISBN-13: 978-1-03-233657-2 (pbk)
DOI: 10.1201/9781351245111

**Visit the Taylor & Francis Web site at
http://www.taylorandfrancis.com**

**and the CRC Press Web site at
http://www.crcpress.com**

Contents

Preface

The purpose of this book is to be a tool for students, with little mathematical background, who aim to study Mathematical Finance. The only prerequisites assumed are one–dimensional differential calculus, infinite series, Riemann integral and elementary linear algebra.

In a sense, it is a sort of intensive course, or crash–course, which allows students, with minimal knowledge in Mathematical Analysis, to reach the level of mathematical expertise necessary in modern Quantitative Finance. These lecture notes concern pure mathematics, but the arguments presented are oriented to Financial applications. The n–dimensional Euclidean space is briefly introduced, in order to deal with multivariable differential calculus. Sequences and series of functions are introduced, in view of theorems concerning the passage to the limit in Measure theory, and their role in the general theory of ordinary differential equations, which is also presented. Due to its importance in Quantitative Finance, the Radon–Nykodim theorem is stated, without proof, since the Von Neumann argument requires notions of Functional Analysis, which would require a dedicated course. Finally, in order to solve the Black–Scholes partial differential equation, basics in ordinary differential equations and in the Fourier transform are provided.

We kept our exposition as short as possible, as the lectures are intended to be a preliminary contact with the mathematical concepts used in Quantitative Finance and provided, often, in a one–semester course. This book, therefore, is not intended for a specialized audience, although the material presented here can be used by both experts and non-experts, to have a clear idea of the mathematical tools used in Finance.

Chapter 1

Euclidean space

This chapter introduces basic, though fundamental notions, on vector spaces and $\mathbb{R}^n$ topology, that are necessary throughout the book; this is done both to keep to a minumum the requirement of familiarity with the concepts presented, here and in the following chapters, and for reasons of completeness.

1.1 Vectors

If $n \in \mathbb{N}$, we use the symbol $\mathbb{R}^n$ to indicate the Cartesian[1] product of n copies of $\mathbb{R}$ with itself, i.e.:

$$\mathbb{R}^n := \{(x_1, x_2, \dots, x_n) \mid x_j \in \mathbb{R} \text{ for } j = 1, 2, \dots, n\}.$$

The concept of Euclidean[2] space is not limited to the set $\mathbb{R}^n$, but it also includes the so–called *Euclidean inner product*, introduced in Definition 1.1. The integer n is called *dimension* of $\mathbb{R}^n$, the elements $\boldsymbol{x} = (x_1, x_2, \dots, x_n)$ of $\mathbb{R}^n$ are called *points*, or vectors or ordered n–tuples, while $x_j, j = 1, \dots, n$, are the coordinates, or *components*, of $\boldsymbol{x}$. Vectors $\boldsymbol{x}$ and $\boldsymbol{y}$ are equal if $x_j = y_j$ for $j = 1, 2, \dots, n$. The zero vector is the vector whose components are null, that is, $\boldsymbol{0} := (0, 0, \dots, 0)$. In low dimension situations, i.e. for $n = 2$ or $n = 3$, we will write $\boldsymbol{x} = (x, y)$ and $\boldsymbol{x} = (x, y, z)$, respectively.

For our purposes, that is extending differential calculus to functions of several variables, we need to define an algebraic structure in $\mathbb{R}^n$. This is done by introducing operations in $\mathbb{R}^n$.

Definition 1.1. Let $\boldsymbol{x} = (x_1, x_2, \dots, x_n), \boldsymbol{y} = (y_1, y_2, \dots, y_n) \in \mathbb{R}^n$ and $\alpha \in \mathbb{R}$.

(i) The sum of $\boldsymbol{x}$ and $\boldsymbol{y}$ is the vector:

$$\boldsymbol{x} + \boldsymbol{y} := (x_1 + y_1, x_2 + y_2, \dots, x_n + y_n);$$

---

[1] Renatus Cartesius (1596–1650), French mathematician and philosopher.
[2] Euclid of Alexandria (350–250 B.C. circa), Greek mathematician.

(ii) The difference of x and y is the vector:

$$x - y := (x_1 - y_1, x_2 - y_2, \ldots, x_n - y_n);$$

(iii) The α–multiple of x is the vector:

$$\alpha x = (\alpha x_1, \alpha x_2, \ldots, \alpha x_n);$$

(iv) The Euclidean inner product of x and y is the real number:

$$x \cdot y := x_1 y_1 + x_2 y_2 + \ldots + x_n y_n.$$

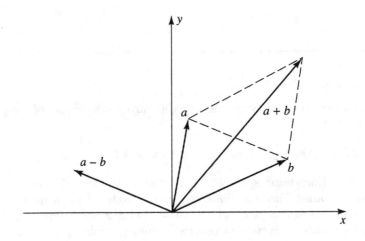

Figure 1.1: Vector operations.

The vector operations of Definition 1.1, illustrated in Figure 1.1, represent the analogues of the algebraic operations in $\mathbb{R}$ and imply algebraic rules in $\mathbb{R}^n$.

Proposition 1.2. Let $x, y, z \in \mathbb{R}^n$ and $\alpha, \beta, \in \mathbb{R}$. Then:

(a) $\alpha \, 0 = 0$;

(b) $0 \, x = 0$;

(c) $1 \, x = x$;

(d) $\alpha \, (\beta \, x) = \beta \, (\alpha \, x) = (\alpha \, \beta) \, x$;

(e) $\alpha \, (x \cdot y) = (\alpha \, x) \cdot y = x \cdot (\alpha \, y)$;

(f) $\alpha \, (x + y) = \alpha \, x + \alpha \, y$;

(g) $0 + x = x$;

(h) $x - x = 0$;

(i) $0 \cdot x = 0$;

(j) $x + (y + z) = (x + y) + z$;

(k) $x + y = y + x$;   (m) $x \cdot (y + z) = x \cdot y + x \cdot z$.

(l) $x \cdot y = y \cdot x$;

Definition 1.3. The standard base of $\mathbb{R}^n$ is the set $\mathbb{E}_n = \{e_1, \ldots, e_n\}$, where:

$$e_1 = (1, 0, \ldots, 0), \quad e_2 = (0, 1, 0, \ldots, 0), \ldots, e_n = (0, \ldots, 0, 1).$$

Note that a generic $x = (x_1, \ldots, x_n) \in \mathbb{R}^n$ can be represented as a linear combination of vectors in $\mathbb{E}_n$:

$$x = \sum_{j=1}^{n} x_j \, e_j = \sum_{j=1}^{n} x \cdot e_j \, e_j .$$

It is worth noting that, when $n = 2$ and $n = 3$, the standard base $\mathbb{E}_n$ is made of pairwise orthogonal vectors. In order to extend orthogonality to n dimensions, consider the main property of the standard base, i.e., $e_j \cdot e_k = 0$ for $j \neq k$.

Definition 1.4. Let $x, y \in \mathbb{R}^n$ be non–zero vectors; then:

(i) x, y are parallel if and only if there exists $t \in \mathbb{R}$ such that $x = t\,y$; this is denoted with $x \| y$;

(ii) x, y are orthogonal if and only if $x \cdot y = 0$; this is denoted with $x \perp y$.

As an example, $a = (3, 5)$ and $b = (-6, -10)$ are parallel, while $c = (1, 1)$ and $d = (1, -1)$ are orthogonal.

The Euclidean inner product allows introducing a metric in $\mathbb{R}^n$, as shown in the following Definition 1.5.

Definition 1.5. Let $x \in \mathbb{R}^n$. The (Euclidean) norm of x is the scalar:

$$\|x\| := \left(\sum_{k=1}^{n} x_k^2 \right)^{\frac{1}{2}} .$$

Remark 1.6. Observe that:

$$\|x\|^2 = x \cdot x .$$

If T is the triangle of vertices $(0, 0)$, (a, b), $(a, 0) \in \mathbb{R}^2$, then, by Pythagora[3] theorem, the hypotenuse $\sqrt{a^2 + b^2}$ of T is exactly the norm of $x = (a, b)$.

We now define the Euclidean distance between two points in $\mathbb{R}^n$.

---

[3]Pythagora of Samos (circa 570–495 B.C.), ancient Ionian Greek philosopher.

Definition 1.7. Given $x, y \in \mathbb{R}^n$, their (Euclidean) distance is defined, and denoted, as:

$$\text{dist}(x, y) := ||x - y||.$$

Theorem 1.8 (Cauchy–Schwarz inequality). Let $x, y \in \mathbb{R}^n$. Then:

$$|x \cdot y| \leq ||x|| \, ||y||.$$

Proof. We only consider the non–trivial situation $x, y \neq 0$. For $t \in \mathbb{R}$, define:

$$f(t) := ||x - t \, y||^2.$$

Since $0 \leq f(t) = ||x||^2 - 2 t \, x \cdot y + t^2 \, ||y||^2$, it must hold:

$$\frac{\Delta}{4} = (x \cdot y)^2 - ||x||^2 \, ||y||^2 \leq 0,$$

and the thesis follows.

The main properties of the Euclidean norm are stated in the following Proposition 1.9, in which inequalities (iii)–(iv) are called *triangular inequalities*.

Proposition 1.9. If $x, y \in \mathbb{R}^n$, then:

(i) $||x|| \geq 0$, with $||x|| = 0$ only when $x = 0$;

(ii) $||\alpha \, x|| = |\alpha| \, ||x||$ for all scalars α;

(iii) $||x + y|| \leq ||x|| + ||y||$;

(iv) $||x - y|| \geq ||x|| - ||y||$.

Proof. Inequality (i) and equality (ii) are trivial. To get (iii), observe that:

$$||x + y||^2 = ||x||^2 + 2 \, x \cdot y + ||y||^2.$$

From the Cauchy–Schwarz[4] inequality of Theorem 1.8, we infer:

$$||x||^2 + 2 \, x \cdot y + ||y||^2 \leq ||x||^2 + 2 \, ||x|| \, ||y|| + ||y||^2 = (||x|| + ||y||)^2.$$

Inequality (iv) follows analogously from:

$$||x - y||^2 = ||x||^2 - 2 \, x \cdot y + ||y||^2,$$

and from the Cauchy–Schwarz inequality.

Remark 1.10. Let $a, b \in \mathbb{R}^2 \setminus \{0\}$ and let T be the triangle determined by points $0, a, b$; then, T has sides of length $||a||, ||b||, ||a - b||$, as shown in Figure 1.2.

---

[4] Augustin–Louis Cauchy (1789–1857), French mathematician, engineer and physicist. Karl Hermann Amandus Schwarz (1843–1921), German mathematician.

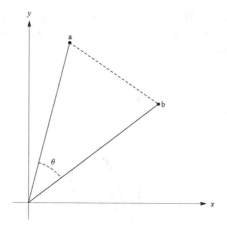

Figure 1.2: Angle between two vectors.

If θ is the angle between the sides of length $||a||$, $||b||$ then, by the Law of Cosines (linked to Carnot[5] theorem), the following equality (1.1) holds:

$$||a - b||^2 = ||a||^2 + ||b||^2 - 2\,||a||\,||b||\cos\theta. \tag{1.1}$$

Recalling that:

$$||a - b||^2 = ||a||^2 + ||b||^2 - 2\,a\bullet b,$$

we conclude that:

$$\cos\theta = \frac{a\bullet b}{||a||\,||b||}. \tag{1.2}$$

Taking into consideration equation (1.2) and the Cauchy–Schwarz inequality, we can formulate the following Definitions 1.11–1.12.

Definition 1.11. Let x, $y \in \mathbb{R}^n$ be two non–zero vectors. Their angle $\vartheta(x, y)$ is defined by:

$$\arccos\vartheta(x, y) = \frac{||x||\,||y||}{x\bullet y}.$$

Observe that, when x, y are orthogonal, then $\vartheta(x, y) = \dfrac{\pi}{2}$.

Definition 1.12. The hyperplane (a plane where $n = 3$) passing through a point $a \in \mathbb{R}^n$, with normal $b \neq 0$, is the set:

$$\Pi_b(a) = \{x \in \mathbb{R}^n \mid (x - a)\bullet b = 0\}.$$

---

[5] Nicolas Leonard Sadi Carnot (1796–1832), French military scientist and physicist.

Note that, by definition, $\Pi_b(a)$ is the set of all points x such that $x - a$ and b are orthogonal; observe that, given a, b, the normal $x - a$ is not unique, as Figure 1.3 illustrates.

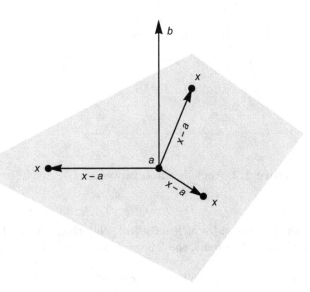

Figure 1.3: Hyperplane.

By definition, the hyperplane $\Pi_b(a)$ is given by:

$$b_1 x_1 + b_2 x_2 + \cdots + b_n x_n = d ,$$

where $b = (b_1, \ldots, b_n)$ is a normal and $d = b \cdot a$ is a constant related to the distance from $\Pi_b(a)$ to the origin. Planes in $\mathbb{R}^3$ have equations of the form:

$$ax + by + cz = d .$$

1.2   Topology of $\mathbb{R}^n$

Topology, that is the description of the relations among subsets of $\mathbb{R}^n$, is based on the concept of open and closed sets, that generalises the notion of open and closed intervals. After introducing these concepts, we state their most basic properties. The first step is the natural generalisation of intervals in $\mathbb{R}^n$.

Definition 1.13. Open and closed balls are defined as follows:

(i) $\forall r > 0$, the open ball, centered at a, of radius r, is the set of points:

$$B_r(a) := \{x \in \mathbb{R}^n \mid ||x - a|| < r\} ;$$

(ii) $\forall r \geq 0$, the closed ball, centered at a, of radius r, is the set of points:

$$\overline{B}_r(a) \{x \in \mathbb{R}^n \mid ||x - a|| \leq r\} .$$

Note that, when $n = 1$, the open ball centered at a of radius r is the open interval $(a - r, a + r)$, and the corresponding closed ball is the closed interval $[a - r, a + r]$. Here we adopt the convention of representing open balls as dashed circumferences, while closed balls are drawn as solid circumferences, as shown in Figure 1.4.

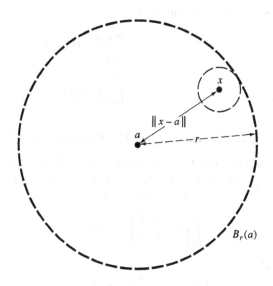

Figure 1.4: Open ball $n = 2$.

To generalise the concept of open and closed intervals even further, observe that each element of an open interval I lies *inside I*, i.e., it is surrounded by other points in I. Although closed intervals do not satisfy this property, their complements do. Accordingly, we give the following Definition 1.14, in which "$\subset$" denotes non–strict inclusion (and analogously for "$\supset$").

Definition 1.14. The open and closed sets are defined as follows:

(i) a set $V \subset \mathbb{R}^n$ is open if and only if, for every $a \in V$, there exists $\varepsilon > 0$ such that $B_\varepsilon(a) \subset V$;

(ii) a set $E \subset \mathbb{R}^n$ is closed if and only if its complement $E^c := \mathbb{R}^n \setminus E$ is open.

It follows that every open ball is an open set. Note that, if $a \in \mathbb{R}^n$, then $\mathbb{R}^n \setminus \{a\}$ is open and $\{a\}$ is closed.

Remark 1.15. For each $n \in \mathbb{N}$, the empty set $\emptyset$ and the whole space $\mathbb{R}^n$ are both open and closed.

We state, without proof, the following Theorem 1.16, which explains the basic properties of open and closed sets. Notions on sets, set operators and Topology are presented in greater detail in Chapter 7, while in this first chapter only strictly necessary concepts are introduced.

Theorem 1.16. Let $\{V_\alpha\}_{\alpha \in A}$ and $\{E_\alpha\}_{\alpha \in A}$ be any collections of respectively open and closed subsets of $\mathbb{R}^n$, where A is any set of indexes. Let further $\{V_k \mid k = 1, \ldots, p\}$ and $\{E_k \mid k = 1, \ldots, p\}$ be finite collections of respectively open and closed subsets of $\mathbb{R}^n$. Then:

(i) $\displaystyle\bigcup_{\alpha \in A} V_\alpha$ is open;   (iii) $\displaystyle\bigcap_{\alpha \in A} E_\alpha$ is closed;

(ii) $\displaystyle\bigcap_{k=1}^{p} V_k$ is open;   (iv) $\displaystyle\bigcup_{k=1}^{p} E_k$ is closed;

(v) If V is open and E is closed, then $V \setminus E$ is open and $E \setminus V$ is closed.

Remark 1.17. In Theorem 1.16, statements (ii) and (iv) are false if arbitrary collections are used in place of finite collections. In the one–dimensional Euclidean space $\mathbb{R}^1 = \mathbb{R}$, in fact, we have that:

$$\bigcap_{k \in \mathbb{N}} \left(-\frac{1}{k}, \frac{1}{k} \right) = \{0\}$$

is a closed set and

$$\bigcup_{k \in \mathbb{N}} \left[\frac{1}{k+1}, \frac{k}{k+1} \right] = (0, 1)$$

is open.

Definition 1.18. Let $E \subset \mathbb{R}^n$.

(i) The interior of E is the set $E^\circ := \bigcup \{V \mid V \subset E, \ V \text{ is open in } \mathbb{R}^n\}$.

(ii) The closure of E is the set $\overline{E} := \bigcap \{B \mid B \supset E, \ B \text{ is closed in } \mathbb{R}^n\}$.

Note that every set E contains the open set $\emptyset$ and is contained in the closed set $\mathbb{R}^n$; hence, E° and $\overline{E}$ are well–defined. Notice further that E° is always open and $\overline{E}$ is always closed: E° is the largest open set contained in E, and $\overline{E}$ is the smallest closed set containing E. The following Theorem 1.19 illustrates the properties of E° and $\overline{E}$.

Theorem 1.19. Let $E \subset \mathbb{R}^n$, then:

(i) $E^\circ \subset E \subset \overline{E}$;

(ii) if V is open and $V \subset E$, then $V \subset E^\circ$;

(iii) if C is closed and $C \supset E$, then $C \supset \overline{E}$.

Let us, now, introduce the notion of *boundary* of a set.

Definition 1.20. The boundary of E is the set:

$$\partial E := \{x \in \mathbb{R}^n \mid \text{ for all } r > 0, B_r(x) \cap E \neq \emptyset \text{ and } B_r(x) \cap E^c \neq \emptyset\}.$$

Given a set E, its boundary ∂E is closely related to E° and $\overline{E}$.

Theorem 1.21. If $E \subset \mathbb{R}^n$ then $\partial E = \overline{E} \setminus E^\circ$.

1.3 Limits of functions

A *vector function* is a function f of the form $f : A \to \mathbb{R}^m$, where $A \subset \mathbb{R}^n$. Since $f(x) \in \mathbb{R}^m$ for each $x \in A$, then there are m functions $f_j : A \to \mathbb{R}$, called *component functions* of f, such that:

$$f(x) = (f_1(x), \ldots, f_m(x)) \qquad \text{for each} \quad x \in A.$$

When $m = 1$, function f has only one component and we call f *real–valued*. If $f = (f_1, \ldots, f_m)$ is a vector function, where the components f_j have intrinsic domains, then the maximal domain of f is defined to be the intersection of the domains of all components f_j.

To set up a notation for the algebra of vector functions, let $E \subset \mathbb{R}^n$ and let $f, g : E \to \mathbb{R}^m$. For each $x \in E$, the following operations can be defined. The scalar multiple of $\alpha \in \mathbb{R}$ by f is given by:

$$(\alpha f)(x) := \alpha f(x).$$

The sum of f and g is obtained as:

$$(f + g)(x) := f(x) + g(x).$$

The (Euclidean) dot product of f and g is constructed as:

$$(f \bullet g)(x) := f(x) \bullet g(x).$$

Definition 1.22. Let n, $m \in \mathbb{N}$ and $\boldsymbol{a} \in \mathbb{R}^n$, let V be an open set containing $\boldsymbol{a}$ and let $f : V \setminus \{\boldsymbol{a}\} \to \mathbb{R}^m$. Then, $f(\boldsymbol{x})$ is said to *converge* to $\boldsymbol{L}$, as $\boldsymbol{x}$ approaches $\boldsymbol{a}$, if and only if for every $\varepsilon > 0$ there exists a positive δ (that in general depends on $\varepsilon, f, V, \boldsymbol{a}$) such that:

$$0 < ||\boldsymbol{x} - \boldsymbol{a}|| < \delta \implies ||f(\boldsymbol{x}) - \boldsymbol{L}|| < \varepsilon.$$

In this case we write:

$$\lim_{\boldsymbol{x} \to \boldsymbol{a}} f(\boldsymbol{x}) = \boldsymbol{L}$$

and call $\boldsymbol{L}$ the *limit* of $f(\boldsymbol{x})$ as $\boldsymbol{x}$ approaches $\boldsymbol{a}$. Using the analogy between the norm on $\mathbb{R}^n$ and the absolute value on $\mathbb{R}$, it is possible to extend a great part of the one–dimensional theory on limits of functions to the Euclidean space setting.

Example 1.23. Show that:

$$\lim_{(x,y) \to (0,0)} \frac{x^2 y}{x^2 + y^2} = 0.$$

Using polar coordinates $x = \rho \cos \theta, y = \rho \sin \theta$, we have:

$$\frac{x^2 y}{x^2 + y^2} = \rho \sin \theta \cos^2 \theta.$$

When $(x,y) \to (0,0)$, then $\rho \to 0$ holds too and, since for any $\theta \in [0, 2\pi]$ the quantity $\sin \theta \cos^2 \theta$ is bounded, the equality to zero follows.

Example 1.24. Let us demonstrate that the following limit does not exist:

$$\lim_{(x,y) \to (0,0)} \frac{x y}{x^2 + y^2}.$$

If we move towards the origin $(0,0)$, along the line $y = mx$, we see that:

$$\frac{x y}{x^2 + y^2} = \frac{m}{1 + m^2},$$

that is, the right–hand side depends explicitly on the slope m, and we have different values for different slopes of the line: this means that the limit does not exist.

Definition 1.25. Let $\emptyset \neq E \subset \mathbb{R}^n$ and let $f : E \to \mathbb{R}^m$.

(i) f is said to be continuous at $\boldsymbol{a} \in E$ if and only if for every $\varepsilon > 0$ there exists a positive δ (that in general depends on $\varepsilon, f, \boldsymbol{a}$) such that:

$$||\boldsymbol{x} - \boldsymbol{a}|| < \delta \text{ and } \boldsymbol{x} \in E \implies ||f(\boldsymbol{x}) - f(\boldsymbol{a})|| < \varepsilon;$$

(ii) f is said to be continuous on E if and only if f is continuous at every $\boldsymbol{x} \in E$.

Example 1.26. Function:

$$f(x,y) = \begin{cases} \dfrac{x^2 y}{x^2 + y^2} & (x,y) \neq \mathbf{0} \\ \\ 0 & (x,y) = \mathbf{0} \end{cases}$$

is continuous at every $x \in \mathbb{R}^2$, while:

$$g(x,y) = \begin{cases} \dfrac{x\,y}{x^2 + y^2} & (x,y) \neq \mathbf{0} \\ \\ 0 & (x,y) = \mathbf{0} \end{cases}$$

is not continuous at $\mathbf{0}$.

We now state the two important Theorems 1.27 and 1.28, that establish the topological properties of continuity.

Theorem 1.27. Let $n, m \in \mathbb{N}$ and $f : \mathbb{R}^n \to \mathbb{R}^m$. Then the following conditions are equivalent:

(i) f is continuous on $\mathbb{R}^n$;

(ii) $f^{-1}(V)$ is open in $\mathbb{R}^n$ for every open subset V of $\mathbb{R}^m$;

(iii) $f^{-1}(E)$ is closed in $\mathbb{R}^n$ for every closed subset E of $\mathbb{R}^m$.

Theorem 1.28. Let $n, m \in \mathbb{N}$, E be open in $\mathbb{R}^n$ and assume $f : E \to \mathbb{R}^m$. Then f is continuous on E if and only if $f^{-1}(V)$ is open in $\mathbb{R}^n$ for every open set V in $\mathbb{R}^m$.

Definition 1.29. A subset $B \subset \mathbb{R}^n$ is bounded if there exists $M > 0$ such that $||x|| \leq M$ for any $x \in B$.

The following Theorem 1.30, due to Weierstrass[6], states the fundamental property that, if a set is both closed and bounded (we call it *compact*), then its image under any continuous function is also compact.

Theorem 1.30 (Weierstrass theorem on compactness). Let $n, m \in \mathbb{N}$. If H is compact in $\mathbb{R}^n$ and $f : H \to \mathbb{R}^m$ is continuous on H, then $f(H)$ is compact in $\mathbb{R}^m$.

In the particular situation of a scalar function, we can state the generalisation of Theorem 1.30 to functions depending on several variables.

---

[6] Karl Theodor Wilhelm Weierstrass (1815–1897), German mathematician.

Theorem 1.31 (Generalisation of Weierstrass theorem). Assume that H is a non–empty subset of $\mathbb{R}^n$ and $f : H \to \mathbb{R}$. If H is compact and f is continuous on H, then:

$$M := \sup\{f(\boldsymbol{x}) \mid \boldsymbol{x} \in H\} \quad \text{and} \quad m := \inf\{f(\boldsymbol{x}) \mid \boldsymbol{x} \in H\}$$

are finite real numbers. Moreover, there exist points $\boldsymbol{x}_M, \boldsymbol{x}_m \in H$ such that $M = f(\boldsymbol{x}_M)$ and $m = f(\boldsymbol{x}_m)$.

Chapter 2

Sequences and series of functions

The notions of sequences and series, of numbers and of functions, are presented in this chapter, with the related concepts of pointwise and uniform convergence. The aim is mainly to minimise assumptions on the mathematical background possessed by the Reader; a purpose of notational introduction is also involved. The Basel problem is described, which will be met again in Chapters 8 and 10.

2.1 Sequences and series of real or complex numbers

A *sequence* is a set of numbers $u_1, u_2, u_3, \ldots$, in a definite order of arrangement, that is, a map $u : \mathbb{N} \to \mathbb{R}$ or $u : \mathbb{N} \to \mathbb{C}$, formed according to a certain rule. Each number in the sequence is called *term*; u_n is called the nth term. The sequence is called *finite* or *infinite*, according to the number of terms. The sequence $u_1, u_2, u_3, \ldots$, when considered as a function, is also designated as $(u_n)_{n \in \mathbb{N}}$ or briefly (u_n).

Definition 2.1. The real or complex number ℓ is called the *limit* of the infinite sequence (u_n) if, for any positive number ε, there exists a positive number n_ε, depending on ε, such that $|u_n - \ell| < \varepsilon$ for all integers $n > n_\varepsilon$. In such a case, we denote:

$$\lim_{n \to \infty} u_n = \ell.$$

Given a sequence (u_n), we say that its associated infinite series $\displaystyle\sum_{n=1}^{\infty} u_n$:

(i) converges, when it exists the limit:

$$\lim_{n \to \infty} \sum_{k=1}^{n} u_k := S = \sum_{n=1}^{\infty} u_n \ ;$$

(ii) diverges, when the limit of the partial sums $\displaystyle\sum_{k=1}^{n} u_k$ does not exist.

2.2   Sequences of functions

Given a real interval $[a,b]$, we denote $\mathscr{F}([a,b])$ the collection of all real functions defined on $[a,b]$:

$$\mathscr{F}([a,b]) = \{f \mid f : [a,b] \to \mathbb{R}\}.$$

Definition 2.2. A sequence of functions with domain $[a,b]$ is a sequence of elements of $\mathscr{F}([a,b])$.

Example 2.3. Functions $f_n(x) = x^n$, where $x \in [0,1]$, form a sequence of functions in $\mathscr{F}([0,1])$.

Let us analyse what happens when $n \to \infty$. It is easy to realise that a sequence of continuous functions may converge to a non–continuous function. Indeed, for the sequence of functions in Example 2.3, it holds:

$$\lim_{n \to \infty} f_n(x) = \lim_{n \to \infty} x^n = \begin{cases} 1 & \text{if } x = 1, \\ 0 & \text{if } 0 \le x < 1. \end{cases}$$

Thus, even if every function of the sequence $f_n(x) = x^n$ is continuous, the limit function $f(x)$, defined below, may not be continuous:

$$f(x) := \lim_{n \to \infty} f_n(x).$$

The convergence of a sequence of functions, like that of Example 2.3, is called *simple convergence*. We now provide its rigorous definition.

Definition 2.4. If (f_n) is a sequence of functions in $I \subset [a,b]$ and f is a real function on I, then f_n *pointwise converges* to f if, for any $x \in I$, there exists the limit of the real sequence $(f_n(x))$ and its value is $f(x)$:

$$\lim_{n \to \infty} f_n(x) = f(x).$$

Pointwise convergence is denoted as follows:

$$f_n \xrightarrow{I} f.$$

Remark 2.5. Definition 2.4 can be reformulated as follows: it holds that $f_n \xrightarrow{I} f$ if, for any $\varepsilon > 0$ and for any $x \in I$, there exists $n_{\varepsilon,x} \in \mathbb{N}$, depending on ε and x, such that:

$$|f_n(x) - f(x)| < \varepsilon$$

for any $n \in \mathbb{N}$ with $n > n_{\varepsilon,x}$.

Example 2.3 shows that the pointwise limit of a sequence of continuous functions may not be continuous.

2.3 Uniform convergence

Pointwise convergence does not allow, in general, interchanging between limit and integral operators, a possibility that we call *passage to the limit* and that we also address in § 8.10. To explain it, consider the sequence of functions:

$$f_n(x) = n\,e^{-n^2 x^2}$$

defined on $[0, \infty)$; it is a sequence that clearly converges to the zero function. Employing the substitution $n\,x = y$, evaluation of the integral of f_n yields:

$$\int_0^\infty f_n(x)\,\mathrm{d}x = \int_0^\infty e^{-y^2}\,\mathrm{d}y\;.$$

We do not have the tools, yet, to evaluate the integral in the left–hand side of the above equality (but we will soon), but it is clear that it is a positive real number, so we have:

$$\lim_{n\to\infty}\int_0^\infty f_n(x)\,\mathrm{d}x = \int_0^\infty e^{-y^2}\,\mathrm{d}y = \alpha > 0 \neq \int_0^\infty \lim_{n\to\infty} f_n(x)\,\mathrm{d}x = 0\;.$$

To establish a 'good' notion of convergence, that allows the passage to the limit, when we take the integral of the considered sequence, and that preserves continuity, we introduce the fundamental notion of uniform convergence.

Definition 2.6. If (f_n) is a sequence of functions defined on the interval I, then f_n *converges uniformly* to the function f if, for any $\varepsilon > 0$, there exists $n_\varepsilon \in \mathbb{N}$ such that, for $n \in \mathbb{N}, n > n_\varepsilon$, it holds:

$$\sup_{x\in I} |f_n(x) - f(x)| < \varepsilon. \tag{2.1}$$

Uniform convergence is denoted by:

$$f_n \overset{I}{\rightrightarrows} f\;.$$

Remark 2.7. Definition 2.6 is equivalent to requesting that, for any $\varepsilon > 0$, there exists $n_\varepsilon \in \mathbb{N}$ such that, for $n \in \mathbb{N}, n > n_\varepsilon$, it holds:

$$|f_n(x) - f(x)| < \varepsilon, \qquad \text{for any } x \in I. \tag{2.2}$$

Proof. Let $f_n \overset{I}{\rightrightarrows} f$. Then, for any $\varepsilon > 0$, there exists $n_\varepsilon \in \mathbb{N}$ such that:

$$\sup_I |f_n(x) - f(x)| < \varepsilon, \qquad \text{for any } n \in \mathbb{N}, \quad n > n_\varepsilon,$$

and this implies (2.2). Vice versa, if (2.2) holds then, for any $\varepsilon > 0$, there exists $n_\varepsilon \in \mathbb{N}$ such that:

$$\sup_{x \in I} |f_n(x) - f(x)| < \varepsilon, \qquad \text{for any } n \in \mathbb{N}, n > n_\varepsilon,$$

that is to say, $f_n \overset{I}{\rightrightarrows} f$. □

Remark 2.8. Uniform convergence implies pointwise convergence. The converse does not hold, as Example 2.3 shows.

In the next Theorem 2.9, we state the so–called Cauchy uniform convergence criterion.

Theorem 2.9. Given a sequence of functions (f_n) in $[a, b]$, the following statements are equivalent:

(i) (f_n) converges uniformly;

(ii) for any $\varepsilon > 0$, there exists $n_\varepsilon \in \mathbb{N}$ such that, for $n, m \in \mathbb{N}$, with $n, m > n_\varepsilon$, it holds:

$$|f_n(x) - f_m(x)| < \varepsilon, \qquad \text{for any } x \in [a, b].$$

Proof. We show that (i) $\implies$ (ii). Assume that (f_n) converges uniformly, i.e., for a fixed $\varepsilon > 0$, there exists $n_\varepsilon > 0$ such that, for any $n \in \mathbb{N}, n > n_\varepsilon$, inequality $|f_n(x) - f(x)| < \dfrac{\varepsilon}{2}$, holds for any $x \in [a, b]$. Using the triangle inequality, we have:

$$|f_n(x) - f_m(x)| \le |f_n(x) - f(x)| + |f(x) - f_m(x)| < \frac{\varepsilon}{2} + \frac{\varepsilon}{2} = \varepsilon$$

for $n, m > n_\varepsilon$.

To show that (ii) $\implies$ (i), let us first observe that, for a fixed $x \in [a, b]$, the numerical sequence $(f_n(x))$ is indeed a Cauchy sequence, thus, it converges to a real number $f(x)$. We prove that such a convergence is uniform. Let us fix $\varepsilon > 0$ and choose $n_\varepsilon \in \mathbb{N}$ such that, for $n, m \in \mathbb{N}, n, m > n_\varepsilon$, it holds:

$$|f_n(x) - f_m(x)| < \varepsilon$$

for any $x \in [a, b]$. Now, taking the limit for $m \to +\infty$, we get:

$$|f_n(x) - f(x)| < \varepsilon$$

for any $x \in [a, b]$. This completes the proof. □

Example 2.10. The sequence of functions $f_n(x) = x(1 + nx)^{-1}$ converges uniformly to $f(x) = 0$ in the interval $[0,1]$. Since $f_n(x) \geq 0$ for $n \in \mathbb{N}$ and for $x \in [0,1]$, we have:

$$\sup_{x \in [0,1]} \frac{x}{1+nx} = \frac{1}{1+n} \to 0 \quad \text{as} \quad n \to \infty.$$

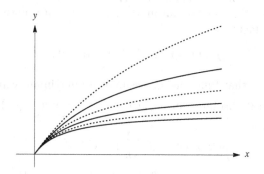

Figure 2.1: $f_n(x) = x(1+nx)^{-1}, \quad n = 1, \ldots, 6.$

Example 2.11. The sequence of functions $f_n(x) = (1+nx)^{-1}$ does not converge uniformly to $f(x) = 0$ in the interval $[0,1]$. In spite of the pointwise limit of f_n for $x \in]0,1]$, we have in fact:

$$\sup_{x \in [0,1]} \frac{1}{1+nx} = 1.$$

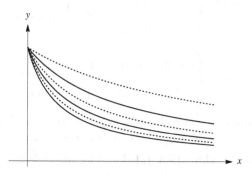

Figure 2.2: $f_n(x) = (1+nx)^{-1}, \quad n = 1, \ldots, 6.$

Example 2.12. If $\alpha \in \mathbb{R}^+$, the sequence of functions, defined on $\mathbb{R}^+$ by $f_n(x) = n^\alpha x e^{-\alpha n x}$, converges pointwise to 0 on $\mathbb{R}^+$, and uniformly if $\alpha < 1$.

For $x > 0$, in fact, by taking the logarithm, we obtain:

$$\ln f_n(x) = \alpha \ln n + \ln x - \alpha n x.$$

It follows that $\lim_{n \to \infty} \ln f_n(x) = -\infty$ and, then, $\lim_{n \to \infty} f_n(x) = 0$. Pointwise convergence is proved.

For uniform convergence, we show that, for any $n \in \mathbb{N}$, the associate function f_n reaches its absolute maximum in $\mathbb{R}^+$. By differentiating with respect to x, we obtain, in fact:

$$f_n'(x) = n^\alpha e^{-\alpha n x} (1 - \alpha n x),$$

from which we see that function f_n assumes its maximum value in $x_n = \dfrac{1}{n}$; such a maximum is absolute, since $f_n(0) = 0$ and $\lim_{x \to +\infty} f_n(x) = 0$. We have thus shown that:

$$\sup_{x \in \mathbb{R}^+} f_n(x) = f_n\left(\frac{1}{n}\right) = e^{-\alpha} n^{\alpha - 1}.$$

Now, $\lim_{n \to \infty} \sup_{x \in \mathbb{R}^+} f_n(x) = \lim_{n \to \infty} e^{-\alpha} n^{\alpha - 1} = 0$, when $\alpha < 1$. Hence, in this case, convergence is indeed uniform.

In the following example, we compare two sequences of functions, apparently very similar, but the first one is pointwise convergent, while the second one is uniformly convergent.

Example 2.13. Consider the sequences of functions (f_n) and (g_n), both defined on $[0, 1]$:

$$f_n(x) = \begin{cases} n^2 x (1 - n x) & \text{if } 0 \leq x < \dfrac{1}{n}, \\ 0 & \text{if } \dfrac{1}{n} \leq x \leq 1, \end{cases} \tag{2.3}$$

and

$$g_n(x) = \begin{cases} n x^2 (1 - n x) & \text{if } 0 \leq x < \dfrac{1}{n}, \\ 0 & \text{if } \dfrac{1}{n} \leq x \leq 1. \end{cases} \tag{2.4}$$

Sequence (f_n) converges pointwise to $f(x) = 0$ for $x \in [0, 1]$; in fact, it is $f_n(0) = 0$ and $f_n(1) = 0$ for any $n \in \mathbb{N}$. When $x \in (0, 1)$, since $n_0 \in \mathbb{N}$ exists such that $\dfrac{1}{n_0} < x$, it follows that $f_n(x) = 0$ for any $n \geq n_0$.

The convergence of (f_n) is not uniform; to show this, observe that $\xi_n = \dfrac{1}{2n}$ maximises f_n, since:

$$f_n'(x) = \begin{cases} n^2 (1 - 2 n x) & \text{if } 0 \leq x < \dfrac{1}{n}, \\ 0 & \text{if } \dfrac{1}{n} \leq x \leq 1. \end{cases}$$

It then follows:

$$\sup_{x\in[0,1]} |f_n(x) - f(x)| = \sup_{x\in[0,1]} f_n(x) = f_n(\xi_n) = \frac{n}{4}$$

which prevents uniform convergence. With similar considerations, we can prove that (g_n) converges pointwise to $g(x) = 0$, and that the convergence is also uniform, since:

$$g_n'(x) = \begin{cases} n\,x\,(2 - 3\,n\,x) & \text{if } 0 \le x < \dfrac{1}{n}, \\ 0 & \text{if } \dfrac{1}{n} \le x \le 1, \end{cases}$$

implying that $\eta_n = \dfrac{2}{3\,n}$ maximises g_n and that:

$$\sup_{x\in[0,1]} |g_n(x) - g(x)| = \sup_{x\in[0,1]} g_n(x) = g_n(\eta_n) = \frac{4}{27\,n},$$

which ensures the uniform convergence of (g_n).

Uniform convergence implies remarkable properties. If a sequence of continuous functions is uniformly convergent, in fact, its limit is also a continuous function.

Theorem 2.14. If (f_n) is a sequence of continuous functions on a closed and bounded interval $[a\,,b]$, which converges uniformly to f, then f is a continuous function.

Proof. Let $f(x)$ be the limit of f_n. Choose $\varepsilon > 0$ and $x_0 \in [a,b]$. Due to uniform convergence, there exists $n_\varepsilon \in \mathbb{N}$ such that, if $n \in \mathbb{N}, n > n_\varepsilon$, then:

$$\sup_{x\in[a\,,b]} |f_n(x) - f(x)| < \frac{\varepsilon}{3}. \tag{2.5}$$

Using the continuity of f_n, we can see that there exists $\delta > 0$ such that:

$$|f_n(x) - f_n(x_0)| < \frac{\varepsilon}{3} \tag{2.6}$$

for any $x \in [a\,,b]$ with $|x - x_0| < \delta$.
To end the proof, we have to show that, given $x_0 \in [a\,,b]$, if $x \in [a\,,b]$ is such that $|x - x_0| < \delta$, then $|f(x) - f(x_0)| < \varepsilon$. By the triangular inequality:

$$|f(x) - f(x_0)| \le |f(x) - f_n(x)| + |f_n(x) - f_n(x_0)| + |f_n(x_0) - f(x_0)|.$$

Observe that:

$$|f(x) - f_n(x)| < \frac{\varepsilon}{3}, \quad |f_n(x_0) - f(x_0)| < \frac{\varepsilon}{3}, \quad |f_n(x) - f_n(x_0)| < \frac{\varepsilon}{3},$$

the first two inequalities being due to (2.5), while the third one is due to (2.6). Hence:

$$|f(x) - f(x_0)| < \varepsilon$$

if $|x - x_0| < \delta$; this concludes the proof. $\qquad\square$

When we are in presence of uniform convergence, for a sequence of continuous functions, defined on the bounded and closed interval $[a, b]$, then the following passage to the limit holds:

$$\lim_{n\to\infty} \int_a^b f_n(x)\,\mathrm{d}x = \int_a^b \lim_{n\to\infty} f_n(x)\,\mathrm{d}x \ . \qquad (2.7)$$

We can, in fact, state the following Theorem 2.15.

Theorem 2.15. If (f_n) is a sequence of continuous functions on $[a, b]$, converging uniformly to $f(x)$, then (2.7) holds true.

Proof. From Theorem 2.14, $f(x)$ is continuous, thus, it is Riemann integrable (see § 8.7.1). Now, choose $\varepsilon > 0$ so that $n_\varepsilon \in \mathbb{N}$ exists such that, for $n \in \mathbb{N}$, $n > n_\varepsilon$:

$$|f_n(x) - f(x)| < \frac{\varepsilon}{b-a} \quad \text{for any } x \in [a, b]. \qquad (2.8)$$

By integration:

$$\left| \int_a^b f_n(x)\,\mathrm{d}x - \int_a^b f(x)\,\mathrm{d}x \right| \le \int_a^b |f_n(x) - f(x)|\,\mathrm{d}x < \frac{\varepsilon}{b-a}(b-a) = \varepsilon \ ,$$

which ends the proof. $\qquad\square$

Remark 2.16. The passage to the limit is sometimes possible under less restrictive hypotheses than Theorem 2.15. In the following example, passage to the limit is possible without uniform convergence. Consider the sequence in $[0, 1]$, given by $f_n(x) = n\,x\,(1 - x)^n$. For such a sequence, it is:

$$\sup_{x\in[0,1]} |f_n(x)| = f_n\left(\frac{1}{n+1}\right) = \frac{n}{n+1}\left(1 - \frac{1}{n+1}\right)^n,$$

thus, f_n is not uniformly convergent, since it holds:

$$\lim_{n\to\infty} \sup_{x\in[0,1]} |f_n(x)| = \frac{1}{e} \ne 0,$$

On the other hand, it holds that $f_n \xrightarrow{[0,1]} 0$. Moreover, we can use integration by part as follows:

$$\lim_{n\to\infty} \int_0^1 f_n(x)\,\mathrm{d}x = \lim_{n\to\infty} \int_0^1 n\,x\,(1 - x)^n\,\mathrm{d}x$$

$$= \lim_{n\to\infty} \int_0^1 n\,x \left(-\frac{1}{n+1}(1 - x)^{n+1}\right)'\,\mathrm{d}x$$

$$= \lim_{n\to\infty} \int_0^1 \frac{n}{n+1}(1 - x)^{n+1}\,\mathrm{d}x = \lim_{n\to\infty} \frac{n}{(n+1)(n+2)} = 0 \ ,$$

and it also holds:

$$\int_0^1 \lim_{n \to \infty} f_n(x)\,\mathrm{d}x = \int_0^1 0\,\mathrm{d}x = 0 \,.$$

Remark 2.17. Consider again the sequences of functions (2.3) and (2.4), defined on $[0,1]$, with $f_n \to 0$ and $g_n \rightrightarrows 0$. Observing that:

$$\int_0^1 f_n(x)\,\mathrm{d}x = \int_0^{\frac{1}{n}} n\,x^2\,(1 - n\,x)\,\mathrm{d}x = \frac{1}{6}$$

and

$$\int_0^1 g_n(x)\,\mathrm{d}x = \int_0^{\frac{1}{n}} n^2\,x\,(1 - n\,x)\,\mathrm{d}x = \frac{1}{12\,n^2} \,,$$

it follows:

$$\lim_{n \to \infty} \int_0^1 f_n(x)\,\mathrm{d}x = \frac{1}{6} \neq \int_0^1 f(x)\,\mathrm{d}x = 0 \,,$$

while:

$$\lim_{n \to \infty} \int_0^1 g_n(x)\,\mathrm{d}x = \lim_{n \to \infty} \frac{1}{12\,n^2} = 0 = \int_0^1 g(x)\,\mathrm{d}x = 0 \,.$$

In other words, the pointwise convergence of (f_n) does not permit the passage to the limit, while the uniform convergence of (g_n) does.

We provide a second example to illustrate, again, that pointwise convergence, alone, does not allow the passage to the limit.

Example 2.18. Consider the sequence of functions (f_n) on $[0,1]$ defined by:

$$f_n(x) = \begin{cases} n^2\,x & \text{if } \ 0 \le x \le \dfrac{1}{n} \,, \\[2mm] 2\,n - n^2 x & \text{if } \ \dfrac{1}{n} < x \le \dfrac{2}{n} \,, \\[2mm] 0 & \text{if } \ \dfrac{2}{n} < x \le 1 \,. \end{cases}$$

Observe that each f_n is a continuous function. Plots of f_n are shown in Figure 2.3, for some values of n; it is clear that, pointwise, $f_n(x) \to 0$ for $n \to \infty$.

By construction, though, each triangle in Figure 2.3 has area equal to 1, thus, for any $n \in \mathbb{N}$:

$$\int_0^1 f_n(x)\,\mathrm{d}x = 1 \,.$$

In conclusion:

$$1 = \lim_{n \to \infty} \int_0^1 f_n(x)\,\mathrm{d}x \neq \int_0^1 \lim_{n \to \infty} f_n(x)\,\mathrm{d}x = 0 \,.$$

In presence of pointwise convergence alone, therefore, swapping between integral and limit is not possible.

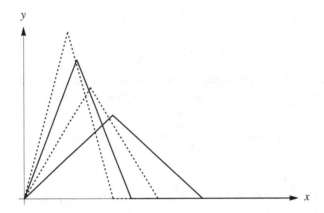

Figure 2.3: Plot of functions $f_n(x)$, $n = 3, \ldots, 6$, in Example 2.18. Solid lines are used for even values of n; dotted lines are employed for odd n.

Uniform convergence leads to a third interesting consequence, connected to the behaviour of sequences of differentiable functions.

Theorem 2.19. Let (f_n) be a sequence of continuous functions, on $[a, b]$. Assume that each f_n is differentiable, with continuous derivative, and that:

(i) $\lim_{n \to \infty} f_n(x) = f(x)$, for any $x \in [a, b]$;

(ii) (f'_n) converges uniformly in $[a, b]$.

Then $f(x)$ is differentiable and

$$\lim_{n \to \infty} f'_n(x) = f'(x).$$

Proof. Define $g(x)$ as:
$$g(x) = \lim_{n \to \infty} f'_n(x),$$

and recall that such a limit is uniform. For Theorems 2.14 and 2.15, $g(x)$ is continuous on $[a, b]$. A classical result from Calculus states that, for any $x \in [a, b]$:

$$\int_a^x g(t)\, dt = \lim_{n \to \infty} \int_a^x f'_n(t)\, dt = \lim_{n \to \infty} (f_n(x) - f_n(a)) = f(x) - f(a)$$

This means that $f(x)$ is differentiable and its derivative is $g(x)$. $\qquad \square$

The hypotheses of Theorem 2.19 are essential, as it is shown by the following example.

Example 2.20. Consider the sequence (f_n) on the open interval $]-1,1[$:

$$f_n(x) = \frac{2x}{1 + n^2 x^2}, \ n \in \mathbb{N}.$$

Observe that f_n converges to 0 uniformly in $]-1,1[$, since:

$$\sup_{x \in]-1,1[} |f_n(x)| = \frac{1}{n} \xrightarrow[n \to \infty]{} 0 .$$

Function f_n is differentiable for any $n \in \mathbb{N}$ and, for any $x \in]-1,1[$ and any $n \in \mathbb{N}$, the derivative of f_n , with respect to x , is:

$$f_n'(x) = \frac{2(1 - n^2 x^2)}{(1 + n^2 x^2)^2} .$$

Now, consider function $g :]-1,1[\to \mathbb{R}$:

$$g(x) = \begin{cases} 0 & \text{if } x \neq 0 , \\ 2 & \text{if } x = 0 . \end{cases}$$

Clearly, $f_n' \xrightarrow{]-1,1[} g$; such a convergence holds pointwise, but not uniformly; by Theorem 2.14, in fact, uniform convergence of (f_n') would imply g to be continuous, which is not, in this case. Here, the hypotheses of Theorem 2.19 are not fulfilled, thus its thesis does not hold.

We end this section with Theorem 2.21, due to Dini[1], and the important Corollary 2.23, a consequence of the Dini theorem, very useful in many applications. Theorem and corollary connect monotonicity and uniform convergence for a sequence of functions; for their proof, we refer the Reader to [16].

Theorem 2.21 (Dini). Let (f_n) be a sequence of continuous functions, converging pointwise to a continuous function f , defined on the interval $[a,b]$. Furthermore, assume that, for any $x \in [a,b]$ and for any $n \in \mathbb{N}$, it holds $f_n(x) \geq f_{n+1}(x)$. Then f_n converges uniformly to f in $[a,b]$.

Remark 2.22. In Theorem 2.21, hypothesis $f_n(x) \geq f_{n+1}(x)$ can be replaced with its reverse monotonicity assumption $f_n(x) \leq f_{n+1}(x)$, obtaining the same thesis.

Corollary 2.23 (Dini). Let (f_n) be sequence of non–negative, continuous and integrable functions, defined on $\mathbb{R}$, and assume that it converges pointwise to f , which is also non–negative, continuous and integrable. Suppose further that it is either $0 \leq f_n(x) \leq f_{n+1} \leq f(x)$ or $0 \leq f(x) \leq f_{n+1} \leq f_n(x)$, for any $x \in \mathbb{R}$ and any $n \in \mathbb{N}$. Then:

$$\lim_{n \to \infty} \int_{-\infty}^{+\infty} f_n(x) \, dx = \int_{-\infty}^{+\infty} f(x) \, dx .$$

---

[1] Ulisse Dini (1845–1918), Italian mathematician and politician.

Example 2.24. Let us consider an application of Theorem 2.21 and Corollary 2.23. Define $f_n(x) = x^n \sin(\pi x)$, $x \in [0,1]$. It is immediate to see that, for any $x \in [0, 1]$:

$$\lim_{n \to \infty} x^n \sin(\pi x) = 0 \ .$$

Moreover, since it is $0 \le f(x) \le f_{n+1}(x) \le f_n(x)$ for any $x \in [0,1]$, the convergence is uniform and, then:

$$\lim_{n \to \infty} \int_0^1 x^n \ \sin(\pi x) \ \mathrm{d}x = 0 \ .$$

2.4   Series of functions

The process of transformation of a sequence of real numbers into an infinite series works, also, when extending sequences of functions into series of functions.

Definition 2.25. The series of functions:

$$\sum_{n=1}^{\infty} f_n(x) = f_1(x) + f_2(x) + \cdots + f_m(x) + \cdots \cdots \tag{2.9}$$

converges in $[a,b]$, if the sequence of its partial sums:

$$s_n(x) = \sum_{k=1}^{n} f_k(x) \tag{2.10}$$

converges in $[a,b]$. The same result applies to uniform convergence, that is, if (2.10) converges uniformly in $[a,b]$, then (2.9) converges uniformly in $[a,b]$.

Remark 2.26. Defining $r_n(x) := f(x) - f_n(x)$, then (2.9) converges uniformly in $[a,b]$ if, for any $\varepsilon > 0$, there exists n_ε such that:

$$\sup_{x \in [a,b]} |r_n(x)| < \varepsilon, \qquad \text{for any } n > n_\varepsilon \ .$$

The following Theorem 2.27, due to Weierstrass, establishes a sufficient condition to ensure the uniform convergence of a series of functions.

Theorem 2.27 (Weierstrass M-Test). Let (f_n) be a sequence of functions defined on $[a,b]$. Assume that for any $n \in \mathbb{N}$, there exists $M_n \in \mathbb{N}$ such that $|f_n(x)| \le M_n$ for any $x \in [a,b]$. Moreover, assume convergence for the numerical series:

$$\sum_{n=1}^{\infty} M_n \ .$$

Then (2.9) converges uniformly in $[a,b]$.

Proof. For the Cauchy criterion of convergence (Theorem 2.9), the series of functions (2.9) converges uniformly if and only if, for any $\varepsilon > 0$, there exists $n_\varepsilon \in \mathbb{N}$ such that:

$$\sup_{x \in [a,b]} \left| \sum_{k=n+1}^{m} f_k \right| < \varepsilon, \qquad \text{for any } m > n > n_\varepsilon .$$

In our case, once $\varepsilon > 0$ is fixed, since the numerical series $\sum\limits_{n=1}^{\infty} M_n$ converges, there exists $n_\varepsilon \in \mathbb{N}$ such that:

$$\sum_{k=n+1}^{m} M_k < \varepsilon, \qquad \text{for any } m > n > n_\varepsilon .$$

Now, we use the triangle inequality:

$$\sup_{x \in [a,b]} \left| \sum_{k=n+1}^{m} f_k \right| \leq \sup_{x \in [a,b]} \sum_{k=n+1}^{m} |f_k| \leq \sum_{k=n+1}^{m} M_k < \varepsilon .$$

This proves the theorem. □

Theorem 2.15 is useful for swapping between sum and integral of a series, and Theorem 2.19 for term–by–term differentiability. We now state three further helpful theorems.

Theorem 2.28. If (f_n) is a sequence of continuous functions on $[a,b]$ and if their series (2.9) converges uniformly on $[a,b]$, then:

$$\sum_{n=1}^{\infty} \left(\int_a^b f_n(x) \, \mathrm{d}x \right) = \int_a^b \left(\sum_{n=1}^{\infty} f_n(x) \right) \mathrm{d}x . \tag{2.11}$$

Proof. Define:

$$f(x) := \sum_{n=1}^{\infty} f_n(x) = \lim_{n \to \infty} \sum_{m=1}^{n} f_m(x) . \tag{2.11a}$$

By Theorem 2.14, function $f(x)$ is continuous and:

$$\int_a^b f(x) \, \mathrm{d}x = \lim_{n \to \infty} \sum_{m=1}^{n} \int_a^b f_m(x) \, \mathrm{d}x . \tag{2.11b}$$

Now, using the linearity of the integral:

$$\sum_{m=1}^{n} \int_a^b f_m(x) \, \mathrm{d}x = \int_a^b \sum_{m=1}^{n} f_m(x) \, \mathrm{d}x = \int_a^b s_n(x) \, \mathrm{d}x .$$

From Theorem 2.15, the thesis (2.11) then follows. □

We state (without proof) a more general result, that is not based on the uniform convergence, but only on simple convergence and few other assumptions.

Theorem 2.29. Let (f_n) be a sequence of functions on an interval $[a, b] \subset \mathbb{R}$. Assume that each f_n is both piecewise continuous and integrable on I, and that (2.9) converges pointwise, on I, to a piecewise continuous function f. Moreover, assume convergence for the numerical (positive terms) series:

$$\sum_{n=1}^{\infty} \int_a^b |f_n(x)| \, dx \ .$$

Then the limit function f is integrable in $[a, b]$ and:

$$\int_a^b f(x) \, dx = \sum_{n=1}^{\infty} \int_a^b f_n(x) \, dx \ .$$

Example 2.30. The series of functions:

$$\sum_{n=1}^{\infty} \frac{\sin(n\,x)}{n^2} \tag{2.12}$$

converges uniformly on any interval $[a, b]$.
It is, in fact, easy to use the Weierstrass Theorem 2.27 and verify that:

$$\left| \frac{\sin(n\,x)}{n^2} \right| \le \frac{1}{n^2} \ .$$

Our statement follows from the convergence of the infinite series $\displaystyle\sum_{n=1}^{\infty} \frac{1}{n^2}$.

Moreover, if $f(x)$ denotes the sum of the series (2.12), then, due to the uniform convergence:

$$\int_0^{\pi} f(x) \, dx = \int_0^{\pi} \sum_{n=1}^{\infty} \frac{\sin(n\,x)}{n^2} \, dx = \sum_{n=1}^{\infty} \frac{1}{n^2} \int_0^{\pi} \sin(n\,x) \, dx = \sum_{n=1}^{\infty} \frac{1 - \cos(n\,\pi)}{n^3} \ .$$

Now, observe that:

$$1 - \cos(n\pi) = \begin{cases} 2 & \text{if } n \text{ is odd,} \\ 0 & \text{if } n \text{ is even.} \end{cases}$$

It is thus possible to infer:

$$\int_0^{\pi} f(x) \, dx = \sum_{n=1}^{\infty} \frac{2}{(2\,n - 1)^3} \ .$$

Theorem 2.31. Assume that (2.9), defined on $[a, b]$, converges uniformly, and assume that each f_n has continuous derivative $f'_n(x)$ for any $x \in [a, b]$; assume further that the series of the derivatives is uniformly convergent. If $f(x)$ denotes the sum of the series (2.9), then $f(x)$ is differentiable and, for any $x \in [a, b]$:

$$f'(x) = \sum_{n=1}^{\infty} f'_n(x) . \tag{2.13}$$

The derivatives at the extreme points a and b are obviously understood as right and left derivatives, respectively.

Proof. We present here the proof given in [33]. Let us denote by $g(x)$, with $x \in [a, b]$, the sum of the series of the derivatives $f'_n(x)$:

$$g(x) = \sum_{n=1}^{\infty} f'_n(x) .$$

By Theorem 2.14, function $g(x)$ is continuous and, by Theorem 2.15, we can integrate term by term in $[a, x]$:

$$\int_a^x g(\xi) \, \mathrm{d}\xi = \sum_{n=1}^{\infty} \int_a^x f'_n(\xi) \, \mathrm{d}\xi = \sum_{n=1}^{\infty} \left(f_n(x) - f_n(a) \right) = \sum_{n=1}^{\infty} f_n(x) - \sum_{n=1}^{\infty} f_n(a) , \tag{2.14}$$

where linearity of the sum is used in the last step of the chain of equalities. Now, recalling definition (2.11a) of $f(x)$, formula (2.14) can be rewritten as:

$$\int_a^x g(\xi) \, \mathrm{d}\xi = f(x) - f(a) . \tag{2.14a}$$

Differentiating both sides of (2.14a), by the Fundamental Theorem of Calculus[2] we obtain $g(x) = f'(x)$, which means that:

$$f'(x) = g(x) = \sum_{n=1}^{\infty} f'_n(x) .$$

Hence, the proof is completed. □

Uniform convergence of series of functions satisfies linearity properties expressed by the following Theorem 2.32, whose proof is left as an exercise.

Theorem 2.32. Given two uniformly convergent power series in $[a, b]$:

$$\sum_{n=1}^{\infty} f_n(x), \qquad \sum_{n=1}^{\infty} g_n(x) ,$$

---

[2]See, for example, mathworld.wolfram.com/FundamentalTheoremsofCalculus.html

the following series converges uniformly for any $\alpha, \beta \in \mathbb{R}$:

$$\sum_{n=1}^{\infty} (\alpha \, f_n(x) + \beta \, g_n(x)) \, .$$

Moreover, if $h(x)$ is a continuous function, defined on $[a, b]$, then the following series is uniformly convergent:

$$\sum_{n=1}^{\infty} h(x) \, f_n(x) \, .$$

2.5   Power series: radius of convergence

The problem dealt with in this section is as follows. Consider a sequence of real numbers $(a_n)_{n \geq 0}$ and the function defined by the so–called *power series*:

$$f(x) = \sum_{n=0}^{\infty} a_n \, (x - x_0)^n \, . \tag{2.15}$$

Given $x_0 \in \mathbb{R}$, it is important to find all the values $x \in \mathbb{R}$ such that the series of functions (2.15) converges.

Example 2.33. With $x_0 = 0$, the power series:

$$\sum_{n=0}^{\infty} \frac{(\,(2\,n)!\,)^2}{16^n \, (\,n!\,)^4} \, x^n$$

converges for $|x| < 1$.

Remark 2.34. It is not restrictive, by using a translation, to consider the following simplified–form power series, obtained from (2.15) with $x_0 = 0$:

$$\sum_{n=0}^{\infty} a_n \, x^n \, . \tag{2.16}$$

Obviously, the choice of x in (2.16) determinates the convergence of the series. The following Lemma 2.35 is of some importance.

Lemma 2.35. If (2.16) converges for $x = r_0$ then, for any $0 \leq r < |r_0|$, it is absolutely and uniformly convergent in $[-r, r]$.

Proof. It is assumed the convergence of the numerical series $\sum_{n=0}^{\infty} a_n \, r_0^n$, that is to say, there exists a positive constant K such that $\left| a_n\, r_0^n \right| \leq K$. Since $\left| \dfrac{r}{r_0} \right| < 1$, then the geometrical series $\displaystyle\sum_{n=0}^{\infty} \left(\dfrac{r}{r_0} \right)^n$ converges. Now, for any $n \geq 0$ and any $x \in [-r, r]$:

$$\left| a_n\, x^n \right| = \left| a_n\, r_0^n \right| \left| \dfrac{x}{r_0} \right|^n \leq K \left| \dfrac{x}{r_0} \right|^n \leq K \left| \dfrac{r}{r_0} \right|^n . \qquad (2.17)$$

By Theorem 2.27, inequality (2.17) implies that (2.16) is uniformly convergent. Due to positivity, the convergence is also absolute. □

From Lemma 2.35 it follows the fundamental Theorem 2.36, due to Cauchy and Hadamard[3], which explains the behaviour of a power series:

Theorem 2.36 (Cauchy–Hadamard). Given the power series (2.16), then only one of the following alternatives holds:

(i) series (2.16) converges for any x ;

(ii) series (2.16) converges only for $x = 0$;

(iii) there exists a positive number r such that series (2.16) converges for any $x \in\,]-r, r[$ and diverges for any $x \in\,]-\infty, -r\,[\, \cup\,]r, +\infty[$.

Proof. Define the set:

$$C := \left\{ x \in [0, +\infty[\ \middle|\ \sum_{n=0}^{\infty} a_n\, x^n \text{ converges} \right\} .$$

If $C = [0, +\infty[$, then (i) holds. Otherwise, C is bounded. If $C = \{0\}$, then (ii) holds. If both (i) and (ii) are not true, then there exists the positive real number $r = \sup C$. Now, choose any $y \in\,]-r, r[$ and form $\bar{y} = \dfrac{|y| + r}{2}$. Since $\bar{y}$ is not an upper bound of C, then a number $z \geq \bar{y}$ exists, for which it converges the series:

$$\sum_{n=0}^{\infty} a_n\, z^n .$$

As a consequence, by Lemma 2.35, series (2.16) converges for any $x \in\,]-z, z[$, and, in particular, it is convergent the series:

$$\sum_{n=0}^{\infty} a_n y^n .$$

---

[3] Jacques Salomon Hadamard (1865–1963), French mathematician.

To end the proof, take $|y| > r$ and assume, by contradiction, that it is still convergent the series:

$$\sum_{n=0}^{\infty} a_n\, y^n \, .$$

If so, using Lemma 2.35, it would follow that series (2.16) converges for any $x \in\,] - |y|\, , |y|\, [$ and, in particular, it would converge for the number:

$$\frac{|y| + r}{2} > r\, ,$$

which contradicts the assumption $r = \sup C$.

$\square$

Definition 2.37. The interval within which (2.16) converges is called *interval of convergence* and r is called *radius of convergence*.

The radius of convergence can be calculated as stated in Theorem 2.38.

Theorem 2.38 (Radius of convergence). Consider the power series (2.16) and assume that the following limit exists:

$$\ell = \lim_{n \to \infty} \left| \frac{a_{n+1}}{a_n} \right| \, .$$

Then:

   (i) if $\ell = \infty$, series (2.16) converges only for $x = 0$;

   (ii) if $\ell = 0$, series (2.16) converges for all x;

   (iii) if $\ell > 0$, series (2.16) converges for $|x| < \dfrac{1}{\ell}$.

Therefore $r = \dfrac{1}{\ell}$ is the radius of convergence of (2.16).

Proof. Consider the series:

$$\sum_{n=0}^{\infty} |a_n\, x^n| = \sum_{n=0}^{\infty} |a_n|\, |x|^n$$

and apply the *ratio test*, that is to say, study the limit of the fraction between the $(n+1)$–th term and the n-th term in the series:

$$\lim_{n \to \infty} \frac{|a_{n+1}|\, |x|^{n+1}}{|a_n|\, |x|^n} = |x| \lim_{n \to \infty} \left| \frac{a_{n+1}}{a_n} \right| = |x|\, \ell \, .$$

If $\ell = 0$, then series (2.16) converges for any $x \in \mathbb{R}$, since it holds:

$$\lim_{n \to \infty} \frac{|a_{n+1}|\, |x|^{n+1}}{|a_n|\, |x|^n} = 0 < 1 \, .$$

If $\ell > 0$, then:

$$\lim_{n \to \infty} \frac{|a_{n+1}| |x|^{n+1}}{|a_n| |x|^n} = |x| \ell < 1 \quad \Longleftrightarrow \quad |x| < \frac{1}{\ell}.$$

Eventually, if $\ell = \infty$, series (2.16) does not converge when $x \neq 0$, since it is:

$$\lim_{n \to \infty} \frac{|a_{n+1}| |x|^{n+1}}{|a_n| |x|^n} > 1,$$

while, for $x = 0$, series (2.16) reduces to the zero series, which converges trivially. □

Example 2.39. The power series (2.18), known as *geometric series*, has radius of convergence $r = 1$.

$$\sum_{n=0}^{\infty} x^n . \tag{2.18}$$

Proof. In (2.18), it is $a_n = 1$ for all $n \in \mathbb{N}$, thus:

$$\lim_{n \to \infty} \left| \frac{a_{n+1}}{a_n} \right| = 1.$$

which means that series (2.18) converges for $-1 < x < 1$. At the boundary of the interval of convergence, namely $x = 1$ and $x = -1$, the geometric series (2.18) does not converge. In conclusion, the interval of convergence of (2.18) is the open interval $] - 1, 1[$. □

Example 2.40. The power series (2.19) has radius of convergence $r = 1$.

$$\sum_{n=1}^{\infty} \frac{x^n}{n} . \tag{2.19}$$

Proof. Here, $a_n = \dfrac{1}{n}$, thus:

$$\lim_{n \to \infty} \left| \frac{a_{n+1}}{a_n} \right| = \lim_{n \to \infty} \frac{n}{n + 1} = 1.$$

that is, (2.19) converges for $-1 < x < 1$.
At the boundary of the interval of convergence, (2.19) behaves as follows; when $x = 1$, it reduces to the divergent *harmonic* series:

$$\sum_{n=0}^{\infty} \frac{1}{n} ,$$

while, when $x = -1$, (2.19) reduces to the convergent *alternate signs* series:

$$\sum_{n=0}^{\infty} (-1)^n \frac{1}{n} .$$

The interval of convergence of (2.19) is, thus, $[-1, 1[$. □

Example 2.41. Series (2.20), given below, has infinite radius of convergence:

$$\sum_{n=0}^{\infty} \frac{x^n}{n!}.$$ (2.20)

Proof. Since it is $a_n = \dfrac{1}{n!}$ for any $n \in \mathbb{N}$, it follows that:

$$\lim_{n\to\infty} \left| \frac{a_{n+1}}{a_n} \right| = \lim_{n\to\infty} \frac{1}{n+1} = 0.$$

□

It is possible to differentiate and integrate power series, as stated in the following Theorem 2.42, which we include for completeness, as it represents a particular case of Theorems 2.28 and 2.31.

Theorem 2.42. Let $f(x)$ be the sum of the power series (2.16), with radius of convergence r. The following results hold.

(i) $f(x)$ is differentiable and, for any $|x| < r$, it is:

$$f'(x) = \sum_{n=1}^{\infty} n\, a_n\, x^{n-1};$$

(ii) if $F(x)$ is the *primitive* of $f(x)$, which vanishes for $x = 0$, then:

$$F(x) = \sum_{n=0}^{\infty} \frac{a_n}{n+1} x^{n+1}.$$

The radius of convergence of both power series $f'(x)$ and $F(x)$ is that of $f(x)$.

Power series behave nicely with respect to the usual arithmetic operations, as shown in Theorem 2.43, which states some useful results.

Theorem 2.43. Consider two power series, with radii of convergence r_1 and r_2 respectively:

$$f_1(x) = \sum_{n=0}^{\infty} a_n\, x^n, \qquad f_2(x) = \sum_{n=0}^{\infty} b_n\, x^n.$$ (2.21)

Then $r = \min\{r_1, r_2\}$ is the radius of convergence of:

$$(f_1 + f_2)(x) = \sum_{n=0}^{\infty} (a_n + b_n)\, x^n.$$

If $\alpha \in \mathbb{R}$, then r_1 is the radius of convergence of:

$$\alpha\, f_1(x) = \sum_{n=0}^{\infty} \alpha\, a_n\, x^n.$$

We state, without proof, Theorem 2.44, concerning the product of two power series.

Theorem 2.44. Consider the two power series in (2.21), with radii of convergence r_1 and r_2 respectively. The product of the two power series is defined by the Cauchy formula:

$$\sum_{n=0}^{\infty} c_n\, x^n, \qquad \text{where} \quad c_n = \sum_{j=0}^{n} a_j\, b_{n-j}, \qquad (2.22)$$

that is:

$$c_0 = a_0\, b_0,$$
$$c_1 = a_0\, b_1 + a_1\, b_0,$$
$$\vdots$$
$$c_n = a_0\, b_n + a_1\, b_{n-1} + \cdots + a_{n-1}\, b_1 + a_n\, b_0.$$

Series (2.22) has interval of convergence given by $|x| < r = \min\{r_1, r_2\}$, and its sum is the pointwise product $f_1(x)\, f_2(x)$.

2.6 Taylor–Maclaurin series

Our starting point, here, is the Taylor[4] formula with Lagrange[5] remainder term. Let $f : I \to \mathbb{R}$ be a function that admits derivatives of any order at $x_0 \in I$. The Taylor–Lagrange theorem states that, if $x \in I$, then there exists a real number ξ, between x and x_0, such that:

$$f(x) = P_n\left(f(x), x_0\right) + R_n\left(f(x), x_0\right)$$
$$= \sum_{k=0}^{n} \frac{f^{(k)}(x_0)}{k!}(x - x_0)^k + \frac{f^{(n+1)}(\xi)}{(n+1)!}(x - x_0)^{n+1}. \qquad (2.23)$$

Since f has derivatives of any order, we may form the limit of (2.23) as $n \to \infty$; a condition is stated in Theorem 2.45 to detect when the passage to the limit is effective.

Theorem 2.45. If f has derivatives of any order in the open interval I, with $x_0, x \in I$, and if:

$$\lim_{n \to \infty} R_n\left(f(x), x_0\right) = \lim_{n \to \infty} \frac{f^{(n+1)}(\xi)}{(n+1)!}(x - x_0)^{n+1} = 0,$$

---

[4] Brook Taylor (1685–1731), English mathematician.
[5] Giuseppe Luigi Lagrange (1736–1813), Italian mathematician.

then:

$$f(x) = \sum_{n=0}^{\infty} y \, \frac{f^{(n)}(x_0)}{n!} \, (x - x_0)^n \,. \tag{2.24}$$

Definition 2.46. A function $f(x)$, defined on an open interval I, is *analytic* at $x_0 \in I$, if its Taylor series about x_0 converges to $f(x)$ in some neighbourhood of x_0.

Remark 2.47. Assuming the existence of the derivatives of any order is not enough to infer that a function is analytic and, thus, it can be represented with a convergent power series. For instance, the function:

$$f(x) = \begin{cases} e^{-\frac{1}{x^2}} & \text{if } x \neq 0 \\ 0 & \text{if } x = 0 \end{cases}$$

has derivatives of any order in $x_0 = 0$, but such derivates are all zero, therefore the Taylor series reduces to the zero function. This happens because the Lagrange remainder does not vanish as $n \to \infty$.

Note that the majority of the functions, that interest us, does not possess the behaviour shown in Remark 2.47. The series expansion of the most important, commonly used, functions can be inferred from Equation (2.23), i.e., from the Taylor–Lagrange theorem. And Theorem 2.45 yields a sufficient condition to ensure that a given function is analytic.

Corollary 2.48. Consider f with derivatives of any order in the interval $I = \,]a, b[$. Assume that there exist $L, M > 0$ such that, for any $n \in \mathbb{N} \cup \{0\}$ and for any $x \in I$:

$$\left| f^{(n)}(x) \right| \leq M L^n. \tag{2.25}$$

Then, for any $x_0 \in I$, function $f(x)$ coincides with its Taylor series in I.

Proof. Assume $x > x_0$. The Lagrange remainder for $f(x)$ is given by:

$$R_n\left(f(x), x_0\right) = \frac{f^{(n+1)}(\xi)}{(n+1)!} \, (x - x_0)^{n+1} \,,$$

where $\xi \in (x_0, x)$, which can be written as $\xi = x_0 + \alpha(x - x_0)$, with $0 < \alpha < 1$. Now, using condition (2.25), it follows:

$$\left| R_n\left(f(x), x_0\right) \right| \leq \frac{\left(L(b-a)\right)^{n+1}}{(n+1)!} M \,.$$

The thesis follows from the limit:

$$\lim_{n \to \infty} \frac{\left(L(b-a)\right)^{n+1}}{(n+1)!} = 0 \,.$$

$\square$

Corollary 2.48, together with Theorem 2.42, allows to find the power series expansion for the most common elementary functions. Theorem 2.49 concerns a first group of power series that converges for any $x \in \mathbb{R}$.

Theorem 2.49. For any $x \in \mathbb{R}$, the *exponential* power series expansion holds:

$$e^x = \sum_{n=0}^{\infty} \frac{x^n}{n!} \,. \qquad (2.26)$$

The *goniometric, hyperbolic,* power series expansions also hold:

$$\sin x = \sum_{n=0}^{\infty} \frac{(-1)^n \, x^{2n+1}}{(2n+1)!}, \quad (2.27) \qquad \cos x = \sum_{n=0}^{\infty} \frac{(-1)^n \, x^{2n}}{(2n)!}, \quad (2.28)$$

$$\sinh x = \sum_{n=0}^{\infty} \frac{x^{2n+1}}{(2n+1)!}, \quad (2.29) \qquad \cosh x = \sum_{n=0}^{\infty} \frac{x^{2n}}{(2n)!}. \quad (2.30)$$

Proof. First, observe that the general term in each of the five series (2.26)–(2.30) comes from the Maclaurin[6] formula.

To show that $f(x) = e^x$ is the sum of the series (2.26), let us use the fact that $f^{(n)}(x) = e^x$ for any $n \in \mathbb{N}$; in this way, it is possible to infer that, in any interval $[a, b]$, the disequality (2.25) is fulfilled if we take $M = 1$ and $L = \max\{e^x \mid x \in [a, b]\}$.

To prove (2.27) and (2.28), in which derivatives of the goniometric functions $\sin x$ and $\cos x$ are considered, condition (2.25) is immediately verified by taking $M = 1$ and $L = 1$.

Finally, (2.29) and (2.30) are a straightforward consequence of the definition of the hyperbolic functions in terms of the exponential:

$$\cosh x = \frac{e^x + e^{-x}}{2}, \quad \sinh x = \frac{e^x - e^{-x}}{2},$$

together with Theorem 2.43.

$\square$

Theorem 2.50 concerns a second group of power series converging for $|x| < 1$.

Theorem 2.50. If $|x| < 1$, the following power series expansions hold:

$$\frac{1}{1-x} = \sum_{n=0}^{\infty} x^n, \qquad (2.31) \qquad \frac{1}{(1-x)^2} = \sum_{n=0}^{\infty} (n+1)x^n, \qquad (2.32)$$

$$\ln(1-x) = -\sum_{n=0}^{\infty} \frac{x^{n+1}}{n+1}, \qquad (2.33) \quad \ln(1+x) = \sum_{n=0}^{\infty} (-1)^n \frac{x^{n+1}}{n+1}, \qquad (2.34)$$

---

[6]Colin Maclaurin (1698–1746), Scottish mathematician.

$$\arctan x = \sum_{n=0}^{\infty}(-1)^n\frac{x^{2n+1}}{2n+1}, \quad (2.35) \quad \ln\left(\frac{1+x}{1-x}\right) = 2\sum_{n=0}^{\infty}\frac{x^{2n+1}}{2n+1}. \quad (2.36)$$

Proof. To prove (2.31), define $f(x) = \dfrac{1}{1-x}$ and build the Maclaurin polynomial of order n, which is $P_n\left(f(x),0\right) = 1 + x + x^2 + \ldots + x^n$; the remainder can be thus estimated directly:

$$\begin{aligned}
R_n\left(f(x),n\right) &= f(x) - P_n\left(f(x),0\right) \\
&= \frac{1}{1-x} - (1+x+x^2+\cdots+x^n) \\
&= \frac{1}{1-x} - \frac{1-x^{n+1}}{1-x} = \frac{x^{n+1}}{1-x}.
\end{aligned} \qquad (2.37)$$

Assuming $|x| < 1$, we see that the remainder vanishes for $n \to \infty$, thus (2.31) follows.

Indentity (2.32) can be proven by employing both formula (2.31) and Theorem 2.44, with $a_n = b_n = 1$.

To obtain (2.33), the geometric series in (2.31) can be integrated term by term, using Theorem 2.42; in fact, letting $|x| < 1$, we can consider the integral:

$$\int_0^x \frac{dt}{1-t} = -\ln(1-x).$$

Now, from Theorem 2.42 it follows:

$$\int_0^x \sum_{n=0}^{\infty} x^n \, dx = \sum_{n=0}^{\infty}\frac{x^{n+1}}{n+1}.$$

Formula (2.33) is then a consequence of formula (2.31). Formula (2.34) can be proven analogously to (2.33), by considering $-x$ instead of x.

To prove (2.35), we use again formula (2.31) with $t = -x$, so that we have:

$$\frac{1}{1+x^2} = \sum_{n=0}^{\infty}(-1)^n x^{2n}.$$

Integrating and invoking Theorem 2.42, we obtain:

$$\arctan x = \int_0^x \frac{dt}{1+t^2} = \sum_{n=0}^{\infty}(-1)^n\frac{x^{2n+1}}{2n+1}.$$

Finally, to prove (2.36), let us consider $x = t^2$ in formula (2.31), so that:

$$\frac{1}{1-t^2} = \sum_{n=0}^{\infty} t^{2n}. \qquad (2.38)$$

Integrating, taking $|x| < 1$, and using Theorem 2.42, the following result is obtained:

$$\int_0^x \frac{dt}{1-t^2} = \frac{1}{2} \int_0^x \left(\frac{1}{1+t} + \frac{1}{1-t} \right) dt = \frac{1}{2} \ln \frac{1+x}{1-x} = \sum_{n=0}^{\infty} \frac{x^{2n+1}}{2n+1}.$$

□

2.6.1 Binomial series

The role of the so–called *binomial series* is pivotal. Let us recall the *binomial formula* (2.39). If $n \in \mathbb{N}$ and $x \in \mathbb{R}$ then:

$$(1+x)^n = \sum_{k=0}^{n} \binom{n}{k} x^k , \tag{2.39}$$

where the *binomial coefficient* is defined as:

$$\binom{n}{k} = \frac{n \cdot (n-1) \cdots \cdots (n-k+1)}{k!}. \tag{2.40}$$

Observe that the left–hand side of (2.40) does not require the numerator to be a natural number. Therefore, if $\alpha \in \mathbb{R}$ and if $n \in \mathbb{N}$, the *generalised binomial coefficient* is defined as:

$$\binom{\alpha}{n} = \frac{\alpha \cdot (\alpha-1) \cdots \cdots (\alpha-n+1)}{n!}. \tag{2.41}$$

From (2.41) a useful property of the generalised binomial coefficient can be inferred, and later used to expand in power series the function $f(x) = (1+x)^\alpha$.

Proposition 2.51. For any $\alpha \in \mathbb{R}$ and any $n \in \mathbb{N}$, the following identity holds:

$$n \binom{\alpha}{n} + (n+1) \binom{\alpha}{n+1} = \alpha \binom{\alpha}{n}. \tag{2.42}$$

Proof. The thesis follows from a straightforward computation:

$$n \binom{\alpha}{n} + (n+1) \binom{\alpha}{n+1} = n \binom{\alpha}{n} + (n+1) \frac{\alpha \cdot (\alpha-1) \cdots \cdots (\alpha-n)}{(n+1)!}$$

$$= n \binom{\alpha}{n} + \frac{\alpha \cdot (\alpha-1) \cdots \cdots (\alpha-n)}{n!}$$

$$= n \binom{\alpha}{n} + (\alpha-n) \frac{\alpha \cdot (\alpha-1) \cdots \cdots (\alpha-n+1)}{n!}$$

$$= n \binom{\alpha}{n} + (\alpha-n) \binom{\alpha}{n} = \alpha \binom{\alpha}{n}.$$

□

By using Proposition 2.51, it is possible to prove the so–called *generalised Binomial theorem* 2.52.

Theorem 2.52 (Generalised Binomial). For any $\alpha \in \mathbb{R}$ and $|x| < 1$, the following identity holds:

$$(1+x)^\alpha = \sum_{n=0}^{\infty} \binom{\alpha}{n} x^n . \tag{2.43}$$

Proof. Let us denote with $f(x)$ the sum of the generalised binomial series:

$$f(x) = \sum_{n=0}^{\infty} \binom{\alpha}{n} x^n ,$$

and introduce function $g(x)$ as follows:

$$g(x) = \frac{f(x)}{(1+x)^\alpha} .$$

To prove the thesis, let us show that $g(x) = 1$ for any $|x| < 1$. Differentiating $g(x)$ we obtain:

$$g'(x) = \frac{(1+x) f'(x) - \alpha f(x)}{(1+x)^{\alpha+1}} . \tag{2.44}$$

Moreover, differentiating $f(x)$ term by term, using Theorem 2.42, we get:

$$(1+x) f'(x) = (1+x) \sum_{n=1}^{\infty} n \binom{\alpha}{n} x^{n-1} = \sum_{n=1}^{\infty} n \binom{\alpha}{n} x^{n-1} + \sum_{n=1}^{\infty} n \binom{\alpha}{n} x^n$$

$$= \sum_{n=0}^{\infty} (n+1) \binom{\alpha}{n+1} x^n + \sum_{n=1}^{\infty} n \binom{\alpha}{n} x^n$$

$$= \sum_{n=0}^{\infty} (n+1) \binom{\alpha}{n+1} x^n + \sum_{n=0}^{\infty} n \binom{\alpha}{n} x^n$$

$$= \sum_{n=0}^{\infty} \left[(n+1) \binom{\alpha}{n+1} + n \binom{\alpha}{n} \right] x^n = \sum_{n=0}^{\infty} \alpha \binom{\alpha}{n} x^n = \alpha f(x) .$$

Thus, $g'(x) = 0$ for any $|x| < 1$, which implies that $g(x)$ is a constant function. It follows that $g(x) = g(0) = f(0) = 1$, which proves thesis (2.43). □

When considering the power series expansion of *arcsin*, the particular value $\alpha = -\frac{1}{2}$ turns out to be important. Let us, then, study the generalised binomial coefficient (2.41) corresponding to such an α.

Proposition 2.53. For any $n \in \mathbb{N}$, the following identity holds true:

$$\binom{-\frac{1}{2}}{n} = (-1)^n \frac{(2n-1)!!}{(2n)!!} , \tag{2.45}$$

in which $n!!$ denotes the *double factorial* function (or *semi–factorial*) of n.

Proof. Evaluation of the binomial coefficient yields, for $\alpha = -\dfrac{1}{2}$:

$$\binom{-\frac{1}{2}}{n} = \frac{-\frac{1}{2}(-\frac{1}{2} - 1)(-\frac{1}{2} - 2) \cdots (-\frac{1}{2} - n + 1)}{n!}$$

$$= (-1)^n \frac{\frac{1}{2}(\frac{1}{2} + 1)(\frac{1}{2} + 2) \cdots (\frac{1}{2} + n - 1)}{n!} = (-1)^n \frac{\frac{1}{2} \cdot \frac{3}{2} \cdot \frac{5}{2} \cdots \frac{2n-1}{2}}{n!}.$$

Recalling that the double factorial $n!!$ is the product of all integers from 1 to n of the same parity (odd or even) as n, we obtain:

$$\frac{1}{2} \cdot \frac{3}{2} \cdot \frac{5}{2} \cdots \frac{2n-1}{2} = \frac{(2n-1)!!}{2^n}.$$

Therefore:

$$\binom{-\frac{1}{2}}{n} = (-1)^n \frac{(2n-1)!!}{2^n \, n!}.$$

Recalling further that $(2n)!! = 2^n \, n!$, thesis (2.45) follows. $\qquad\square$

The following Corollary 2.54 is a consequence of Proposition 2.53.

Corollary 2.54. For any $|x| < 1$, using the convention $(-1)!! = 1$, it holds:

$$\frac{1}{\sqrt{1-x}} = \sum_{n=0}^{\infty} \frac{(2n-1)!!}{(2n)!!} x^n, \tag{2.46}$$

$$\frac{1}{\sqrt{1+x}} = \sum_{n=0}^{\infty} (-1)^n \frac{(2n-1)!!}{(2n)!!} x^n. \tag{2.47}$$

Formula (2.46) is implied by Theorem 2.42 and yields the Maclaurin series for $\arcsin x$, while (2.47) gives the series for $\operatorname{arcsinh} x$, as expressed in the following Theorem 2.55.

Theorem 2.55. Considering $|x| < 1$ and letting $(-1)!! = 1$, then:

$$\arcsin x = \sum_{n=0}^{\infty} \frac{(2n-1)!!}{(2n)!!} \frac{x^{2n+1}}{2n+1}, \tag{2.48}$$

$$\operatorname{arcsinh} x = \sum_{n=0}^{\infty} (-1)^n \frac{(2n-1)!!}{(2n)!!} \frac{x^{2n+1}}{2n+1}. \tag{2.49}$$

Proof. For $|x| < 1$, we can write:

$$\arcsin x = \int_0^x \frac{dt}{\sqrt{1-t^2}}.$$

Using (2.46) with $x = t^2$ and applying Theorem 2.42, it follows:

$$\arcsin x = \int_0^x \sum_{n=0}^{\infty} \frac{(2n-1)!!}{(2n)!!} t^{2n} \, dt = \sum_{n=0}^{\infty} \frac{(2n-1)!!}{(2n)!!} \int_0^x t^{2n} \, dt$$

$$= \sum_{n=0}^{\infty} \frac{(2n-1)!!}{(2n)!!} \frac{x^{2n+1}}{2n+1} \, .$$

Equation (2.49) can be proved analogously. □

Using the power series (2.46) and the *Central Binomial Coefficient* formula:

$$\binom{2n}{n} = \frac{2^n (2n-1)!!}{n!} \tag{2.50}$$

provable by induction, a result can be obtained, due to Lehmer[7] [38].

Theorem 2.56 (Lehmer). If $|x| < \dfrac{1}{4}$, then:

$$\frac{1}{\sqrt{1-4x}} = \sum_{n=0}^{\infty} \binom{2n}{n} x^n \, . \tag{2.51}$$

Proof. Formula (2.50) yields:

$$\sum_{n=0}^{\infty} \binom{2n}{n} x^n = \sum_{n=0}^{\infty} \frac{2^n (2n-1)!!}{n!} x^n = \sum_{n=0}^{\infty} \frac{4^n (2n-1)!!}{2^n \, n!} x^n \, .$$

Using again the relation $(2n)!! = 2^n \, n!$, it follows:

$$\sum_{n=0}^{\infty} \binom{2n}{n} x^n = \sum_{n=0}^{\infty} \frac{4^n (2n-1)!!}{(2n)!!} x^n = \sum_{n=0}^{\infty} \frac{(2n-1)!!}{(2n)!!} (4x)^n \, .$$

Finally, equality (2.51) follows from (2.46). □

### 2.6.2	The error function

We present here an example on how to deal with power series expansion of a function, which has great importance in Probability theory. In Statistics, it is fundamental to deal with the following definite integral:

$$F(x) = \int_0^x e^{-t^2} \, dt \, . \tag{2.52}$$

The main issue with the integral (2.52) is that it is not possible to express it by means of the known elementary functions [39]. On the other hand, some

---

[7]Derrick Henry Lehmer (1905–1991), American mathematician.

probabilistic applications require to know, at least numerically, the values of the function introduced in (2.52). A way to achieve this goal is integrating by series. Using the power series for the exponential function, it is possible to write:

$$e^{-t^2} = \sum_{n=0}^{\infty} (-1)^n \frac{t^{2n}}{n!} .$$

Since the power series is uniformly convergent, we can invoke Theorem 2.15 and transform the integral (2.52) into a series as:

$$\int_0^x e^{-t^2} \, dt = \sum_{n=0}^{\infty} (-1)^n \int_0^x \frac{t^{2n}}{n!} \, dt = \sum_{n=0}^{\infty} (-1)^n \frac{1}{n!} \frac{x^{2n+1}}{2n+1} . \qquad (2.53)$$

The error function, used in Statistics, is defined in the following way:

$$\mathrm{erf}(x) = \frac{2}{\sqrt{\pi}} \int_0^x e^{-t^2} \, dt . \qquad (2.54)$$

Our previous argument, which led to equation (2.53), shows that the power series expansion of the error function, introduced in (2.54), is

$$\boxed{\mathrm{erf}(x) = \frac{2}{\sqrt{\pi}} \sum_{n=0}^{\infty} (-1)^n \frac{1}{n!} \frac{x^{2n+1}}{2n+1} .} \qquad (2.55)$$

Notice that from Theorem 2.38 it follows that the radius of convergence of the power series (2.55) is infinite.

2.6.3 Abel theorem and series summation

We present here an important theorem, due to Abel[8], which explains the behaviour of a given power series, with positive radius of convergence, at the boundary of the interval of convergence. In the previous Examples 2.39 and 2.40, we observed different behaviours at the boundary of the convergence interval: they can be explained by Abel Theorem 2.57, for the proof of which we refer to [9].

Theorem 2.57 (Abel). Denote by $f(x)$ the sum of the power series (2.16), in which we assume that the radius of convergence is $r > 0$. Assume further that the numerical series $\sum_{n=0}^{\infty} r^n a_n$ converges. Then:

$$\lim_{x \to r^-} f(x) = \sum_{n=0}^{\infty} a_n r^n . \qquad (2.56)$$

---

[8]Niels Henrik Abel (1802–1829), Norwegian mathematician.

Proof. The generality of the proof is not affected by the choice $r = 1$, as different radii can be achieved with a straightforward change of variable. Let:

$$s_n = \sum_{m=0}^{n-1} a_m;$$

then:

$$s = \lim_{n \to \infty} s_n = \sum_{n=0}^{\infty} a_n.$$

Now, observe that $a_0 = s_1$ and $a_n = s_{n+1} - s_n$ for any $n \in \mathbb{N}$. If $|x| < 1$, then 1 is the radius of convergence of the power series:

$$\sum_{n=0}^{\infty} s_{n+1} x^n. \tag{2.57}$$

To show it, notice that:

$$\lim_{n \to \infty} \left| \frac{s_{n+2}}{s_{n+1}} \right| = \lim_{n \to \infty} \left| \frac{a_{n+1} + s_{n+1}}{s_{n+1}} \right| = 1.$$

When $|x| < 1$, series (2.57) can be multiplied by $1 - x$, yielding:

$$
\begin{aligned}
(1-x) \sum_{n=0}^{\infty} s_{n+1} x^n &= \sum_{n=0}^{\infty} s_{n+1} x^n - \sum_{n=0}^{\infty} s_{n+1} x^{n+1} \\
&= \sum_{n=0}^{\infty} s_{n+1} x^n - \sum_{n=1}^{\infty} s_n x^n \\
&= s_1 + \sum_{n=1}^{\infty} (s_{n+1} - s_n) x^n \\
&= a_0 + \sum_{n=1}^{\infty} a_n x^n = f(x).
\end{aligned}
\tag{2.58}
$$

To obtain thesis (2.56) we have to show that, for any $\varepsilon > 0$, there exists $\delta_\varepsilon > 0$ such that $|f(x) - s| < \varepsilon$, for any x such that $1 - \delta_\varepsilon < x < 1$. From (2.58) and using formula (2.31) for the sum of the geometric series, we have:

$$
\begin{aligned}
f(x) - s &= (1-x) \sum_{n=0}^{\infty} s_{n+1} x^n - s \\
&= (1-x) \sum_{n=0}^{\infty} s_{n+1} x^n - s(1-x) \sum_{n=0}^{\infty} x^n \\
&= (1-x) \sum_{n=0}^{\infty} s_{n+1} x^n - (1-x) \sum_{n=0}^{\infty} s x^n \\
&= (1-x) \sum_{n=0}^{\infty} (s_{n+1} - s) x^n.
\end{aligned}
\tag{2.59}
$$

Now, fixed $\varepsilon > 0$, there exists $n_\varepsilon \in \mathbb{N}$ such that $|s_{n+1} - s| < \dfrac{\varepsilon}{2}$ for any $n \in \mathbb{N}$, $n > n_\varepsilon$; therefore, using the triangle inequality in (2.59), the following holds for $x \in]-1, 1[$:

$$
\begin{aligned}
|f(x) - s| &= (1 - x)\left| \sum_{n=0}^{n_\varepsilon}(s_{n+1} - s)x^n + \sum_{n=n_\varepsilon+1}^{\infty}(s_{n+1} - s)x^n \right| \\
&\leq (1 - x)\left| \sum_{n=0}^{n_\varepsilon}(s_{n+1} - s)x^n \right| + (1 - x)\left| \sum_{n=n_\varepsilon+1}^{\infty}(s_{n+1} - s)x^n \right| \\
&\leq (1 - x)\sum_{n=0}^{n_\varepsilon}|s_{n+1} - s|\,|x|^n + (1 - x)\frac{\varepsilon}{2}\sum_{n=n_\varepsilon+1}^{\infty}|x|^n \\
&\leq (1 - x)\sum_{n=0}^{n_\varepsilon}|s_{n+1} - s|\,|x|^n + \frac{\varepsilon}{2} \leq (1 - x)\sum_{n=0}^{n_\varepsilon}|s_{n+1} - s| + \frac{\varepsilon}{2}\,.
\end{aligned}
$$
$$(2.60)$$

Observing that the function:

$$
x \mapsto (1 - x)\sum_{n=0}^{n_\varepsilon}|s_{n+1} - s|
$$

is continuous and vanishes for $x = 1$, it is possible to choose $\delta \in]0, 1[$ such that, if $1 - \delta < x < 1$, we have:

$$
(1 - x)\sum_{n=0}^{n_\varepsilon}|s_{n+1} - s| < \frac{\varepsilon}{2}\,.
$$

Thesis (2.56) thus follows. $\qquad\square$

Theorem 2.57 allows to compute, exactly, the sum of many interesting series.

Example 2.58. Recalling the power series expansion (2.33), from Theorem 2.57, with $x = 1$, it follows:

$$
\ln 2 = \sum_{n=1}^{\infty}\frac{(-1)^{n+1}}{n}\,.
$$

Example 2.59. Recalling the power series expansion (2.35), Theorem 2.57, with $x = 1$, allows finding the sum of the Leibniz–Gregory[9] series:

$$
\frac{\pi}{4} = \sum_{n=0}^{\infty}(-1)^n\,\frac{1}{2n + 1}\,.
$$

---

[9] James Gregory (1638–1675), Scottish mathematician and astronomer.
Gottfried Wilhelm von Leibniz (1646–1716), German mathematician and philosopher.

Example 2.60. Recalling the particular binomial expansion (2.46), Abel Theorem 2.57 implies that, for $x = 1$, the following holds:

$$\frac{1}{\sqrt{2}} = \sum_{n=0}^{\infty} (-1)^n \frac{(2n-1)!!}{(2n)!!} .$$

Using the fact that $\arccos x = \frac{\pi}{2} - \arcsin x$, it is possible to obtain a second series, which gives π.

Example 2.61. Recalling the arcsin expansion (2.48), from Theorem 2.57 it follows:

$$\frac{\pi}{2} = \sum_{n=0}^{\infty} \frac{(2n-1)!!}{(2n)!!} \frac{1}{2n+1} .$$

Example 2.62. We show here two summation formulæ connecting π to the central binomial coefficients:

$$\sum_{n=0}^{\infty} \frac{\binom{2n}{n}}{4^n (2n+1)} = \frac{\pi}{2} ; \tag{2.61}$$

$$\sum_{n=0}^{\infty} \frac{\binom{2n}{n}}{16^n (2n+1)} = \frac{\pi}{3} . \tag{2.62}$$

The key to show (2.61) and (2.62) lies in the representation of the central binomial coefficient (2.50), whose insertion in the left–hand side of (2.61) leads to the infinite series:

$$\sum_{n=0}^{\infty} \frac{(2n+1)!!}{2^n \, n! \, (2n+1)} . \tag{2.63}$$

We further notice that, from the power expansion of the arcsin function (2.48), it is possible to infer the following equality:

$$\frac{\arcsin \sqrt{x}}{\sqrt{x}} = \sum_{n=0}^{\infty} \frac{(2n+1)!!}{2^n \, n! \, (2n+1)} x^n . \tag{2.64}$$

The radius of convergence of the power series (2.64) is $r = 1$; Abel Theorem 2.57 can thus be applied to arrive to (2.61). It is worth noting that (2.61) can also be obtained using the Lehmer series (2.51), via the change of variable $y = 4x$ and integrating term by term.

A similar argument leads to (2.62); here, the starting point is the following power series expansion, which has, again, radius of convergence $r = 1$:

$$\frac{\arcsin x}{x} = \sum_{n=0}^{\infty} \frac{\binom{2n}{n}}{4^n (2n+1)} x^{2n} . \tag{2.65}$$

Equality (2.62) follows by evaluating formula (2.65) at $x = \frac{1}{2}$.

2.7 Basel problem

One of the most celebrated problems in Classical Analysis is the *Basel Problem*, which consists in determining the exact value of the infinite series:

$$\sum_{n=1}^{\infty} \frac{1}{n^2}. \tag{2.66}$$

Mengoli[10] originally posed, in 1644, this problem that takes its name from Basel, birthplace of Euler[11] who first provided the correct solution $\frac{\pi^2}{6}$ in [19]. There exist several solutions of the Basel problem; here we present the solution of Choe [11], based on the power series expansion of $f(x) = \arcsin x$, shown in Formula (2.48), as well as on the Abel Theorem 2.57 and on the following integral Formula (2.67), which can be proved by induction on $m \in \mathbb{N}$:

$$\int_0^{\frac{\pi}{2}} \sin^{2m+1} t \, dt = \frac{(2\,m)!!}{(2\,m+1)!!} \,. \tag{2.67}$$

The first step towards solving the Basel problem is to observe that, in the sum (2.66), the attention can be confined to odd indexes only. Namely, if E denotes the sum of the series (2.66), then E can be computed by considering, separately, the sums on even and odd indexes:

$$\sum_{n=1}^{\infty} \frac{1}{(2\,n)^2} + \sum_{n=0}^{\infty} \frac{1}{(2\,n+1)^2} = E \,.$$

On the other hand:

$$\sum_{n=1}^{\infty} \frac{1}{(2\,n)^2} = \sum_{n=1}^{\infty} \frac{1}{4\,n^2} = \frac{E}{4} \,,$$

yielding:

$$\sum_{n=0}^{\infty} \frac{1}{(2\,n+1)^2} = \frac{3}{4} E \,. \tag{2.68}$$

Now, observe that $E = \frac{\pi^2}{6} \iff \frac{3}{4} E = \frac{\pi^2}{8}$. In other words, the Basel problem is equivalent to show that:

$$\sum_{n=0}^{\infty} \frac{1}{(2\,n+1)^2} = \frac{\pi^2}{8} \,, \tag{2.69}$$

whose proof can be found in [11].

---

[10]Pietro Mengoli (1626–1686), Italian mathematician and clergyman from Bologna.
[11]Leonhard Euler (1707–1783), Swiss mathematician and physicist.

Abel Theorem 2.57 applies to the power series (2.48), since we can prove that (2.48) converges for $x = 1$, using *Raabe*[12] test, that is to say, forming:

$$\rho = \lim_{n \to \infty} n \left(\frac{a_n}{a_{n+1}} - 1 \right),$$

in which a_n is the n-th series term, and proving that $\rho > 1$. In the case of (2.48), with $x = 1$:

$$\rho = \lim_{n \to \infty} n \left(\frac{(2n-1)!!}{(2n)!!} \frac{x^{2n+1}}{2n+1} \frac{(2n+2)!!}{(2n+1)!!} \frac{2n+3}{x^{2n+3}} - 1 \right)$$

$$= \lim_{n \to \infty} n \left(\frac{2(n+1)(2n+3)}{(2n+1)^2} - 1 \right) = \lim_{n \to \infty} \frac{n(6n+5)}{(2n+1)^2} = \frac{3}{2}.$$

This implies, also, that the series (2.48) converges uniformly. The change of variable $x = \sin t$ in both sides of (2.48) yields, when $-\dfrac{\pi}{2} < t < \dfrac{\pi}{2}$:

$$t = \sin t + \sum_{n=1}^{\infty} \frac{(2n-1)!!}{(2n)!!} \frac{\sin^{2n+1} t}{2n+1}. \tag{2.70}$$

Integrating (2.70) term by term, on the interval $\left[0, \dfrac{\pi}{2}\right]$, and using (2.67), we obtain:

$$\frac{\pi^2}{8} = 1 + \sum_{n=1}^{\infty} \frac{(2n-1)!!}{(2n)!!} \int_0^{\frac{\pi}{2}} \frac{\sin^{2n+1} t}{2n+1} \, dt$$

$$= 1 + \sum_{n=1}^{\infty} \frac{(2n-1)!!}{(2n)!!} \frac{(2n)!!}{(2n+1)!!} \frac{1}{2n+1}$$

$$= 1 + \sum_{n=1}^{\infty} \frac{1}{(2n+1)^2} = \sum_{n=0}^{\infty} \frac{1}{(2n+1)^2}.$$

This shows (2.69) and, thus, the Euler summation formula:

$$\sum_{n=1}^{\infty} \frac{1}{n^2} = \frac{\pi^2}{6}. \tag{2.71}$$

2.8   Extension of elementary functions to the complex field

The set $\mathbb{C}$ of complex numbers, as well as the set $\mathbb{R}$ of reals, in force of the triangle inequality, possesses the topological structure of metric space. The

---

[12] Joseph Ludwig Raabe (1801–1859), Swiss mathematician.

theory of convergence of sequences and sequences of functions with complex values is, therefore, analogous to that of real–valued sequences and sequences of functions. As a consequence, it is possible to extend the domain of the elementary functions, that are representable in terms of convergent power series, to the complex domain.

2.8.1 Complex exponential

Let us start considering the complex exponential. In $\mathbb{C}$, the exponential function is defined in terms of the usual power series, which is thought, here, as a function of a variable $z \in \mathbb{C}$.

Definition 2.63.

$$e^z := \sum_{n=0}^{\infty} \frac{z^n}{n!}. \qquad (2.72)$$

Equations (2.26) and (2.72) only differ in the fact that, in the latter, the argument can be a complex number. Almost all the familiar properties of the exponential still hold, with the one exception of positivity, which has no sense in the unordered field $\mathbb{C}$. The fundamental property of the complex exponential is stated in the following Theorem 2.64, due to Euler.

Theorem 2.64 (Euler). For any $z = x + iy \in \mathbb{C}$, with $x, y \in \mathbb{R}$, it holds:

$$e^{x+iy} = e^x \left(\cos y + i \sin y \right). \qquad (2.73)$$

Proof. Let $z = x + iy \in \mathbb{C}$; then:

$$
\begin{aligned}
e^z = e^{x+iy} &= e^x \cdot e^{iy} \\
&= e^x \cdot \left(1 + \frac{iy}{1!} + \frac{(iy)^2}{2!} + \frac{(iy)^3}{3!} + \frac{(iy)^4}{4!} + \cdots \right) \\
&= e^x \cdot \left\{ \left(1 - \frac{y^2}{2!} + \frac{y^4}{4!} + \cdots \right) + i \left(y - \frac{y^3}{3!} + \frac{y^5}{5!} + \cdots \right) \right\} \\
&= e^x \cdot \left(\cos y + i \sin y \right).
\end{aligned}
$$

The last step, above, exploits the real power series expansion for the sine and cosine functions given in (2.27) and (2.28) respectively. □

The first beautiful consequence of Theorem 2.64 is the famous *Euler identity*.

Corollary 2.65 (Euler identity).

$$e^{i\pi} + 1 = 0. \qquad (2.74)$$

Proof. First observe that, if $x = 0$ in (2.73), then it holds, for any $y \in \mathbb{R}$:

$$e^{iy} = \cos y + i \sin y. \qquad (2.75)$$

Now, with $y = \pi$, equation (2.75) follows. □

The $\mathbb{C}$–extension of the exponential has an important consequence since the exponential function, when considered as a function $\mathbb{C} \to \mathbb{C}$, is no longer one–to–one, but it is a periodic function. In fact, if $z, w \in \mathbb{C}$, then:

$$e^z = e^w \iff z = w + 2n\pi i \qquad \text{with } n \in \mathbb{Z}.$$

2.8.2 Complex goniometric hyperbolic functions

Equality (2.75) implies the following formulæ (2.76), again due to Euler and valid for any $y \in \mathbb{R}$:

$$\sin y = \frac{e^{iy} - e^{-iy}}{2i}, \qquad \cos y = \frac{e^{iy} + e^{-iy}}{2}. \tag{2.76}$$

It is thus possible to use (2.76) to extend to $\mathbb{C}$ the goniometric functions.

Definition 2.66. For any $z \in \mathbb{C}$, define:

$$\sin z = \frac{e^{iz} - e^{-iz}}{2i}, \qquad \cos z = \frac{e^{iz} + e^{-iz}}{2}. \tag{2.77}$$

In essence, for the sine and cosine functions, both in their goniometric and hyperbolic versions, the power series expansions (2.27), (2.28), (2.29) and (2.30) are understood as functions of a complex variable.

2.8.3 Complex logarithm

To define the complex logarithm, it must be taken into account that the $\mathbb{C}$–exponential function is periodic, with period $2\pi i$, thus the $\mathbb{C}$–logarithm is not univocally determined. With this in mind, we formulate the following definition.

Definition 2.67. If $w \in \mathbb{C}$, the logarithm of w is any complex number $z \in \mathbb{C}$ such that $e^z = w$.

Remark 2.68. In $\mathbb{C}$, as well as in $\mathbb{R}$, the logarithm of zero is undefined, since, from (2.73), it follows $e^z \neq 0$, for any $z \in \mathbb{C}$.

Using the polar representation of a complex number, we can represent its logarithms as shown below.

Theorem 2.69. If $w = \rho e^{i\vartheta}$ is a non–zero complex number, the logarithms of w are defined as:

$$\log w = \ln \rho + i(\vartheta + 2n\pi), \qquad n \in \mathbb{Z}. \tag{2.78}$$

Proof. Let $w = e^z$ and let $z = x + iy$; then, we have to solve the system:

$$e^z = \rho e^{i\vartheta},$$

with

$$e^z = e^{x+iy} = e^x\, e^{iy} = e^x\, (\cos y + i\, \sin y)\,, \qquad \rho\, e^{i\vartheta} = \rho\, (\cos \vartheta + i\, \sin \vartheta)\,,$$

from which the real and imaginary components of z are obtained:

$$x = \ln \rho\,, \qquad \rho \geq 0\,, \qquad y = \vartheta + 2n\pi\,.$$

Since $\log w = z$, thesis (2.78) follows. □

Among the infinite logarithms of a complex number, we pin down one, corresponding to the most convenient argument.

Definition 2.70. Consider $w = \rho\, e^{i\vartheta}$, $w \neq 0$. The main argument of w is ϑ, with $-\pi < \vartheta \leq \pi$, and is referred to as $\arg(w)$. Note, also, that $\rho = |w|$. Then, the principal determination of the logarithm of w is:

$$\operatorname{Log} w = \ln \rho + i\vartheta = \ln |w| + i\, \arg(w)\,.$$

Example 2.71. Compute $\operatorname{Log}(-1)$. Here, $w = \rho\, e^{i\vartheta}$, with $\rho = |-1|$ and $\vartheta = \arg(-1)$. Since $\ln 1 = 0$ and $\arg(-1) = \pi$, we obtain $\operatorname{Log}(-1) = i\pi$.

In other words, for a non–zero complex w, the principal determination (or principal value) $\operatorname{Log} w$ is the logarithm whose imaginary part lies in the interval $(-\pi, \pi]$.

We end this section introducing the complex power.

Definition 2.72. Given $z \in \mathbb{C}, z \neq 0$, and $w \in \mathbb{C}$, the complex power function is defined as:

$$z^w = e^{w\,\operatorname{Log} z}$$

Example 2.73. Compute i^i. Applying Definition 2.72: $i^i = e^{i\,\operatorname{Log} i}$. Since $\arg(i) = \dfrac{\pi}{2}$ and $|i| = 1$, then $\operatorname{Log} i = i\dfrac{\pi}{2}$. Finally, $i^i = e^{i\,i\,\frac{\pi}{2}} = e^{-\frac{\pi}{2}}$.

Example 2.74. In $\mathbb{C}$, it is possible to solve equations like $\sin z = 2$, obviously finding complex solutions. From the sin definition (2.77), in fact, we obtain:

$$e^{2iz} - 4i\, e^{iz} - 1 = 0\,.$$

Thus:

$$e^{iz} = \left(2 \pm \sqrt{3}\right) i\,.$$

Evaluating the logarithms, the following solutions can be found:

$$z = \frac{\pi}{2} + 2n\pi - i \ln\left(2 \pm \sqrt{3}\right)\,, \qquad n \in \mathbb{Z}\,.$$

2.9 Exercises

2.9.1 Solved exercises

1. Given the following sequence of functions, establish whether it is pointwise and/or uniformly convergent:

$$f_n(x) = \frac{n\,x + x^2}{n^2}, \qquad x \in [0, 1],$$

2. Evaluate the pointwise limit of the sequence of functions:

$$f_n(x) = \sqrt[n]{1 + x^n}, \qquad x \geq 0.$$

3. Show that the following sequence of functions converges pointwise, but not uniformly, to $f(x) = 0$:

$$f_n(x) = n\,x\,e^{-n\,x}, \qquad x > 0,$$

4. Show that the following sequence of functions converges uniformly to $f(x) = 0$:

$$f_n(x) = \frac{\sqrt{1 - x^n}}{n^2}, \qquad x \in [-1, 1],$$

5. Show that:

$$\sum_{n=1}^{\infty} \frac{1}{n\,2^n} = \ln 2.$$

6. Evaluate:

$$\lim_{n \to \infty} \int_{1}^{\infty} \frac{n\,e^{-n\,x}}{1 + n\,x}\, dx\, .$$

7. Use the definite integral $\displaystyle \int_{0}^{1} \frac{x\,(1 - x)}{1 + x}\, dx$ to prove that:

$$\sum_{n=1}^{\infty} \frac{(-1)^n}{(n + 1)\,(n + 2)} = \frac{3}{2} - \ln 4\, .$$

8. Let $f_n(x) = \left(1 + \dfrac{x^2}{n} \right)^{-n}, \quad x \geq 0.$

   a. Show that (f_n) is pointwise convergent to a function $f(x)$ to be determined.

   b. Show that:

$$\lim_{n \to \infty} \int_{0}^{\infty} f_n(x)\, dx = \int_{0}^{\infty} f(x)\, dx\, .$$

Solutions to Exercises 2.9.1

1. Sequence $f_n(x)$ converges pointwise to zero for any $x \in [0,1]$ since:

$$\lim_{n \to \infty} f_n(x) = \lim_{n \to \infty} \left(\frac{x}{n} + \frac{x^2}{n^2} \right) = \lim_{n \to \infty} \frac{x}{n} + \lim_{n \to \infty} \frac{x^2}{n^2} = 0 + 0 = 0 \,.$$

To establish if such a convergence is also uniform, we evaluate:

$$\sup_{x \in [0,1]} |f_n(x) - 0| = \sup_{x \in [0,1]} \left(\frac{nx + x^2}{n^2} \right) = \frac{n+1}{n^2} \,.$$

Observe that:

$$\lim_{n \to \infty} \sup_{x \in [0,1]} |f_n(x) - 0| = \lim_{n \to \infty} \frac{n+1}{n^2} = 0 \,.$$

The uniform convergence on the interval $[0,1]$ follows.

2. If $x = 0$ then $f_n(0) = 1$ for any $n \in \mathbb{N}$.

   If $x \leq 1$, then $x^n \to 0$, and we can infer that there exists $n_0 \in \mathbb{N}$ such that $\frac{3}{2} < 1 + x^n < \frac{5}{2}$; therefore, for any $n > n_0$, we have:

$$\sqrt[n]{\frac{3}{2}} < f_n(x) < \sqrt[n]{\frac{5}{2}} \,.$$

Since $\lim_{n \to \infty} \sqrt[n]{\frac{3}{2}} = \lim_{n \to \infty} \sqrt[n]{\frac{5}{2}} = 1$, we can use the Sandwich theorem (or Squeeze theorem[13]), for $x \leq 1$, to prove the limit relation:

$$\lim_{n \to \infty} f_n(x) = 1 \,.$$

Now, examine what happens when $x > 1$. First, notice that:

$$f_n(x) = \sqrt[n]{1 + x^n} = \sqrt[n]{x^n \left(\frac{1}{x^n} + 1 \right)} = x \sqrt[n]{\frac{1}{x^n} + 1} \,.$$

Recalling that, here, $\frac{1}{x} < 1$, $x \neq 0$, we consider a change of variable $t = \frac{1}{x}$ and repeat the previous argument (that we followed in the case of a variable $t < 1$) to obtain $\lim_{n \to \infty} \sqrt[n]{t^n + 1} = 1$, that is:

$$\lim_{n \to \infty} \sqrt[n]{\frac{1}{x^n} + 1} = 1 \,.$$

---

[13] See, for example, mathworld.wolfram.com/SqueezingTheorem.html

In other words, for $x > 1$, we have shown that:

$$\lim_{n \to \infty} f_n(x) = \lim_{n \to \infty} \sqrt[n]{1 + x^n} = x.$$

Putting everything together, we have proven that:

$$\lim_{n \to \infty} f_n(x) = f(x) \qquad \text{where} \quad f(x) = \begin{cases} 1 & \text{if } x \leq 1, \\ x & \text{if } x > 1. \end{cases}$$

3. The pointwise limit of sequence $f_n(x) = n\,x\,e^{-n\,x}$ is $f(x) = 0$, due to the exponential decay of the factor $e^{-n\,x}$. To investigate the possible uniform convergence, we consider:

$$\sup_{x>0} |f_n(x) - f(x)| = \sup_{x>0} n\,x\,e^{-n\,x}.$$

Differentiating we find:

$$\frac{d}{dx}\left(n\,x\,e^{-n\,x}\right) = n\,e^{-n\,x}(1 - n\,x),$$

showing that $x = \dfrac{1}{n}$ is a local maximiser and the corresponding extremum is:

$$f_n\left(\frac{1}{n}\right) = \frac{1}{e}.$$

But this implies that the found convergence cannot be uniform, since:

$$\lim_{n \to \infty} \sup_{x>0} |f_n(x) - f(x)| = \frac{1}{e} \neq 0.$$

4. For any $x \in [-1, 1]$ and any $n \in \mathbb{N}$, it holds that $\sqrt{1 - x^n} \leq \sqrt{2}$, thus:

$$f_n(x) = \frac{\sqrt{1 - x^n}}{n^2} \leq \frac{\sqrt{2}}{n^2}. \tag{2.79}$$

Now, observe that inequality (2.79) is independent of $x \in [-1, 1]$: this fact, taking the supremum with respect to $x \in [-1, 1]$, ensures uniform convergence.

5. Consider, for any $n \in \mathbb{N}$, the definite integral:

$$\int_0^{\frac{1}{2}} x^{n-1}\, dx = \frac{1}{n\,2^n}.$$

Summing up for all positive integers between 1 and n, we get:

$$\sum_{n=1}^{\infty} \frac{1}{n\,2^n} = \sum_{n=1}^{\infty} \int_0^{\frac{1}{2}} x^{n-1}\, dx.$$

Since the geometric series, in the right–hand side above, converges uniformly, we can swap series and integral, obtaining:

$$\sum_{n=1}^{\infty} \frac{1}{n\,2^n} = \int_0^{\frac{1}{2}} \left(\sum_{n=1}^{\infty} x^{n-1} \right) dx \ .$$

Therefore

$$\sum_{n=1}^{\infty} \frac{1}{n\,2^n} = \int_0^{\frac{1}{2}} \frac{1}{1-x}\,dx \ .$$

The thesis follows by integrating:

$$\int_0^{\frac{1}{2}} \frac{1}{1-x}\,dx = [-\ln(1-x)]_{x=0}^{x=\frac{1}{2}} \ .$$

6. Define the (decreasing) function $h_n(x) = \dfrac{n}{1+n\,x}$, with $x \in [1,+\infty)$. Then:

$$h_n'(x) = -\frac{n^2}{(1+n\,x)^2} < 0 \ .$$

Since:

$$\lim_{x \to \infty} \frac{n}{1+n\,x} = 0$$

and

$$\sup_{x \in [1,\infty)} \frac{n}{1+n\,x} = h_n(1) = \frac{n}{1+n} \ ,$$

we can infer that:

$$|h_n(x)| \le \frac{n}{1+n} < 1 \ .$$

Therefore:

$$\left| \frac{n\,e^{-n\,x}}{1+n\,x} \right| < e^{-n\,x} \ .$$

This shows uniform convergence for f_n. We can now invoke Theorem 2.15, to obtain:

$$\lim_{n \to \infty} \int_1^{\infty} \frac{n\,e^{-n\,x}}{1+n\,x}\,dx = \int_1^{\infty} \lim_{n \to \infty} \frac{n\,e^{-n\,x}}{1+n\,x}\,dx = \int_1^{\infty} 0\,dx = 0 \ .$$

7. First, evaluate the definite integral:

$$\int_0^1 \frac{x\,(1-x)}{1+x}\,dx = \int_0^1 \left(2 - x - \frac{2}{x+1} \right) dx = \frac{3}{2} - \ln 4 \ .$$

Then, recall that, in $[0,1]$, it holds true the geometric series expansion:

$$\frac{1}{1+x} = \sum_{m=0}^{\infty} (-x)^m = \sum_{m=0}^{\infty} (-1)^m\,x^m \ .$$

It is thus possible to integrate term by term, obtaining:

$$\int_0^1 \frac{x\,(1-x)}{1+x}\,dx = \int_0^1 x\,(1-x)\sum_{n=0}^{\infty}(-1)^m\,x^m\,dx$$

$$= \sum_{m=0}^{\infty}(-1)^m\int_0^1 x^{m+1}\,(1-x)\,dx\ .$$

Now, evaluating the last right–hand side integral, we get:

$$\int_0^1 \frac{x\,(1-x)}{1+x}\,dx = \sum_{m=0}^{\infty}(-1)^m\left(\frac{1}{m+2}-\frac{1}{m+3}\right)$$

$$= \sum_{m=0}^{\infty}\frac{(-1)^m}{(m+2)\,(m+3)}\ .$$

Our statement follows using the change of index $n = m+1$.

8. Observe that:

$$\left(1+\frac{x^2}{n}\right)^{-n} = e^{-n\,\ln\left(1+\frac{x^2}{n}\right)} = e^{-n\left(\frac{x^2}{n}+o\left(\frac{x^2}{n}\right)\right)} = e^{-x^2\,(1+o(1))}$$

where we have used $\ln(1+t) = t + o(t)$ when $t \simeq 0$. This means that:

$$\lim_{n\to\infty}\left(1+\frac{x^2}{n}\right)^{-n} = e^{-x^2}\ .$$

We have thus shown point (a). The second statement follows from Corollary 2.23.

2.9.2   Unsolved exercises

1. Show that the sequence of functions $f_n(x) = \dfrac{x}{n}$, with $x \in \mathbb{R}$, converges pointwise to $f(x) = 0$, but the convergence is not uniform.

   Show also that, on the other hand, when $a > 0$, the sequence (f_n) converges uniformly to $f(x) = 0$ for $x \in [-a,a]$.

2. Let $f_n(x) = \left(\cos\dfrac{x}{\sqrt{n}}\right)^n$, $x \in \mathbb{R}$. Show that:

   a. f_n converges pointwise to a non–zero function $f(x)$ to be determined;

   b. if $a > 0$, the sequence (f_n) converges uniformly on $[-a,a]$.

   Hint. Consider the sequence $g_n(x) = \ln f_n(x)$ *and use the power series* (2.33) *and* (2.28).

3. Establish if the sequence of functions (f_n), defined by $f_n(x) = \dfrac{x + x^2 \, e^{n\,x}}{1 + e^{n\,x}}$,
for $x \in \mathbb{R}$, converges pointwise and/or uniformly.

4. Show that $\displaystyle\sum_{n=1}^{\infty} \dfrac{1}{n\,3^n} = \ln \dfrac{3}{2}$.

5. Show that $\displaystyle\lim_{n \to \infty} \int_0^1 e^{\frac{x+1}{n}} \, dx = 1$.

6. Consider the following equality and say if (and why) it is true or false:

$$\lim_{n \to \infty} \int_0^1 \frac{x^4}{x^2 + n^2} \, dx = \int_0^1 \lim_{n \to \infty} \frac{x^4}{x^2 + n^2} \, dx \; .$$

7. Let $f_n(x) = \dfrac{n(x^3 + x)\,e^{-x}}{1 + n\,x}$, with $x \in [0, 1]$.

    a. Show that (f_n) is pointwise convergent to a function $f(x)$ to be determined.

    b. Show that, for any $x \in [0, 1]$ and for any $n \in \mathbb{N}$:

$$|f_n(x) - f(x)| \le \frac{2}{1 + n\,x} \; .$$

    c. Show that, for any $a > 0$, sequence (f_n) converges uniformly to f on $[a, 1]$, but the convergence is not uniform on $[0, 1]$.

    d. Evaluate $\displaystyle\lim_{n \to \infty} \int_0^1 f_n(x) \, dx$.

8. Use the definite integral:

$$\int_0^1 \frac{1 + \sqrt{x}}{1 + x} \, dx$$

to show that:

$$\sum_{n=1}^{\infty} (-1)^{n-1} \frac{4\,n + 1}{n\,(2\,n + 1)} = 2 + \ln 2 - \frac{\pi}{2} \; .$$

9. Show that $\displaystyle\int_0^{\infty} \frac{x^5}{e^{x^2} - 1} \, dx = \sum_{n=1}^{\infty} \frac{1}{n^3}$.

10. Show that $\cos z = 2 \iff z = 2\,n\,\pi - i \ln \left(2 \pm \sqrt{3} \right)$, $\quad n \in \mathbb{Z}$.

Chapter 3

Multidimensional differential calculus

This chapter presents the concept of differentiability of vector–valued functions, which is crucial for studying the theory of ordinary and partial differential equations, that will be presented in the following chapters of this book, and in particular in Chapters 4, 5 and 13. The notion of critical point, and the related definitions of gradient vector, Jacobian and Hessian matrices, and Lagrange multipliers, are recalled. The important Implicit function theorem is also stated and proved.

3.1 Partial derivatives

The most natural way to define derivatives of functions of several variables is to allow only one variable at a time to move, while freezing the others. Thus, if $f : V \to \mathbb{R}$ is a function of n variables, whose domain is the open set V, we define the set $\{x_1\} \times \cdots \times \{x_{j-1}\} \times [a, b] \times \{x_{j+1}\} \times \cdots \times \{x_n\}$, where $[a, b]$ is chosen so to have $\{x_1\} \times \cdots \times \{x_{j-1}\} \times \{t\} \times \{x_{j+1}\} \times \cdots \times \{x_n\} \subset V$ for any $t \in [a, b]$. We shall denote the function:

$$g(t) := f(x_1, \dots, x_{j-1}, t, x_{j+1}, \dots, x_n)$$

by

$$f(x_1, \dots, x_{j-1}, \cdot, x_{j+1}, \dots, x_n).$$

If g is differentiable (see Definition 3.7) at some $t_0 \in (a, b)$, then the *first–order partial derivative* of f at $(x_1, \dots, x_{j-1}, t_0, x_{j+1}, \dots, x_n)$, with respect to x_j, is defined by:

$$f_{x_j}(x_1, \dots, x_{j-1}, t_0, x_{j+1}, \dots, x_n) := \frac{\partial f}{\partial x_j}(x_1, \dots, x_{j-1}, t_0, x_{j+1}, \dots, x_n)$$

$$:= g'(t_0).$$

Therefore, the partial derivative f_{x_j} exists at a point a if and only if the following limit exists:

$$\frac{\partial f}{\partial x_j}(a) := \lim_{h \to 0} \frac{f(a + h\,e_j) - f(a)}{h}.$$

Partial derivatives of order higher than one are defined by iteration. For example, when it exists, the second–order partial derivative of f, with respect to x_j and x_k, is defined by:

$$f_{x_j x_k} := \frac{\partial^2 f}{\partial x_k \, \partial x_j} = \frac{\partial}{\partial x_k}\left(\frac{\partial f}{\partial x_j}\right).$$

Second–order partial derivatives are called *mixed* when $j \neq k$.

Definition 3.1. Let V be a non–empty open subset of $\mathbb{R}^n$, let $f : V \to \mathbb{R}$ and $p \in \mathbb{N}$.

(i) f is said to be $\mathcal{C}^p$ on V if and only if every k–th order partial derivative of f, with $k \leq p$, exists and is continuous on V.

(ii) f is said to be $\mathcal{C}^\infty$ on V if and only if f is $\mathcal{C}^p$ for all $p \in \mathbb{N}$.

If f is $\mathcal{C}^p$ on V and $q < p$, then f is $\mathcal{C}^q$ on V. The symbol $\mathcal{C}^p(V)$ denotes the set of functions that are $\mathcal{C}^p$ on an open set V.

For simplicity, in the following we shall state all results for the case $m = 1$ and $n = 2$, denoting x_1 with x and x_2 with y. With appropriate changes in notation, the same results hold for any $m, n \in \mathbb{N}$.

Example 3.2. By the Product Rule[1], if f_x and g_x exist, then:

$$\frac{\partial f}{\partial x}(f\,g) = f\,\frac{\partial g}{\partial x} + g\,\frac{\partial f}{\partial x}.$$

Example 3.3. By the Mean–Value theorem[2], if $f(\cdot\,,y)$ is continuous on $[a, b]$ and the partial derivative $f_x(\cdot\,,y)$ exists on (a, b), then there exists a point $c \in (a, b)$ (which may depend on y as well as on a and b) such that:

$$f(b\,,y) - f(a\,,y) = (b - a)\,\frac{\partial f}{\partial x}(c\,,y).$$

In most situations, when dealing with high–order partial derivatives, the order of computation of the derivatives is, in some sense, arbitrary. This is expressed by the Clairaut[3]–Schwarz theorem.

Theorem 3.4 (Clairaut–Schwarz). Assume that V is open in $\mathbb{R}^2$, that $(a, b) \in V$ and $f : V \to \mathbb{R}$. Assume further that f is $\mathcal{C}^2$ on V and that one of the two second–order mixed partial derivatives of f exists on V and is continuous at the point (a, b). Then, the other second–order mixed partial derivative exists at (a, b) and the following equality is verified:

$$\frac{\partial^2 f}{\partial y \, \partial x}(a\,,b) = \frac{\partial^2 f}{\partial x \, \partial y}(a\,,b).$$

---

[1] See, for example, mathworld.wolfram.com/ProductRule.html
[2] See, for example, mathworld.wolfram.com/MeanValueTheorem.html
[3] Alexis Claude Clairaut (1713–1765), French mathematician, astronomer, geophysicist.

The hypotheses of Theorem 3.4 are met if $f \in C^2(V)$ on $V \subset \mathbb{R}^2$, V open. For functions of n variables, the following Theorem 3.5 holds.

Theorem 3.5. If f is C^2 on an open subset V of $\mathbb{R}^n$, if $a \in V$, and if $j \neq k$, then:

$$\frac{\partial^2 f}{\partial x_j \, \partial x_k}(a) = \frac{\partial^2 f}{\partial x_k \, \partial x_j}(a).$$

Remark 3.6. Existence of partial derivatives does not ensure continuity. As an example, consider:

$$f(x,y) = \begin{cases} \dfrac{x\,y}{x^2 + y^2} & \text{if} \quad (x,y) \neq (0,0), \\ 0 & \text{if} \quad (x,y) = (0,0). \end{cases}$$

This function is not continuous at $(0,0)$, but admits partial derivatives at any $(x,y) \in \mathbb{R}^2$, since:

$$\lim_{\Delta x \to 0} \frac{f(\Delta x, 0) - f(0,0)}{\Delta x} = \lim_{\Delta x \to 0} 0 = 0,$$

and

$$\lim_{\Delta y \to 0} \frac{f(0, \Delta y) - f(0,0)}{\Delta y} = \lim_{\Delta y \to 0} 0 = 0.$$

3.2 Differentiability

In this section, we define what it means for a vector function f to be differentiable at a point a. Whatever our definition, if f is differentiable at a, then we expect two things:

(1) f will be continuous at a;

(2) all first–order partial derivatives of f will exist at a.

To appreciate the following Definition 3.7 of *total derivative* of a function of n variables, we consider one peculiar aspect of differentiable functions of one variable. Recall that $f : \mathbb{R} \to \mathbb{R}$ is differentiable at $x \in \mathbb{R}$ if the following limit is finite, i.e., it is a real number:

$$\lim_{h \to 0} \frac{f(x + h) - f(x)}{h} := f'(x).$$

The definition above is equivalent to the following: f is differentiable at $x \in \mathbb{R}$ if there exist $\alpha \in \mathbb{R}$ and a function $\omega : (-\delta, \delta) \to \mathbb{R}$, with $\omega(0) = 0$ and $\lim_{h \to 0} \dfrac{\omega(h)}{h} = 0$, such that:

$$f(x + h) = f(x) + \alpha h + \omega(h) h. \tag{3.1}$$

The definition of differentiability for functions of several variables extends Property (3.1).

Definition 3.7. Let f be a real function of n variables. f is said to be differentiable, at a point $a \in \mathbb{R}^n$, if and only if there exists an open set $V \subset \mathbb{R}^n$, such that $a \in V$ and $f : V \to \mathbb{R}$, and there exists $d \in \mathbb{R}^n$ such that:

$$\lim_{h \to 0} \frac{f(a+h) - f(a) - d \cdot h}{\|h\|} = 0$$

d is called *total derivative* of f at a

Theorem 3.8. If f is differentiable at a, then:

   (i) f is continuous at a;

   (ii) all first–order partial derivatives of f exist at a;

   (iii) $d = \nabla f(a) := \left(\dfrac{\partial f}{\partial x_1}(a), \dots, \dfrac{\partial f}{\partial x_n}(a) \right).$

$\nabla f(a)$ is called the *gradient* (or *nabla*) of f at a.

A reverse implication to Theorem 3.8 also holds true.

Theorem 3.9. Let V be open in $\mathbb{R}^n$, let $a \in V$ and suppose that $f : V \to \mathbb{R}$. If all first–order partial derivatives of f exist in V and are continuous at a, then f is differentiable at a.

The hypotheses of Theorem 3.9 are met if $f \in \mathcal{C}^1(V)$ on $V \subset \mathbb{R}^n$, V open.

Theorem 3.10. Let $\alpha \in \mathbb{R}$, $a \in \mathbb{R}^n$, and assume that $f, g : V \to \mathbb{R}$ are differentiable at a, being $V \subset \mathbb{R}^n$ an open set. Then, the functions $f + g$ and αf are differentiable at a, and the following equalities are verified:

   (i) $\nabla (f + g)(a) = \nabla f(a) + \nabla g(a)$;

   (ii) $\nabla (\alpha f)(a) = \alpha \nabla f(a)$;

   (iii) $\nabla (f g)(a) = g(a) \nabla f(a) + f(a) \nabla g(a)$.

Moreover, if $g(a) \neq 0$, then f/g is differentiable at a, and it holds:

   (iv) $\nabla \left(\dfrac{f}{g} \right)(a) = \dfrac{g(a) \nabla f(a) - f(a) \nabla g(a)}{g^2(a)}.$

The composition of functions follows similar rules, for differentiation, as in the one–dimensional case. For instance, the Chain Rule [4] holds in the following way. Consider a vector function $g : I \to \mathbb{R}^n$, $g = (g_1, \dots, g_n)$, defined on an open interval $I \subset \mathbb{R}$, and consider $f : g(I) \subset \mathbb{R}^n \to \mathbb{R}$. If each of the

---

[4]See, for example, mathworld.wolfram.com/ChainRule.html

components g_j of g is differentiable at $t_0 \in I$, and if f is differentiable at $a = (g_1(t_0),\dots,g_n(t_0))$, then the composition $\varphi(t) = f(g(t))$ is differentiable at t_0, and we have:

$$\varphi'(t_0) = \nabla f(a) \bullet g'(t_0),$$

where the symbol " $\bullet$ " denotes the dot (inner) product in $\mathbb{R}^n$, introduced in Definition 1.1, and:

$$g'(t_0) := (g_1'(t_0),\dots,g_n'(t_0)).$$

In order to extend the notion of gradient, we introduce the *Jacobian*[5] matrix associated to a vector–valued function.

Definition 3.11. Let $f : \mathbb{R}^n \to \mathbb{R}^m$ be a function from the Euclidean n–space to the Euclidean m–space. f has m real–valued component functions:

$$f_1(x_1,\dots,x_n),\ \dots,\ f_m(x_1,\dots,x_n).$$

If the partial derivatives of the component functions exist, they can be organized in an m–by–n matrix, namely the Jacobian matrix J of f :

$$J = \begin{bmatrix} \dfrac{\partial f_1}{\partial x_1} & \dfrac{\partial f_1}{\partial x_2} & \cdots & \dfrac{\partial f_1}{\partial x_n} \\ \vdots & \vdots & \ddots & \vdots \\ \dfrac{\partial f_m}{\partial x_1} & \dfrac{\partial f_m}{\partial x_2} & \cdots & \dfrac{\partial f_m}{\partial x_n} \end{bmatrix} := \dfrac{\partial(f_1,\dots f_m)}{\partial(x_1,\dots,x_n)}.$$

The i–th row of J corresponds to the gradient ∇f_i of the i–th component function f_i, for $i = 1,\dots,m$.

We introduce, now, the class of *positive homogeneous* functions.

Definition 3.12. A function $f : \mathbb{R}^n \setminus \{0\} \to \mathbb{R}$ is positive homogeneous, of degree k, if for any $x \in \mathbb{R}^n \setminus \{0\}$:

$$f(\alpha x) = \alpha^k f(x).$$

The following Theorem 3.13 is known as Euler theorem on homogeneous functions.

Theorem 3.13. If $f : \mathbb{R}^n \setminus \{0\} \to \mathbb{R}$ is continuously differentiable, then f is positive homogeneous, of degree k, if and only if:

$$x \bullet \nabla f(x) = k\, f(x).$$

---

[5] Carl Gustav Jacob Jacobi (1804–1851), German mathematician.

3.3   Maxima and Minima

Definition 3.14. Let V be an open set in $\mathbb{R}^n$, let $a \in V$ and suppose that $f : V \to \mathbb{R}$. Then:

(i) $f(a)$ is called a *local minimum* of f if and only if there exists $r > 0$ such that $f(a) \leq f(x)$ for all $x \in B_r(a)$, an open ball neighbourhood of a (recall Definition 1.13);

(ii) $f(a)$ is called a *local maximum* of f if and only if there exists $r > 0$ such that $f(a) \geq f(x)$ for all $x \in B_r(a)$;

(iii) $f(a)$ is called a *local extremum* of f if and only if $f(a)$ is a local maximum or a local minimum of f.

Remark 3.15. If the first–order partial derivatives of f exist at a, and if $f(a)$ is a local extremum of f, then $\nabla f(a) = \mathbf{0}$.

In fact, the one–dimensional function:

$$g(t) = f(a_1, \ldots, a_{j-1}, t, a_{j+1}, \ldots, a_n)$$

has a local extremum at $t = a_j$ for each $j = 1, \ldots, n$. Hence, by the one–dimensional theory:

$$\frac{\partial f}{\partial x_j}(a) = g'(a_j) = 0 .$$

As in the one–dimensional case, condition $\nabla f(a) = \mathbf{0}$ is necessary but not sufficient for $f(a)$ to be a local extremum.

Example 3.16. There exist continuously differentiable functions satisfying $\nabla f(a) = \mathbf{0}$ and such that $f(a)$ is neither a local maximum nor a local minimum.

Consider, for instance, in the case $n = 2$, the following function:

$$f(x, y) = y^2 - x^2 .$$

It is easy to check that $\nabla f(0) = \mathbf{0}$, but the origin is a *saddle point*, as shown in Figure 3.1.

Let us give a formal definition to such a situation.

Definition 3.17. Let V be open in $\mathbb{R}^n$, let $a \in V$, and let $f : V \to \mathbb{R}$ be differentiable at a.

Point a is called a *saddle point* of f if $\nabla f(a) = \mathbf{0}$ and there exists $r_0 > 0$ such that, given any $\rho \in (0, r_0)$, there exist points $x, y \in B_\rho(a)$ satisfying:

$$f(x) < f(a) < f(y) .$$

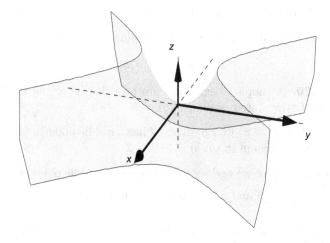

Figure 3.1: Saddle point of the function $z = y^2 - x^2$.

3.4 Sufficient conditions

To establish sufficient conditions for optimization, we introduce the notion of *Hessian*[6] matrix.

Definition 3.18. Let $V \subset \mathbb{R}^n$ be an open set and let $f : V \to \mathbb{R}$ be a $\mathcal{C}^2$ function. The Hessian matrix of f at $x \in V$ (or, simply, the Hessian) is the symmetric square matrix formed by the second–order partial derivatives of f, evaluated at point x :

$$H(f)(x) := \left[\frac{\partial^2 f}{\partial x_i \, \partial x_j}(x) \right] , \qquad \text{for} \quad i, j = 1, \dots, n .$$

Tests for extrema and saddle points, in the simplest situation of $n = 2$, are stated in Theorem 3.19.

Theorem 3.19. Let V be open in $\mathbb{R}^2$, consider $(a, b) \in V$, and suppose that $f : V \to \mathbb{R}$ satisfies $\nabla f(a, b) = \mathbf{0}$. Suppose further that $f \in \mathcal{C}^2$ and set:

$$D := f_{xx}(a, b) \, f_{yy}(a, b) - f_{xy}^2(a, b) .$$

(i) If $D > 0$ and $f_{xx}(a, b) > 0$, then $f(a, b)$ is a local minimum.

(ii) If $D > 0$ and $f_{xx}(a, b) < 0$, then $f(a, b)$ is a local maximum.

---

[6]Ludwig Otto Hesse (1811–1874), German mathematician.

(iii) If $D < 0$, then (a, b) is a saddle point.

Notice that D is the determinant of the Hessian of f evaluated at (a, b) :

$$D = \det[H(f)(a, b)].$$

Example 3.20. A couple of examples are provided here, and the Reader is invited to verify the stated results.

(1) Function $f(x, y) = x^3 + 6xy - 3y^2 + 2$ has a saddle point in $(a, b) = (0, 0)$ and a local maximum in $(a, b) = (-2, -2)$.

(2) Function $f(x, y) = x^2 + y^3 - 2xy - y$ admits a saddle point of coordinates $(a, b) = (-\frac{1}{3}, -\frac{1}{3})$ and a local minimum in $(a, b) = (1, 1)$.

Tests for extrema and saddle points, in the general situation of n variables, are stated in Theorem 3.21.

Theorem 3.21. In n variables, a *critical* point x_0 :

(i) is a local minimum for $f \in C^2$ if, for each $k = 1, \ldots, n$:

$$\det[H_k(f)(x_0)] > 0,$$

(ii) is a local maximum for $f \in C^2$ if, for each $k = 1, \ldots, n$:

$$\det[(-1)^k H_k(f)(x_0)] > 0,$$

where $H_k(f)$ denotes the *principal minor* of order k of $H(f)$.

3.5   Lagrange multipliers

In applications, it is often necessary to optimize functions under some constraints. The Lagrange multipliers theorem 3.22 provides necessary optimum conditions for a problem of the following kind:

$$\max f(x) \qquad \text{subject to} \quad g(x) = 0$$

or

$$\min f(x) \qquad \text{subject to} \quad g(x) = 0.$$

Theorem 3.22 (Lagrange multipliers – general case). Let $m < n$, let V be open in $\mathbb{R}^n$, and let $f, g_j : V \to \mathbb{R}$ be C^1 on V, for $j = 1, 2 \ldots, m$. Suppose that:

$$\frac{\partial(g_1, \ldots, g_m)}{\partial(x_1, \ldots, x_n)}$$

has rank m at $\boldsymbol{x}_0 \in V$, where $g_j(\boldsymbol{x}_0) = 0$ for $j = 1, 2, \ldots, m$. Assume further that $\boldsymbol{x}_0$ is a local extremum for f in the set:

$$M = \{\boldsymbol{x} \in V \mid g_j(\boldsymbol{x}) = 0\}.$$

Then, there exist scalars $\lambda_1, \ldots, \lambda_m$, such that:

$$\nabla \left(f(\boldsymbol{x}_0) - \sum_{k=1}^{m} \lambda_k \, g_k(\boldsymbol{x}_0) \right) = \boldsymbol{0}.$$

We will limit the proof of the Lagrange multipliers theorem 3.22 in a two–dimensional context. To this aim, it is first necessary to consider some preliminary results; we will resume the proof in §3.8.

3.6 Mean–Value theorem

We begin with recalling the definition of a segment in the Euclidean space.

Definition 3.23. Given $\boldsymbol{x}, \boldsymbol{y} \in \mathbb{R}^n$, the *segment* joining $\boldsymbol{x}$ and $\boldsymbol{y}$ is defined as:

$$[\boldsymbol{x}, \boldsymbol{y}] := \{\boldsymbol{z} \in \mathbb{R}^n \mid \boldsymbol{z} = t\boldsymbol{x} + (1 - t)\boldsymbol{y}, \quad 0 \le t \le 1\}.$$

The one–dimensional Mean–Value theorem (already met in Example 3.3), also called Lagrange Mean–Value theorem or First Mean–Value theorem, can be extended to the Euclidean space $\mathbb{R}^n$.

Theorem 3.24 (Mean–Value). Let $A \subset \mathbb{R}^n$, and $f : A \to \mathbb{R}$. Consider $\boldsymbol{x}, \boldsymbol{y} \in \mathbb{R}^n$ such that $[\boldsymbol{x}, \boldsymbol{y}] \subset A^\circ$, the interior of A (see Definition 1.18). Assume that $f(\boldsymbol{x})$ is differentiable in $[\boldsymbol{x}, \boldsymbol{y}]$. Then, there exists $\boldsymbol{z} \in [\boldsymbol{x}, \boldsymbol{y}]$ such that:

$$f(\boldsymbol{x}) - f(\boldsymbol{y}) = \nabla f(\boldsymbol{z}) \bullet (\boldsymbol{x} - \boldsymbol{y}).$$

Proof. Define $\varphi : [0, 1] \to \mathbb{R}^n$, $\varphi(t) = \boldsymbol{y} + t(\boldsymbol{x} - \boldsymbol{y})$. Observe that $\varphi \in \mathcal{C}$ and it is differentiable for any $t \in (0, 1)$. Moreover, $\varphi'(t) = \boldsymbol{x} - \boldsymbol{y}$. It follows that $g = f \circ \varphi : [0, 1] \to \mathbb{R}$ is continuous and differentiable in $(0, 1)$. We can thus apply the one–dimensional version of the Mean–Value theorem, to infer the existence of $\eta \in (0, 1)$ such that:

$$f(\boldsymbol{x}) - f(\boldsymbol{y}) = g(1) - g(0) = g'(\eta).$$

On the other hand, the Chain Rule implies:

$$g'(\eta) = \nabla f(\varphi(\eta)) \bullet \varphi'(\eta) = \nabla f(\varphi(\eta)) \bullet (\boldsymbol{x} - \boldsymbol{y}).$$

Since $\boldsymbol{z} = \varphi(\eta) \in [\boldsymbol{x}, \boldsymbol{y}]$, Theorem 3.24 is proved. □

3.7 Implicit function theorem

The fundamental step that follows is represented by the *Implicit function theorem*, proved by Dini in 1878. For simplicity, we provide its proof only in the $\mathbb{R}^2$ case, presented in Theorem 3.25, but its generalisation to $\mathbb{R}^n$ is quite straightforward, and we state it in Theorem 3.26.

Theorem 3.25 (Implicit function – case $n = 2$). Let Ω be an open set in $\mathbb{R}^2$, and let $f : \Omega \to \mathbb{R}$ be a $\mathcal{C}^1$ function. Suppose there exists $(x_0, y_0) \in \Omega$ such that $f(x_0, y_0) = 0$ and $f_y(x_0, y_0) \neq 0$.
Then, there exist $\delta, \varepsilon > 0$ such that, for any $x \in (x_0 - \delta, x_0 + \delta)$ there exists a unique $y = \varphi(x) \in (y_0 - \varepsilon, y_0 + \varepsilon)$ such that:

$$f(x, y) = 0.$$

Moreover, function $y = \varphi(x)$ is $\mathcal{C}^1$ in $(x_0 - \delta, x_0 + \delta)$ and it holds that, for any $x \in (x_0 - \delta, x_0 + \delta)$:

$$\varphi'(x) = -\frac{f_x(x, \varphi(x))}{f_y(x, \varphi(x))} \, .$$

Proof. Let us assume that $f(x_0, y_0) > 0$. Since function $f_y(x, y)$ is continuos, it is possibile to find a ball $B_{\delta_1}(x_0, y_0)$ in which it is verified that $(x, y) \in B_{\delta_1}(x_0, y_0) \implies f_y(x, y) > 0$.
This means that, with an appropriate narrowing of parameters ε and δ, function $y \mapsto f(x, y)$ can be assumed to be an increasing function, for any $x \in (x_0 - \delta, x_0 + \delta)$.
In particular, $y \mapsto f(x_0, y)$ is increasing and, since $f(x_0, y_0) = 0$ by assumption, the following disequalities are verified, for ε small enough:

$$f(x_0, y_0 + \varepsilon) > 0 \qquad \text{and} \qquad f(x_0, y_0 - \varepsilon) < 0.$$

Using, again, continuity of f and an appropriate narrowing of δ, we infer that, for any $x \in (x_0 - \delta, x_0 + \delta)$:

$$f(x, y_0 + \varepsilon) > 0 \qquad \text{and} \qquad f(x, y_0 - \varepsilon) < 0.$$

In conclusion, using continuity of $y \mapsto f(x, y)$ and the Bolzano theorem[7] on the existence of zeros, we have shown that, for any $x \in (x_0 - \delta, x_0 + \delta)$, there is a unique $y = \varphi(x) \in (y_0 - \varepsilon, y_0 + \varepsilon)$ such that:

$$f(x, y) = f(x, \varphi(x)) = 0 \, .$$

To prove the second part of Theorem 3.25, we need to show that $\varphi(x)$ is

---

[7] Bernard Placidus Johann Nepomuk Bolzano (1781–1848), Czech mathematician, theologian and philosopher. For the theorem of Bolzano see, for example, mathworld.wolfram.com/BolzanoTheorem.html

differentiable. To this aim, consider $h \in \mathbb{R}$ such that $x + h \in (x_0 - \delta, x_0 + \delta)$. In this way, from the Mean–Value Theorem 3.24, there exist $\theta \in (0, 1)$ such that:

$$
\begin{aligned}
0 &= f\big(x + h, \varphi(x + h)\big) - f\big(x, \varphi(x)\big) \\
&= f_x\Big(x + \theta\, h\, \varphi(x) + \theta\, \big(\varphi(x + h) - \varphi(x)\big)\Big)\, h + \\
&\quad + f_y\Big(x + \theta\, h\, \varphi(x) + \theta\, \big(\varphi(x + h) - \varphi(x)\big)\Big)\big(\varphi(x + h) - \varphi(x)\big),
\end{aligned}
$$

thus:

$$
\frac{\varphi(x + h) - \varphi(x)}{h} = -\frac{f_x\Big(x + \theta\, h\, \varphi(x) + \theta\, \big(\varphi(x + h) - \varphi(x)\big)\Big)}{f_y\Big(x + \theta\, h\, \varphi(x) + \theta\, \big(\varphi(x + h) - \varphi(x)\big)\Big)}.
$$

The thesis follows by taking, in the equality above, the limit for $h \to 0$, observing that $h \to 0 \implies \theta \to 0$, and recalling that $f(x, y)$ is $\mathcal{C}^1$. □

We are now ready to state the Implicit function Theorem 3.26 in the general n–dimensional case; here, Ω is an open set in $\mathbb{R}^n \times \mathbb{R}$, thus $(x, y) \in \Omega$ means that $x \in \mathbb{R}^n$ and $y \in \mathbb{R}$.

Theorem 3.26 (Implicit function – general case). Let $\Omega \subset \mathbb{R}^n \times \mathbb{R}$ be open, and let $f \in \mathcal{C}^1(\Omega, \mathbb{R})$. Assume that there exists $(x_0, y_0) \in \Omega$ such that $f(x_0, y_0) = 0$ and $f_y(x_0, y_0) \neq 0$.
Then, there exist an open ball $B_\delta(x_0)$, an open interval $(y_0 - \varepsilon, y_0 + \varepsilon)$ and a function $\varphi : (y_0 - \varepsilon, y_0 + \varepsilon) \to \mathbb{R}$, such that:

(i) $B_\delta(x_0) \times (y_0 - \varepsilon, y_0 + \varepsilon) \subset \Omega$;

(ii) $(x, y) \in B_\delta(x_0) \times (y_0 - \varepsilon, y_0 + \varepsilon) \implies f_y(x, y) \neq 0$;

(iii) for any $(x, y) \in B_\delta(x_0) \times (y_0 - \varepsilon, y_0 + \varepsilon)$ it holds:

$$
f(x, y) = 0 \iff y = \varphi(x) ;
$$

(iv) $\varphi \in \mathcal{C}^1\big(B_\delta(x_0)\big)$ and

$$
\varphi_{x_j}(x) = -\frac{f_{x_j}\big(x, \varphi(x)\big)}{f_y\big(x, \varphi(x)\big)}.
$$

3.8   Proof of Theorem 3.22

We can now prove the multipliers Theorem 3.22; as said before, the proof is given only for the $n = 2$ case, presented in Theorem 3.27.

Theorem 3.27 (Lagrange multipliers – case $n = 2$). Let $A \subset \mathbb{R}^2$ be open, and let $f, g : A \to \mathbb{R}$ be C^1 functions. Consider the subset of A :

$$M = \{(x, y) \in A \mid g(x, y) = 0\} .$$

Assume that $\nabla g(x, y) \neq 0$ for any $(x, y) \in M$. Assume further that $(x_0, y_0) \in M$ is a maximum or a minimum of $f(x, y)$ for any $(x, y) \in M$. Then, there exists $\lambda \in \mathbb{R}$ such that:

$$\nabla f(x_0, y_0) = \lambda \nabla g(x_0, y_0) .$$

Proof. Since $\nabla g(x_0, y_0) \neq 0$, we can assume that $g_y(x_0, y_0) \neq 0$. Thus, from the Implicit function Theorem 3.25, there exist $\varepsilon, \delta > 0$ such that, for $x \in (x_0 - \delta, x_0 + \delta)$, $y \in (y_0 - \varepsilon, y_0 + \varepsilon)$, it holds:

$$g(x, y) = g(x, \varphi(x)) = 0 .$$

Consider the function $x \mapsto f(x, \varphi(x)) := h(x)$, for $x \in (x_0 - \delta, x_0 + \delta)$. By assumption, $h(x)$ admits an extremum in $x = x_0$, therefore its derivative in x_0 vanishes. Using the Chain Rule, it follows:

$$0 = h'(x_0) = f_x(x_0, \varphi(x_0)) + f_y(x_0, \varphi(x_0)) \varphi'(x_0) . \tag{3.2}$$

Again, we use the Implicit function Theorem 3.25, which gives:

$$\varphi'(x_0) = -\frac{g_x(x_0, \varphi(x_0))}{g_y(x_0, \varphi(x_0))} .$$

Substituting into (3.2), recalling that $\varphi(x_0) = y_0$, we get:

$$f_x(x_0, y_0) \, g_y(x_0, y_0) - f_y(x_0, y_0) \, g_x(x_0, y_0) = 0 ,$$

which can be rewritten as:

$$\det \begin{vmatrix} f_x(x_0, y_0) & f_y(x_0, y_0) \\ g_x(x_0, y_0) & g_y(x_0, y_0) \end{vmatrix} = 0 .$$

Since the above determinant is zero, it follows that its rows are proportional, implying that there exists $\lambda \in \mathbb{R}$ such that:

$$(f_x(x_0, y_0), f_y(x_0, y_0)) = \lambda (g_x(x_0, y_0), g_y(x_0, y_0)) .$$

$\square$

3.9   Sufficient conditions

The multiplier Theorem 3.22 expresses necessary conditions for the existence of an optimal solution. Stating sufficient conditions is also important; to such an aim, in the two–dimensional case, the main tool is the so–called *Bordered Hessian*.

Suppose we are dealing with the simplest case of constrained optimization, that is, find the maximum value (max) or the minimum value (min) of $f(x,y)$ under the constraint $g(x,y) = 0$. We form the Lagrangian functional $L(x,y,\lambda) = f(x,y) - \lambda g(x,y)$ and, after solving the critical point system:

$$\begin{cases} f'_x(x,y) - \lambda g'_x(x,y) = 0 , \\ f'_y(x,y) - \lambda g'_y(x,y) = 0 , \\ g(x,y) = 0 , \end{cases}$$

we evaluate:

$$\Lambda = \det \begin{bmatrix} L''_{xx} & L''_{xy} & g_x \\ L''_{xy} & L''_{yy} & g_y \\ g_x & g_y & 0 \end{bmatrix} .$$

Then:

(a) $\Lambda > 0$ indicates a maximum value;

(b) $\Lambda < 0$ indicates a minimum value.

Example 3.28. An example of interest in Economics concerns the maximization of a production function of Cobb–Douglas[8] kind. The mathematical problem can be modelled as:

$$\begin{aligned} \max \ & f(x,y) = x^a \, y^{1-a} \\ \text{subject to} \ & px + qy - c = 0 \end{aligned} \tag{3.3}$$

where $0 < a < 1$, and $p, q, c > 0$.

In a problem like (3.3), $f(x,y)$ is referred to as *objective function*, while , by defining the function $w(x,y) = px + qy - c$, the *constraint* is given by $w(x,y) = 0$.

The Lagrangian is $L(x,y;m) = f(x,y) - m \, w(x,y)$. The critical point equations are:

$$\begin{cases} L_x(x,y;m) = a \, x^{a-1} \, y^{1-a} - mp = 0 , \\ L_y(x,y;m) = (1-a) \, x^a \, y^{-a} - mq = 0 , \\ L_m(x,y;m) = px + qy - c = 0 . \end{cases}$$

Eliminating m from the first two equations, by subtraction, we obtain the two–by–two linear system in the variables x, y :

$$\begin{cases} (1-a) \, px - a \, qy = 0 , \\ px + qy - c = 0 . \end{cases}$$

---

[8]Charles Wiggins Cobb (1875–1949), American mathematician and economist.
Paul Howard Douglas (1892–1976), American politician and Georgist economist.

Solving the 2×2 system and recovering m, from $m = (a\, x^{a-1}\, y^{1-a})\, p^{-1}$, we find the critical point:

$$\begin{cases} x = \dfrac{a\,c}{p} \\[2mm] y = \dfrac{c\,(1-a)}{q} \\[2mm] m = (1-a)^{1-a}\, a^a\, p^{-a}\, q^{a-1} \end{cases}$$

which is a maximum; the Bordered Hessian is, in fact:

$$\begin{bmatrix} (a-1)\,a\left(\dfrac{a\,c}{p}\right)^{a-2}\left(\dfrac{c\,(1-a)}{q}\right)^{1-a} & (1-a)\,a\left(\dfrac{a\,c}{p}\right)^{a-1}\left(\dfrac{c\,(1-a)}{q}\right)^{-a} & p \\[4mm] (1-a)\,a\left(\dfrac{a\,c}{p}\right)^{a-1}\left(\dfrac{c\,(1-a)}{q}\right)^{-a} & (a-1)\,a\left(\dfrac{a\,c}{p}\right)^{a}\left(\dfrac{c\,(1-a)}{q}\right)^{-1-a} & q \\[4mm] p & q & 0 \end{bmatrix}$$

and its determinant is positive:

$$\det = \frac{a^{a-1}\, p^{2-a}\, q^{a+1}}{c\,(1-a)^a} > 0 .$$

Example 3.29. In this example, we reverse the point of view between constraint and objective function in a problem like (3.3). Here, the idea is to minimise the *total cost*, fixing the *level of production*. For the sake of simplicity, we treat the particular two–dimensional problem of finding maxima and minima of $f(x,y) = 2\,x + y$, subject to the constraint $\sqrt[4]{x}\,\sqrt[4]{y^3} = 1$, $x > 0$, $y > 0$. The critical point equations are:

$$\begin{cases} 2 - \dfrac{m\,\sqrt[4]{y^3}}{4\,\sqrt[4]{x^3}} = 0 , \\[3mm] 1 - \dfrac{3\,m\,\sqrt[4]{x}}{4\,\sqrt[4]{y}} = 0 , \\[3mm] \sqrt[4]{x}\,\sqrt[4]{y^3} = 1 . \end{cases}$$

Eliminating m from the first two equations, by substitution, we obtain the two–by–two linear system in the variables x, y :

$$\begin{cases} \dfrac{y}{x} = 6 , \\[2mm] \sqrt[4]{x}\,\sqrt[4]{y^3} = 1 . \end{cases}$$

Solving the 2×2 system and recovering m from $m = \dfrac{4\,\sqrt[4]{y}}{3\,\sqrt[4]{x}}$, the critical point is found:

$$x = \sqrt[4]{6^{-3}} , \qquad y = \sqrt[4]{6} , \qquad m = \frac{4\,\sqrt[4]{2}}{\sqrt[4]{3^3}} .$$

The Bordered Hessian is:

$$\Lambda = \begin{bmatrix} \dfrac{3\,m\,\sqrt[4]{y^3}}{16\,x\,\sqrt[4]{x^3}} & -\dfrac{3\,m}{16\,\sqrt[4]{x^3}\,y} & \dfrac{\sqrt[4]{y^3}}{4\,\sqrt[4]{x^3}} \\[2ex] -\dfrac{3\,m}{16\,\sqrt[4]{x^3}\,y} & \dfrac{3\,m\,\sqrt[4]{x}}{16\,y\,\sqrt[4]{y}} & \dfrac{3\,\sqrt[4]{x}}{4\,\sqrt[4]{y}} \\[2ex] \dfrac{\sqrt[4]{y^3}}{4\,\sqrt[4]{x^3}} & \dfrac{3\,\sqrt[4]{x}}{4\,\sqrt[4]{y}} & 0 \end{bmatrix}.$$

Evaluating Λ at the critical point and computing its determinant:

$$\det \Lambda = -\frac{3\sqrt[4]{3}}{\sqrt[4]{2^3}},$$

we see that we found a minimum.

Example 3.30. The problem presented here is typical in the determination of an optimal investment portfolio in Corporate Finance.
We seek to minimise $f(x, y) = x^2 + 2\,y^2 + 3\,z^2 + 2\,xz + 2\,y\,z$, with the constraints:

$$x + y + z = 1, \qquad 2\,x + y + 3\,z = 7.$$

The Lagrangian is:

$$L(x\,,y\,,z\,;m\,,n) = x^2 + 2\,y^2 + 3\,z^2 + 2\,x\,z + 2\,y\,z - m\,(x+y+z-1) - n\,(2\,x+y+3\,z-7),$$

hence the optimality conditions are:

$$\begin{cases} 2\,x + 2\,z = m + 2\,n\,, \\ 4\,y + 2\,z = m + n\,, \\ 2\,x + 2\,y + 6\,z = m + 3\,n\,, \\ x + y + z = 1\,, \\ 2\,x + y + 3\,z = 7\,. \end{cases}$$

The solution to this 5×5 linear system is

$$x = 0, \quad y = -2, \quad z = 3, \quad m = -10, \quad n = 8.$$

The convexity of the objective function ensures that the found solution is the absolute minimum. Though this statement should be proved rigorously, we do not treat it here.

Chapter 4

Ordinary differential equations of first order: general theory

Our goal, in introducing ordinary differential equations, is to provide a brief account on methods of explicit integration, for the most common types of ordinary differential equations. However, it is not taken for granted the main theoretical problem, concerning existence and uniqueness of the solution of the *Initial Value Problem*, modelled by (4.3). Indeed, the proof of the Picard–Lindelöhf Theorem 4.17 is presented in detail: to do this, we will use some notions from the theory of uniform convergence of sequences of functions, already discussed in Theorem 2.15. An abstract approach followed, for instance, in Chapter 2 of [60], is avoided here.

In Chapter 5, that follows, we present some classes of ordinary differential equations for which, using suitable techniques, the solution can be described in terms of known functions: in this case, we say that we are able to find an *exact solution* of the given ordinary differential equation.

4.1 Preliminary notions

Let x be an independent variable, moving on the real axis, and let y be a dependent variable, that is $y = y(x)$. Let further $y', y'', \ldots, y^{(n)}$ represent successive derivatives of y with respect to x. An *ordinary differential equation* (ODE) is any relation of equality involving at least one of those derivatives and the function itself. For instance, the equation below:

$$\frac{\mathrm{d}y}{\mathrm{d}x}(x) := y'(x) = 2\,x\,y(x) \tag{4.1}$$

states that the first derivative of the function y equals the multiplication of $2\,x$ and y. An additional, implicit statement is that (4.1) holds only for all those x for which both the function and its first derivative are defined.

The term *ordinary* distinguishes this kind of equation from a *partial* differential equation, which would involve two or more independent variables, a dependent variable and the corresponding partial derivatives, i.e., for example:

$$\frac{\partial f(x,y)}{\partial x} + 4\,x\,y\,\frac{\partial f(x,y)}{\partial y} = x + y\,.$$

We will present partial differential equations, used in Quantitative Finance, in Chapter 13.

The general ordinary differential equation of first order has the form:

$$F(x, y, y') = 0\,. \tag{4.2}$$

A function $y = y(x)$ is called a *solution* of (4.2), on an interval J, if $y(x)$ is differentiable on J and if the following equality holds for all $x \in J$:

$$F(x, y(x), y'(x)) \equiv 0\,.$$

In general, we would like to know whether, under certain circumstances, a differential equation has a unique solution. To accomplish this property, it is usual to consider the so–called *Initial Value Problem* (or *IVP*) which, in the simplest scalar case, takes the form presented in Definition 4.1.

Definition 4.1. Given $f : \Omega \subset \mathbb{R}^2 \to \mathbb{R}$, being Ω an open set, the initial value problem (also called *Cauchy problem*) takes the form:

$$\begin{cases} y' = f(x, y)\,, & x \in I\,, \\ y(x_0) = y_0\,, & x_0 \in I,\ y_0 \in J\,, \end{cases} \tag{4.3}$$

where I, J are intervals such that $I \times J \subset \Omega$, and where we have simply denoted y in place of $y(x)$.

Remark 4.2. We say that differential equations are studied by *quantitative* or *exact* methods when they can be solved completely, that is to say, all their solutions are known and could be written in *closed form*, in terms of elementary functions or, at times, in terms of special functions (or in terms of inverses of elementary and special functions).

We now provide some examples of ordinary differential equations.

Example 4.3. Let us consider the differential equation:

$$y' = \frac{1}{x^2}\,, \tag{4.4}$$

If we rewrite equation (4.4) as:

$$\frac{\mathrm{d}}{\mathrm{d}x}\left(y(x) + \frac{1}{x}\right) = 0\,,$$

we see that we are dealing with a function whose derivative is zero. If we seek solutions defined on an interval, then, we can exploit a consequence of the Mean–Value Theorem 3.24 (namely, a function that is continuous and differentiable on $[a,b]$ and has null first–derivative on (a,b), is constant on (a,b)), to see that:

$$y(x) + \frac{1}{x} = C,$$

for some constant C and for all $x \in I$, where I is an interval not containing zero. In other words, as long as we consider the domain of solutions to be an interval like I, any solution of the differential equation (4.4) takes the form:

$$y(x) = C - \frac{1}{x}, \qquad \text{for } x \in I.$$

By choosing an *initial condition*, for example $y(1) = 5$, a particular value $C = 6$ is determined, so that:

$$y(x) = 6 - \frac{1}{x}, \qquad \text{for } x \in I.$$

We can also follow a reverse approach, in the sense that, as illustrated in Example 4.4, given a geometrical locus, we obtain its ordinary differential equation.

Example 4.4. Consider the family of parabolas of equation:

$$y = \alpha x^2. \tag{4.5}$$

Any parabola in the family has the y-axis as common axis, with vertex in the origin. Differentiating, we get:

$$y' = 2\alpha x. \tag{4.6}$$

Eliminating α from (4.5) and (4.6), we obtain the differential equation:

$$y' = \frac{2y}{x}. \tag{4.7}$$

This means that any parabola in the family is solution to the differential equation (4.7).

4.1.1 Systems of ODEs: equations of high order

It is possible to consider differential equations of order higher than one, or systems of many differential equations of first order.

Example 4.5. The following ordinary differential equations are, respectively, of order 2 and of order 3 :

$$x\,y'' + 2\,y' + 3\,y - e^x = 0,$$

$$(y^{(3)})^2 + y'' + y = x \,.$$

The second equation is quadratic in the highest derivative $y^{(3)}$, therefore we say, also, that it has **degree** 2.

A system of first–order differential equations is, for example, the following one:

$$\begin{cases} y_1' = y_1 \, (a - b \, y_2) \,, \\ y_2' = y_2 \, (c \, y_1 - d) \,, \end{cases} \tag{4.8}$$

in which $y_1 = y_1(x)$ and $y_2 = y_2(x)$ are functions of a variable x that, in most applications, takes the meaning of time. System (4.8) of ordinary differential equations is very famous, as it represents the Lotka–Volterra[1] predator–prey system; see, for instance, [23]. Notice that the left–hand sides in (4.8) are not dependent on x : in this particular case, the system is called *autonomous*.

We now state, formally, the definition of initial value problem for a system of n ordinary differential equations, each of first order, and for a differential equation of order n, with integer $n \geq 1$ in both cases.

Definition 4.6. Consider Ω, open set in $\mathbb{R} \times \mathbb{R}^n$, with integer $n \geq 1$, and let $\boldsymbol{f} : \Omega \to \mathbb{R}^n$ be a vector–valued continuous function of $(n+1)$–variables. Let further $(x_0, \boldsymbol{y}) \in \Omega$ and I be an open interval such that $x_0 \in I$. Then, a vector–valued function $\boldsymbol{s} : I \to \mathbb{R}^n$ is a solution of the initial value problem:

$$\begin{cases} \boldsymbol{y}' = \boldsymbol{f}(x, \boldsymbol{y}) \\ \boldsymbol{y}(x_0) = \boldsymbol{y}_0 \end{cases} \tag{4.9}$$

if the following conditions are verified:

(i) $\boldsymbol{s} \in C^1(I)$;   (iii) $\boldsymbol{s}(x_0) = \boldsymbol{y}_0$;

(ii) $(x, \boldsymbol{s}(x)) \in \Omega$ for any $x \in I$;   (iv) $\boldsymbol{s}'(x) = \boldsymbol{f}(x, \boldsymbol{s}(x))$.

Remark 4.7. In the Lotka–Volterra case (4.8), it is $n = 2$, thus $\boldsymbol{y} = (y_1, y_2)$, the open set is $\Omega = \mathbb{R} \times ((0, +\infty) \times (0, +\infty))$ and the continuous function is $\boldsymbol{f}(x, \boldsymbol{y}) = \boldsymbol{f}(x, y_1, y_2) = (y_1 (a - b \, y_2), y_2 (c \, y_1 - d))$.

The rigorous definition of initial value problem for a differential equation of order n is provided below.

Definition 4.8. Consider an open set $\Omega \subset \mathbb{R} \times \mathbb{R}^n$, where $n \geq 1$ is integer. Let $F : \Omega \to \mathbb{R}$ be a scalar continuous function of $(n+1)$–variables. Let further $(x_0, \boldsymbol{b}) \in \Omega$ and I be an open interval such that $x_0 \in I$. Finally, denote $\boldsymbol{b} = (b_1, \ldots, b_n)$.

---

[1]Vito Volterra (1860–1940), Italian mathematician and physicist.
Alfred James Lotka (1880–1949), American mathematician, physical chemist, statistician.

Then, a real function $s : I \to \mathbb{R}$ is a solution of the initial value problem:

$$\begin{cases} y^{(n)} = F(x, y, y', y'', \cdots, y^{(n-1)}) \\ y(x_0) = b_1 \\ y'(x_0) = b_2 \\ \quad \cdots \\ y^{(n-1)}(x_0) = b_n \end{cases} \tag{4.10}$$

if:

(i) $s \in C^n(I)$;

(ii) $\big(x, s(x), s'(x), \ldots, s^{(n-1)}(x)\big) \in \Omega$ for any $x \in I$;

(iii) $s^{(j)}(x_0) = b_{j+1}$, $j = 0, 1, \ldots, n-1$;

(iv) $s^{(n)}(x) = F\big(x, s(x), s'(x), \cdots, s^{(n-1)}(x)\big)$.

Definition 4.9. Consider a family of functions $y(x; c_1, \ldots, c_n)$, depending on x and on n parameters $c_1, \ldots, c_n$, which vary within a set $M \subset \mathbb{R}^n$. Such a family is called a *complete integral*, or a *general solution*, of the n-th order equation:

$$y^{(n)} = F(x, y, y', y'', \cdots, y^{(n-1)}), \tag{4.11}$$

if it satisfies two requirements:

(1) each function $y(x; c_1, \ldots, c_n)$ is a solution to (4.11)

(2) all solutions to (4.11) can be expressed as functions of the family itself, i.e., they take the form $y(x; c_1, \ldots, c_n)$.

Remark 4.10. Systems of first–order differential equations like (4.9) and equations of order n like (4.10) are intimately related. Given the n-th order equation (4.10), in fact, an equivalent system can be built, that has form (4.9), by introducing a new vector variable $z = (z_1, \ldots, z_n)$ and considering the system of differential equations:

$$\begin{cases} z_1' = z_2 \\ z_2' = z_3 \\ \quad \cdots \\ z_{n-1}' = z_n \\ z_n' = F(x, z_1, z_2, \ldots, z_n) \end{cases} \tag{4.12}$$

with the set of initial conditions:

$$\begin{cases} z_1(x_0) = b_1, \\ \quad \cdots \\ z_n(x_0) = b_n. \end{cases} \tag{4.13}$$

System (4.12) can be represented in the vectorial form (4.9), simply by setting $z' = (z'_1, \ldots, z'_n)$, $b = (b_1, \ldots, b_n)$ and:

$$f(x, z) = \begin{pmatrix} z_2 \\ z_3 \\ \cdots \\ z_n \\ F(x, z_1, z_2, \ldots, z_n) \end{pmatrix}.$$

Form Remark 4.10, the following Theorem 4.11 can be inferred, whose straightforward proof is omitted.

Theorem 4.11. Function s is solution of the n–th order initial value problem (4.10) if and only if the vector function z solves system (4.12), with the initial conditions (4.13).

Remark 4.12. It is also possible to go in the reverse way, that is to say, any system of n differential equations, of first order, can be transformed into a scalar differential equation of order n. We illustrate this procedure with the Lotka-Volterra system (4.8). The first step consists in computing the second derivative, with respect to x, of the first equation in (4.8):

$$y'_1 = y_1 (a - b y_2) \quad \Longrightarrow \quad y''_1 = y'_1 (a - b y_2) - b y_1 y'_2 . \tag{4.8a}$$

Then, the values of y'_1 and y'_2 from (4.8) are inserted in (4.8a), yielding:

$$y''_1 = y_1 \left((a - b y_2)^2 + b y_2 (d - c y_1) \right) . \tag{4.8b}$$

Thirdly, using again the first equation in (4.8), y_2 is expressed in terms of y_1 and y'_1, namely:

$$y'_1 = y_1 (a - b y_2) \quad \Longrightarrow \quad y_2 = \frac{a y_1 - y'_1}{b y_1} . \tag{4.8c}$$

Finally, (4.8c) is inserted into (4.8b), which provides the second–order differential equation for y_1 :

$$y''_1 = (a y_1 - y'_1)(d - c y_1) + \frac{y'^2_1}{y_1} . \tag{4.8d}$$

4.2   Existence of solutions: Peano theorem

In this section, we briefly deal with the problem of the existence of solutions for ordinary differential equations, for which continuity is the only essential hypothesis. The Peano[2] Theorem 4.14 on existence is stated, but not demonstated; the interested Reader is referred to Chapter 2 of [29].

We first state Peano theorem in the scalar case.

Theorem 4.13. Consider the rectangle $\mathcal{R} = [x_0 - a, x_0 + a] \times [y_0 - b, y_0 + b]$, and let $f : \mathcal{R} \to \mathbb{R}$ be continuous. Then, the initial value problem (4.3) admits at least a solution in a neighbourhood of x_0.

To extend Theorem 4.13 to systems of ordinary differential equations, the rectangle $\mathcal{R}$ is replaced by a parallelepiped, obtained as the Cartesian product of a real interval with an n–dimensional closed ball.

Theorem 4.14 (Peano). Let us consider the $n+1$–dimensional parallelepiped $\mathcal{P} = [x_0 - a, x_0 + a] \times \overline{B}(y_0, r)$, and let $f : \mathcal{P} \to \mathbb{R}^n$ be a continuous function. Then, the initial value problem (4.9) admits at least a solution in a neighbourhood of x_0.

Remark 4.15. Under the sole continuity assumption, a solution needs not to be unique. Consider, for example, the initial value problem:

$$\begin{cases} y'(x) = 2\sqrt{|y(x)|}, \\ y(0) = 0. \end{cases} \tag{4.14}$$

The zero function $y(x) = 0$ is a solution of (4.14), which is solved, though, by function $y(x) = x|x|$ as well. Moreover, for each pair of real numbers $\alpha < 0 < \beta$, the following $\varphi_{\alpha,\beta}(x)$ function solves (4.14) too:

$$\varphi_{\alpha,\beta}(x) = \begin{cases} -(x - \alpha)^2 & \text{if} \quad x < \alpha, \\ 0 & \text{if} \quad \alpha \leq x \leq \beta, \\ (x - \beta)^2 & \text{if} \quad x > \beta. \end{cases}$$

In other words, the considered initial value problem admits infinite solutions. This phenomenon is known as *Peano funnel*.

4.3 Existence and uniqueness: Picard–Lindelöhf theorem

To ensure existence and uniqueness of the solution to the initial value problem (4.9), a more restrictive condition than continuity needs to be considered and is presented in Theorem 4.17. Given the importance of such a theorem, we provide here its proof, though in the scalar case only; notice that the proof is

---

[2]Giuseppe Peano (1858–1932), Italian mathematician and glottologist.

constructive and turns out useful when trying to evaluate the solution of the given ordinary differential equation.

The key notion to be introduced is *Lipschitz continuity*, which may be considered as a kind of intermediate property, between continuity and differentiability.

For simplicity, we work in a scalar situation; the extension to systems of differential equations is only technical; some details are provided in § 4.3.2.

We use again $\mathcal{R}$ to denote the rectangle:

$$\mathcal{R} = [x_0, x_0 + a] \times [y_0 - b, y_0 + b].$$

Definition 4.16. Function $f : \mathcal{R} \to \mathbb{R}$ is called uniformly Lipschitz continuous in y, with respect to x, if there exists $L > 0$ such that:

$$|f(x, y_1) - f(x, y_2)| < L|y_1 - y_2|, \quad \text{for any } (x, y_1), (x, y_2) \in \mathcal{R}. \quad (4.15)$$

Using the Lipschitz[3] continuity property, we prove the Picard–Lindelöhf[4] theorem.

Theorem 4.17 (Picard–Lindelöhf). Let $f : \mathcal{R} \to \mathbb{R}$ be uniformly Lipschitz continuous in y, with respect to x, and define:

$$M = \max_{\mathcal{R}} |f|, \qquad \alpha = \min\left\{a, \frac{b}{M}\right\}. \qquad (4.16)$$

Then, problem (4.3) admits unique solution $u \in \mathcal{C}^1([x_0, x_0 + \alpha], \mathbb{R})$.

Proof. The proof is somewhat long, so we present it splitted into four steps.
First step. Let $n \in \mathbb{N}$. Define the sequence of functions (u_n) by recurrence:

$$\begin{cases} u_0(x) = y_0, \\ u_{n+1}(x) = y_0 + \int_{x_0}^{x} f(X, u_n(X)) \, \mathrm{d}X. \end{cases}$$

We want to show that $(x, u_n(x)) \in \mathcal{R}$ for any $x \in [x_0, x_0 + \alpha]$. To this aim, it is enough to prove that, for $n \geq 0$, the following inequality is verified:

$$|u_n(x) - y_0| \leq b, \qquad \text{for any } x \in [x_0, x_0 + \alpha]. \qquad (4.17)$$

In the particular case $n = 0$, inequality (4.17) is satisfied, since:

$$|u_0(x) - y_0| = |y_0 - y_0| = 0 \leq b.$$

---

[3]Rudolf Otto Sigismund Lipschitz (1832–1903), German mathematician.
[4]Charles Émile Picard (1856–1941), French mathematician.
Ernst Leonard Lindelöf (1870–1946), Finnish mathematician.

In the general case, we have:

$$|u_{n+1}(x) - y_0| \leq \left| \int_{x_0}^x f(X, u_n(X)) \, dX \right|$$

$$\leq \int_{x_0}^x |f(X, u_n(X))| \, dX \leq M |x - x_0| \leq M \alpha \leq b \,.$$

It is precisely here that we can understand the reason of the peculiar definition (4.16) of the number α, as such a choice turns out appropriate in *correctly* defining each (and any) term u_n in the sequence. It also highlights the local nature of the solution of the initial value problem (4.3).

Second step. We now show that (u_n) converges uniformly on $[x_0, x_0 + \alpha]$. The identity:

$$u_n = u_0 + (u_1 - u_0) + \cdots + (u_n - u_{n-1}) = u_0 + \sum_{k=1}^n (u_k - u_{k-1})$$

suggests that any sequence (u_n) can be thought of as an infinite series: its uniform convergence, thus, can be proved by showing that the following series (4.18) converges totally on $[x_0, x_0 + \alpha]$:

$$\sum_{k=1}^\infty (u_k - u_{k-1}) \,. \tag{4.18}$$

To prove total convergence, we need to prove, for $n \in \mathbb{N}$, the following bound:

$$|u_n(x) - u_{n-1}(x)| \leq M \frac{L^{n-1} |x - x_0|^n}{n!}, \qquad \text{for any } x \in [x_0, x_0 + \alpha]. \tag{4.19}$$

We proceed by induction. For $n = 1$, the bound is verified, since:

$$|u_1(x) - u_0(x)| = |u_1(x) - y_0|$$

$$= \left| \int_{x_0}^x f(X, y_0) \, dX \right| \leq \int_{x_0}^x |f(X, y_0)| \, dX \leq M |x - x_0| \,.$$

We now prove (4.19) for $n + 1$, assuming that it holds true for n. Indeed:

$$|u_{n+1}(x) - u_n(x)| = \left| \int_{x_0}^x \Big(f(X, u_n(X)) - f(X, u_{n-1}(X)) \Big) \, dX \right|$$

$$\leq \int_{x_0}^x |f(X, u_n(X)) - f(X, u_{n-1}(X))| \, dX$$

$$\leq L \int_{x_0}^x |u_n(X) - u_{n-1}(X)| \, dX$$

$$\leq M \frac{L^{n-1}}{n!} \int_{x_0}^x |X - x_0|^n \, dX = M \frac{L^n |x - x_0|^{n+1}}{(n+1)!} \,.$$

Therefore (4.19) is proved and implies that series (4.18) is totally convergent for $[x_0, x_0 + \alpha]$; in fact:

$$\sum_{n=1}^{\infty} (u_n - u_{n-1}) \leq \sum_{n=1}^{\infty} \sup_{[x_0, x_0 + \alpha]} |u_n - u_{n-1}| \leq M \sum_{n=1}^{\infty} \frac{L^{n-1} \alpha^n}{n!}$$

$$= \frac{M}{L} \sum_{n=1}^{\infty} \frac{(L\alpha)^n}{n!} = \frac{M}{L} \left(e^{\alpha L} - 1 \right) < +\infty.$$

Third step. We show that the limit of the sequence of functions (u_n) solves the initial value problem (4.3). From the equality:

$$\lim_{n \to \infty} u_{n+1}(t) = \lim_{n \to \infty} \left(y_0 + \int_{x_0}^{x} f(X, u_n(X)) \, dX \right),$$

we obtain, when $u = \lim_{n \to \infty} u_n$, the fundamental relation:

$$u(x) = y_0 + \int_{x_0}^{x} f(X, u(X)) \, dX, \tag{4.20}$$

since $|f(X, u_n(X))| \leq M$ ensures uniform convergence for $\left(f(X, u_n(X)) \right)_{n \in \mathbb{N}}$.

Now, differentiating both sides of (4.20), we see that $u(x)$ is solution of the initial value problem (4.3).

Fourth step. We have to prove uniqueness of the solution of (4.3). By contradiction, assume that $v \in C^1([x_0, x_0 + \alpha], \mathbb{R})$ solves (4.3) too. Thus:

$$v(x) = y_0 + \int_{x_0}^{x} f(X, v(X)) \, dX.$$

As before, it is possible to show that, for any $n \in \mathbb{N}$ and any $x \in [x_0, x_0 + \alpha]$, the following inequality holds true:

$$|u(x) - v(x)| \leq K \frac{L^n |x - x_0|^n}{n!}, \tag{4.21}$$

where K is given by:

$$K = \max_{x \in [x_0, x_0 + \alpha]} |u(x) - v(x)|.$$

Indeed:

$$|u(x) - v(x)| \leq \int_{x_0}^{x} |f(X, u(X)) - f(x, v(X))| \, dX \leq K L |x - x_0|,$$

which proves (4.21) for $n = 1$. Using induction, if we assume that (4.21) is satisfied for some $n \in \mathbb{N}$, then:

$$|u(x) - v(x)| \leq \int_{x_0}^{x} |f(X, u(X)) - f(X, v(X))| \, dX$$

$$\leq L \int_{x_0}^{x} |u(X) - v(X)| \, dX \leq L \int_{x_0}^{x} K \frac{L^n (X - x_0)^n}{n!} \, dX.$$

After calculating the last integral in the above inequality chain, we arrive at:

$$|u(x) - v(x)| \leq K \frac{L^{n+1}(x - x_0)^{n+1}}{(n+1)!} \,,$$

which proves (4.21) for index $n+1$. By induction, (4.21) holds true for any $n \in \mathbb{N}$.

We can finally end our demonstration of Theorem 4.17. In fact, by taking the limit $n \to \infty$ in (4.21), we obtain that, for any $x \in [x_0 , x_0 + a]$, the following inequality is verified:

$$|u(x) - v(x)| \leq 0 \,,$$

which shows that $u(x) = v(x)$ for any $x \in [x_0 , x_0 + a]$. $\square$

Remark 4.18. Let us go back to Remark 4.15. In such a situation, where the initial value problem (4.14) has multiple solutions, function $f(x , y) = 2\sqrt{|y|}$ does not fulfill the Lipschitz continuity property. In fact, taking, for istance, $y_1 , y_2 > 0$ yields:

$$\frac{|f(y_1) - f(y_2)|}{|y_1 - y_2|} = \frac{2}{\left| \sqrt{y_1} - \sqrt{y_2} \right|} \,,$$

which is unbounded.

The proof of Theorem 4.17, based on successive Picard–Lindelöhf iterates, is also useful in some simple situations, where it allows to compute an approximate solution of the initial value problem (4.3). This is illustrated by Example 4.19.

Example 4.19. Construct the Picard–Lindelhöf iterates for:

$$\begin{cases} y'(x) = -2\,x\,y(x), \\ y(0) = 1. \end{cases} \tag{4.22}$$

The first iterate is $y_0(x) = 1$, while subsequent iterates are:

$$y_1(x) = y_0(x) + \int_0^x f(t , y_0(t))\,dt = 1 - 2\int_0^x t\,dt = 1 - x^2 \,,$$

$$y_2(x) = y_0(x) + \int_0^x f(t , y_1(t))\,dt = 1 - 2\int_0^x t\,(1 - t^2)\,dt = 1 - x^2 + \frac{x^4}{2} \,,$$

$$y_3(x) = 1 - 2\int_0^x t\left(1 - t^2 + \frac{t^4}{4}\right)dt = 1 - x^2 + \frac{x^4}{2} - \frac{x^6}{6} \,,$$

$$y_4(x) = 1 - 2\int_0^x t\left(1 - t^2 + \frac{t^4}{4} - \frac{t^6}{6}\right)dt = 1 - x^2 + \frac{x^4}{2} - \frac{x^6}{6} + \frac{x^8}{24} \,,$$

and so on. A pattern emerges:

$$y_n(x) = 1 - \frac{x^2}{1!} + \frac{x^4}{2!} - \frac{x^6}{3!} + \frac{x^8}{4!} + \cdots + \frac{(-1)^n\,x^{2n}}{n!} \,.$$

The sequence of Picard–Lindelhöf iterates converges only if it also converges the series:

$$y(x) := \lim_{m \to \infty} \sum_{n=0}^{m} \frac{(-1)^n x^{2n}}{n!}.$$

Now, recalling the Taylor series for the exponential function:

$$e^x = \sum_{n=0}^{\infty} \frac{x^n}{n!},$$

it follows:

$$y(x) = \sum_{n=0}^{\infty} \frac{(-x^2)^n}{n!} = e^{-x^2}.$$

We will show later, in Example 5.4, that this function is indeed the solution to (4.22).

We leave it to the Reader, as an exercise, to determine the successive approximations of the IVP:

$$\begin{cases} y'(x) = y(x), \\ y(0) = 1. \end{cases}$$

and to recognise that the solution is the exponential function $y(x) = e^x$.

4.3.1   Interval of existence

The interval of existence of an initial value problem can be defined as the largest interval where the solution is well defined. This means that the initial point x_0 must be within the interval of existence. In the following, we discuss how to detect such an interval with a theoretical approach. When the exact solution is available, as in the case of the following Example 4.20, the determination of the interval of existence is straightforward.

Example 4.20. The initial value problem:

$$\begin{cases} y' = 1 - y^2 \\ y(0) = 0 \end{cases}$$

is solved by the function $y(x) = \tanh x$, so that its interval of existence is $\mathbb{R}$. It is worth noting that the similar initial value problem:

$$\begin{cases} y' = 1 + y^2 \\ y(0) = 0 \end{cases}$$

behaves differently, since it is solved by $y(x) = \tan x$ and has, therefore, interval of existence given by $\left]-\frac{\pi}{2},\frac{\pi}{2}\right[$. Moreover, in the latter case, taking the limit at the boundary of the interval of existence yields:

$$\lim_{x\to\pm\frac{\pi}{2}} y(x) = \lim_{x\to\pm\frac{\pi}{2}} \tan x = \pm\infty.$$

This is not a special situation. Even when the interval of existence is bounded, for some theoretical reason that we present later, in detail, the solution can be unbounded; this case is referred to as a *blow-up* phenomenon.

4.3.2 Vector–valued differential equations

It is not difficult to adapt the argument presented in Theorem 4.17 to the vector–valued situation of a function $\boldsymbol{f} : \Omega \to \mathbb{R}^n$ defined on an open set $\Omega \subset \mathbb{R} \times \mathbb{R}^n$. In this case, the Lipschitz continuity condition is:

$$\|\boldsymbol{f}(x,\boldsymbol{y}_1) - \boldsymbol{f}(x,\boldsymbol{y}_2)\| \leq L\,\|\boldsymbol{y}_1 - \boldsymbol{y}_2\|$$

for any $(x,\boldsymbol{y}_1)$, $(x,\boldsymbol{y}_2) \in \mathcal{R}$, where the rectangle $\mathcal{R}$ is replaced by a cylinder:

$$\mathcal{R} = [x_0,x_0+a] \times \overline{B}_b(\boldsymbol{y}_0).$$

The vector–valued version of the Picard–Lindelöhf Theorem 4.17 is represented by the following Theorem 4.21, whose proof is omitted, as it is very similar to that of Theorem 4.17.

Theorem 4.21 (Picard–Lindelöhf, vector–valued case). Let $f : \mathcal{R} \to \mathbb{R}^n$ br uniformly Lipschitz continuous in $\boldsymbol{y}$, with respect to x, and define:

$$M = \max_{\mathcal{R}} \|\boldsymbol{f}\|\,, \qquad \alpha = \min\left\{a,\frac{b}{M}\right\}. \qquad (4.23)$$

Then, problem (4.9) admits unique solution $\boldsymbol{u} \in \mathcal{C}^1\left([x_0,x_0+\alpha],\mathbb{R}^n\right)$.

4.3.3 Solution continuation

To detect the interval of existence, we start by observing that the Picard–Lindelhöf Theorem 4.17 leads to a solution of the IVP (4.3) which is, by construction, *local*, i.e., it is a solution defined within a neighbourhood of the initial independent data x_0. The radius of this neighbourhood depends on function $\boldsymbol{f}(x,\boldsymbol{y})$ in different ways: to understand them, we introduce the notion of *joined* solutions, as a first (technical) step.

Remark 4.22. Let $\boldsymbol{f} : \Omega \to \mathbb{R}$ be a continuous function, defined on an open set $\Omega \subset \mathbb{R} \times \mathbb{R}^n$. Consider two solutions, $\boldsymbol{y}_1 \in \mathcal{C}^1\left([a,b],\mathbb{R}^n\right)$ and $\boldsymbol{y}_2 \in \mathcal{C}^1\left([b,c],\mathbb{R}^n\right)$, of the differential equation $\boldsymbol{y}' = \boldsymbol{f}(x,\boldsymbol{y})$, such that $\boldsymbol{y}_1(b) = \boldsymbol{y}_2(b)$. Then, function $\boldsymbol{y} : [a,c] \to \mathbb{R}^n$ defined as:

$$\boldsymbol{y}(x) = \begin{cases} \boldsymbol{y}_1(x) & \text{if } x \in [a,b] \\ \boldsymbol{y}_2(x) & \text{if } x \in (b,c] \end{cases}$$

is also a solution of $y' = f(x, y)$.

Function f represents a vector field.

With the Picard–Lindelöhf Theorem 4.21, we can build solutions to initial value problems associated to $y' = f(x, y)$, choosing the initial data in Ω. In other words, given a point $(x_0, y_0) \in \Omega$, we form the IVP (4.9), for which Theorem 4.21 ensures existence of a solution $u(x)$ in a neighbourhood of x_0. If now, in the rectangle $\mathcal{R} = [x_0, x_0 + a] \times \overline{B}_b(y_0)$, we choose $a, b > 0$ so that $\mathcal{R} \subset \Omega$ (which is always possible, since Ω is open), then the solution of IVP (4.9) is defined at least up to the point $x_1 = x_0 + \alpha_1$, where the constant $\alpha_1 > 0$ is given by (4.23).

This allows us to *continue* and consider a new initial value problem:

$$\begin{cases} y' = f(x, y) \\ y(x_1) = u(x_1) := y_1 \end{cases} \tag{4.9a}$$

which is defined at least up to point $x_2 = x_1 + \alpha_2$, where constant $\alpha_2 > 0$ is again given by (4.23).

This procedure can be iterated, leading to the formal Definition 4.23 of maximal domain solution. The idea of *continuation* of a solution may be better understood by looking at Figure 4.1.

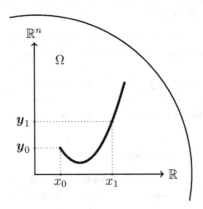

Figure 4.1: Continuation of a solution.

Definition 4.23 (Maximal domain solution). If $u \in \mathcal{C}^1(I, \mathbb{R}^n)$ solves the initial value problem (4.9), we say that u has *maximal domain* (or that u does not admit a continuation) if there exists no function $v \in \mathcal{C}^1(J, \mathbb{R}^n)$ which also solves (4.9) and such that $I \subset J$.

The existence of the maximal domain solution to IVP (4.9) can be understood heuristically, as it comes from indefinitely repeating the continuation procedure. Establishing it with mathematical rigor is beyond the aim of these lecture notes, since it would require notions from advanced theoretical Set theory, such as Zorn's Lemma[5].

We end this section stating, in Theorem 4.24, a result on the asymptotic behaviour of a solution with maximal domain, in the particular case where $\Omega = I \times \mathbb{R}^n$, being I an open interval. Such a result explains what observed in Example 4.20, though we do not provide a proof of Theorem 4.24.

Theorem 4.24. Let f be defined on an open set $I \times \mathbb{R}^n \subset \mathbb{R} \times \mathbb{R}^n$. Given $(x_0, y_0) \in I \times \mathbb{R}^n$, assume that function y is a maximal domain solution of the initial value problem:

$$\begin{cases} y' = f(x, y), \\ y(x_0) = y_0. \end{cases}$$

Denote (α, ω) the maximal domain of y. Then, one of two possibility holds respectively for α and for ω :

(1) it is either $\alpha = \inf I$,

   or $\alpha > \inf I$,   implying $\lim\limits_{x \to \alpha^+} |y(x)| = +\infty$;

(2) it is either $\omega = \sup I$,

   or $\omega < \sup I$,   implying $\lim\limits_{x \to \omega^-} |y(x)| = +\infty$.

---

[5]Max August Zorn (1906–1993), German mathematician.
For his lemma, see, for example, mathworld.wolfram.com/ZornsLemma.html

Chapter 5

Ordinary differential equations of first order: methods for explicit solutions

In the previous Chapter 4 we exposed the general theory, concerning conditions for existence and uniqueness of an initial value problem. Here, we consider some important particular situations, in which, due to the structure of certain kind of scalar ordinary differential equations, it is possible to establish methods to determine their explicit solution.

5.1 Separable equations

Definition 5.1. A differential equation is *separable* if it has the form:

$$\begin{cases} y'(x) = a(x)\ b\big(y(x)\big)\,, \\ y(x_0) = y_0\,, \end{cases} \tag{5.1}$$

where $a(x)$ and $b(y)$ are continuous functions, respectively defined on intervals I_a and I_b, such that $x_0 \in I_a$ and $y_0 \in I_b$.

Theorem 5.2. If, for any $y \in I_b$, it holds:

$$b(y) \neq 0\,, \tag{5.2}$$

then the unique solution to (5.1) is function $y(x)$, defined implicitly by:

$$\int_{y_0}^{y} \frac{dz}{b(z)} = \int_{x_0}^{x} a(s)\,ds\,. \tag{5.3}$$

Remark 5.3. The hypothesis (5.2) cannot be removed, as shown, for instance, in Remark 4.15, where $b(y) = 2\sqrt{|y|}$, which means that $b(0) = 0$.

Proof. Introduce the two-variable function:

$$F(x,y) := \int_{y_0}^{y} \frac{dz}{b(z)} - \int_{x_0}^{x} a(s)\,ds \,, \qquad (5.3a)$$

for which $F(x_0,y_0) = 0$. Recalling (5.2) and since:

$$\frac{\partial F(x,y)}{\partial y} = \frac{1}{b(y)} \,,$$

it follows:

$$\frac{\partial F(x_0,y_0)}{\partial y} = \frac{1}{b(y_0)} \neq 0 \,.$$

We can thus invoke the Implicit function theorem 3.25 and infer the existence of $\delta, \varepsilon > 0$ for which, given any $x \in (x_0 - \delta, x_0 + \delta)$, there is a unique C^1 function $y = y(x) \in (y_0 - \varepsilon, y_0 + \varepsilon)$ such that:

$$F(x,y) = 0$$

and such that, for any $x \in (x_0 - \delta, x_0 + \delta)$:

$$y'(x) = -\frac{F_x\big(x,y(x)\big)}{F_y\big(x,y(x)\big)} = \frac{a(x)}{\dfrac{1}{b(y)}} = a(x)\,b(y) \,.$$

Function y, implicitly defined by (5.3), is thus a solution of (5.1); to complete the proof, we still have to show its uniqueness. Assume that $y_1(x)$ and $y_2(x)$ are both solutions of (5.1), and define:

$$B(y) := \int_{y_0}^{y} \frac{dz}{b(z)} \,;$$

then:

$$\frac{d}{dx}\Big(B\big(y_1(x)\big) - B\big(y_2(x)\big)\Big) = \frac{y_1'(x)}{b\big(y_1(x)\big)} - \frac{y_2'(x)}{b\big(y_2(x)\big)}$$

$$= \frac{a(x)\,b\big(y_1(x)\big)}{b\big(y_1(x)\big)} - \frac{a(x)\,b\big(y_2(x)\big)}{b\big(y_2(x)\big)} = 0 \,.$$

Notice that we used the fact that both $y_1(x)$ and $y_2(x)$ are assumed to solve (5.1). Thus $B\big(y_1(x)\big) - B\big(y_2(x)\big)$ is a constant function, and its constant value is zero, since $y_1(x_0) = y_2(x_0) = y_0$. In other words, we have shown that, for any $x \in I_a$:

$$B\big(y_1(x)\big) - B\big(y_2(x)\big) = 0 \,,$$

which means, recalling the definition of B :

$$0 = \int_{y_0}^{y_1(x)} \frac{dz}{b(z)} - \int_{y_0}^{y_2(x)} \frac{dz}{b(z)} = \int_{y_2(x)}^{y_1(x)} \frac{dz}{b(z)} \,.$$

At this point, using the Mean–Value Theorem 3.24, we infer the existence of a number $X(x)$ between the integration limits $y_2(x)$ and $y_1(x)$, such that:

$$\frac{1}{b(X(x))}\left(y_1(x) - y_2(x)\right) = 0.$$

But, from (5.2), it holds:

$$\frac{1}{b(X(x))} \neq 0,$$

thus:

$$y_1(x) - y_2(x) = 0,$$

and the theorem proof is completed. $\qquad\Box$

Example 5.4. Consider once more the IVP studied, using successive approximations, in Example 4.19:

$$\begin{cases} y'(x) = -2\,x\,y(x), \\ y(0) = 1. \end{cases}$$

Setting $a(x) = -2\,x$, $b(y) = y$, $x_0 = 0$, $y_0 = 1$ in (5.3) leads to:

$$\int_1^y \frac{1}{z}\,dz = \int_0^x (-2\,z)\,dz \qquad \Longleftrightarrow \qquad \ln y = -x^2 \qquad \Longleftrightarrow \qquad y(x) = e^{-x^2}.$$

In the next couple of examples, some interesting, particular cases of separable equations are considered.

Example 5.5. The choice $b(y) = y$ in (5.3) yields the particular separable equation:

$$\begin{cases} y'(x) = a(x)\,y(x), \\ y(x_0) = y_0, \end{cases} \tag{5.4}$$

where $a(x)$ is a given continuous function. Using (5.3), we get:

$$\int_{y_0}^y \frac{1}{z}\,dz = \int_{x_0}^x a(s)\,ds \qquad \Longrightarrow \qquad \ln\frac{y}{y_0} = \int_{x_0}^x a(s)\,ds,$$

thus:

$$y = y_0\,e^{\displaystyle\int_{x_0}^x a(s)\,ds}.$$

For instance, if $a(x) = -\dfrac{x}{2}$, the initial value problem:

$$\begin{cases} y'(x) = -\dfrac{x}{2}\,y(x) \\ y(0) = 1 \end{cases}$$

has solution:

$$y(x) = e^{-\frac{x^2}{4}}.$$

Example 5.6. In (5.3), let $b(y) = y^2$, which leads to the separable equation:

$$\begin{cases} y'(x) = a(x)\, y^2(x)\,, \\ y(x_0) = y_0\,, \end{cases} \tag{5.5}$$

with $a(x)$ continuous function. Using (5.3), we find:

$$\int_{y_0}^{y} \frac{1}{z^2}\, dz = \int_{x_0}^{x} a(s)\, ds \quad \Longrightarrow \quad -\frac{1}{y} + \frac{1}{y_0} = \int_{x_0}^{x} a(s)\, ds\,,$$

and, solving with respect to y:

$$y = \frac{1}{\dfrac{1}{y_0} - \displaystyle\int_{x_0}^{x} a(s)\, ds}\,.$$

For instance, if $a(x) = -2\,x$, the initial value problem:

$$\begin{cases} y'(x) = -2\,x\,y^2(x) \\ y(0) = 1 \end{cases}$$

has solution

$$y = \frac{1}{1 + x^2}\,.$$

We now provide some practical examples, recalling that a complete treatment needs, both, finding the analytical expression of the solution and determining the maximal solution domain.

Example 5.7. Consider equation:

$$\begin{cases} y'(x) = (x+1)\left(1 + y^2(x)\right)\,, \\ y(-1) = 0\,. \end{cases}$$

Using (5.3), we find:

$$\int_{0}^{y(x)} \frac{1}{1 + z^2}\, dz = \int_{-1}^{x} (s+1)\, ds\,,$$

and, evaluating the integrals:

$$\int_{0}^{y(x)} \frac{1}{1 + z^2}\, dz = \arctan y(x)\,, \qquad \int_{-1}^{x} (s+1)\, ds = \frac{1}{2}\,(x+1)^2\,,$$

the solution is obtained:

$$y(x) = \tan \frac{(x+1)^2}{2}\,.$$

Observe that the solution $y(x)$ is only well–defined for x in a neighbourhood of $x_0 = -1$ and such that:

$$-\frac{\pi}{2} < \frac{(x+1)^2}{2} < \frac{\pi}{2},$$

that is, $-1 - \pi < x < -1 + \pi$.

Example 5.8. Solve the initial value problem:

$$\begin{cases} y'(x) = \dfrac{x-1}{y(x)+1}, \\ y(0) = 0. \end{cases}$$

From (5.3):

$$\int_0^{y(x)} (z+1)\,dz = \int_0^x (s-1)\,ds,$$

performing the relevant computations, we get:

$$\frac{1}{2} y^2(x) + y(x) = \frac{1}{2} x^2 - x,$$

so that:

$$y(x) = -1 \pm \sqrt{(x-1)^2} = \begin{cases} x-2, \\ -x. \end{cases}$$

Now, recall that x lies in a neighbourhood of zero and that the initial condition requires $y(0) = 0$; it can be inferred, therefore, that $y(x) = -x$ must be chosen. To establish the maximal domain, observe that $x - 1$ vanishes for $x = 1$; thus, we infer that $x < 1$.

Example 5.9. As a varies in $\mathbb{R}$, investigate the maximal domain of the solutions to the initial value problem:

$$\begin{cases} u'(x) = a\left(1 + u^2(x)\right) \cos x, \\ u(0) = 0. \end{cases}$$

Form (5.3), to obtain:

$$\int_0^{u(x)} \frac{dz}{1+z^2} = \int_0^x a \, \cos s \, ds.$$

After performing the relevant computations, we get:

$$\arctan u(x) = a \, \sin x. \tag{5.6}$$

It is clear that the Range of the right–hand side of (5.6) is $[-a, a]$. To obtain a solution defined on $\mathbb{R}$, we have to impose that $a < \dfrac{\pi}{2}$. In such a case, solving with respect to u yields:

$$u(x) = \tan\left(a \sin x\right).$$

Vice versa, when $a \geq \dfrac{\pi}{2}$, since there exists $\bar{x} \in \mathbb{R}^+$ for which $a \sin \bar{x} = \dfrac{\pi}{2}$, then, the obtained solution is defined in $(-\bar{x}, \bar{x})$, and $\bar{x}$ is the minimum positive number verifying the equality $a \sin \bar{x} = \dfrac{\pi}{2}$.

5.1.1   Exercises

1. Solve the following separable equations:

(a) $\begin{cases} y'(x) = \dfrac{2\,x}{\sin y(x)}, \\[2mm] y(0) = \dfrac{\pi}{2}, \end{cases}$

(b) $\begin{cases} y'(x) = \dfrac{2\,x}{\cos y(x)}, \\[2mm] y(0) = 0, \end{cases}$

(c) $\begin{cases} y'(x) = \dfrac{2\,x}{\cot y(x)}, \\[2mm] y(0) = \dfrac{\pi}{4}, \end{cases}$

(d) $\begin{cases} y'(x) = \dfrac{e^x}{\cosh y(x)}, \\[2mm] y(0) = 0, \end{cases}$

(e) $\begin{cases} y'(x) = \dfrac{e^x}{\cos y(x)}, \\[2mm] y(0) = 0, \end{cases}$

(f) $\begin{cases} y'(x) = \dfrac{x^2}{1+x}\,y^2(x), \\[2mm] y(0) = 1. \end{cases}$

Solutions:

(a) $y_a(x) = \arccos(-x^2)$,

(b) $y_f(x) = \arcsin(x^2)$,

(c) $y_c(x) = \arcsin\left(\dfrac{e^{x^2}}{\sqrt{2}}\right)$,

(d) $y_d(x) = \operatorname{arcsinh}(e^x - 1)$,

(e) $y_e(x) = \arcsin(e^x - 1)$,

(f) $y_b(x) = \dfrac{2}{2(1 + x - \ln(1 + x)) - x^2}$.

2. Show that the solution to the initial value problem:

$$\begin{cases} y'(x) = y(x)\,\dfrac{e^{2x}}{4 + e^{2x}} \\[2mm] y(0) = \sqrt{5} \end{cases}$$

is $y(x) = \sqrt{4 + e^{2x}}$. What is the maximal domain of such a solution?

3. Show that the solution to the initial value problem:

$$\begin{cases} y'(x) = \dfrac{x \sin x}{1 + y(x)} \\[2mm] y(0) = 0 \end{cases}$$

is $y(x) = \sqrt{2 \sin x - 2x \cos x + 1} - 1$. Find the maximal domain of the solution.

4. Show that the solution to the initial value problem:

$$\begin{cases} y'(x) = \dfrac{y^3(x)}{1+x^2} \\ y(0) = 2 \end{cases}$$

is $y(x) = \dfrac{2}{\sqrt{1 - 8\arctan x}}$. Find the maximal domain of the solution.

5. Show that the solution to the initial value problem:

$$\begin{cases} y'(x) = (\sin x + \cos x)\, e^{-y(x)} \\ y(0) = 1 \end{cases}$$

is $y(x) = \ln(1+e+\sin x - \cos x)$. Find the maximal domain of the solution.

6. Show that the solution to the initial value problem:

$$\begin{cases} y'(x) = (1 + y^2(x))\ln(1 + x^2) \\ y(0) = 1 \end{cases}$$

is $y(x) = \tan\left(x\ln(x^2+1) - 2x + 2\arctan x + \dfrac{\pi}{4}\right)$. Find the maximal domain of the solution.

7. Show that the solution to the initial value problem:

$$\begin{cases} y'(t) = -\dfrac{1}{x}\, y(x) - \dfrac{1}{x}\, y^4(x) \\ y(1) = 1 \end{cases}$$

is $y(x) = -\dfrac{1}{\sqrt[3]{1 - 2x^3}}$. Find the maximal domain of the solution.

8. Solve the initial value problem:

$$\begin{cases} y''(x) = (\sin x)\, y'(x) \\ y(0) = 1 \\ y'(1) = 0 \end{cases}$$

Hint. Set $z(x) = y'(x)$ *and solve the equation* $z'(x) = (\sin x)\, z(x)$.

5.2   Singular integrals

Given a differential equation, one may want to describe its general solution, without fixing a set of initial conditions. Consider, for instance, the differential equation:

$$y' = (y - 1)(y - 2).\tag{5.7}$$

Equation (5.7) is separable, so we can easily adapt formula (5.3), using indefinite integrals and adding a constant of integration:

$$\ln\frac{y-2}{y-1} = x + c_1.$$

Solving for y, the general solution to (5.7) is obtained:

$$y(x) = \frac{2 - e^{x+c_1}}{1 - e^{x+c_1}} = \frac{2 - c\,e^x}{1 - c\,e^x},\tag{5.8}$$

where we set $c = e^{c_1}$.

Observe that the two constant functions $y = 1$ and $y = 2$ are solutions of equation (5.7). Observe further that $y = 2$ is obtained from (5.8) taking $c = 0$, thus such a solution is a *particular* solution to (5.7). Vice versa, solution $y = 1$ cannot be obtained using the general solution (5.8); for this reason this solution is called *singular*.

Singular solutions of a differential equation can be found with a computational procedure, illustrated in Remark 5.10.

Remark 5.10. Given the differential equation (4.2), suppose that its general solution is given by $\Phi(x, y, c) = 0$. When there exists a singular integral of (4.2), it can be detected eliminating c from the system:

$$\begin{cases} \Phi(x, y, c) = 0, \\ \dfrac{\partial \Phi}{\partial c}(x, y, c) = 0. \end{cases}\tag{5.9}$$

In the case of equation (5.7), system (5.9) becomes:

$$\begin{cases} c\,e^x(1 - y) + y - 2 = 0, \\ e^x(1 - y) = 0, \end{cases}$$

which confirms that $y = 1$ is a singular integral of (5.7).

Remark 5.11. When the differential equation (5.3) is given in implicit form, uniqueness of the solution does not hold; this generates the occurrence of a singular integral, which can be detected without solving the differential equation (4.2), eliminating y' from system 5.10:

$$\begin{cases} F(x,y,y') = 0 \ , \\ \dfrac{\partial F}{\partial y'}(x,y,y') = 0 \ . \end{cases} \tag{5.10}$$

A detailed discussion can be found in § 23 of [13].

5.3 Homogeneous equations

To obtain exact solutions of non–separable differential equations, it is possible, in some specific situations, to use some ansatz and transform the given equations. A few examples are provided in the following. The first kind of transformable equations, that we consider here, are the so–called homogeneous equations.

Theorem 5.12. Given $f : [0,\infty) \times [0,\infty) \to \mathbb{R}$, if $f(\alpha x, \alpha y) = f(x,y)$, and $x_0, y_0 \in \mathbb{R}, x_0 \neq 0$, are such that $f\left(1, \dfrac{y_0}{x_0}\right) \neq \dfrac{y_0}{x_0}$, then the change of variable $y(x) = x\, u(x)$ can be employed to transform the differential equation:

$$\begin{cases} y'(x) = f\big(x,y(x)\big) \\ y(x_0) = y_0 \end{cases} \tag{5.11}$$

into the separable equation:

$$\begin{cases} u'(x) = \dfrac{f\big(1,u(x)\big) - u(x)}{x} \ , \\ u(x_0) = \dfrac{y_0}{x_0} \ . \end{cases} \tag{5.12}$$

Proof. We represent the solution to (5.11) in the form $y(x) = x\, u(x)$ and look for the auxiliary unknown $u(x)$; this is a change of variable that, in force of homogeneity conditions, transforms (5.11) into a separable equation, expressed in terms of $u(x)$. Differentiating $y(x) = x\, u(x)$ yields, in fact, $y'(x) = x\, u'(x) + u(x)$. Now, imposing the equality $y'(x) = f\big(x,y(x)\big)$ leads to $x\, u'(x) + u(x) = f\big(x, x\, u(x)\big) = f\big(1, u(x)\big)$, where we used the fact that $f(\alpha x, \alpha y) = f(x,y)$. Therefore, $u(x)$ solves the differential equation:

$$u'(x) = \frac{f(1,u(x)) - u(x)}{x} \ .$$

Observe further that the initial condition $y(x_0) = y_0$ is changed into

$x_0\, u(x_0) = y_0$. Recalling that $x_0 \neq 0$, equation (5.11) is changed into the separable problem (5.12), whose solution $u(x)$ is defined by:

$$\int_{\frac{y_0}{x_0}}^{u(x)} \frac{1}{f(1,s)-s}\, ds = \ln|x| - \ln|x_0| \, .$$

□

Example 5.13. Consider the initial value problem:

$$\begin{cases} y'(x) = \dfrac{x^2 + y^2(x)}{x\, y(x)}\,, \\ y(2) = 2\,. \end{cases}$$

In this case, $f(x,y)$ is an homogeneous function:

$$f(x,y) = \frac{x^2 + y^2}{x\, y}\,.$$

Using $y(x) = x\, u(x)$:

$$f(1,u(x)) = \frac{1 + u^2(x)}{u(x)} = \frac{1}{u(x)} - u(x)\,.$$

Thus, the transformed problem turns out to be in separable form:

$$\begin{cases} u'(x) = \dfrac{1}{x\, u(x)}\,, \\ u(2) = 1\,, \end{cases}$$

and its solution can be found by integration:

$$\int_{1}^{u(x)} s\, ds = \ln|x| - \ln 2\,,$$

yielding:

$$u(x) = \sqrt{2\ln|x| + 1 - \ln 4}\,.$$

Observe that the solution is defined on the interval $x > \sqrt{e^{\ln 4 - 1}} = \dfrac{2}{\sqrt{e}}\,.$

At this point, going back to our original initial value problem, we arrive at the solution of the homogeneous problem:

$$y(x) = x\,\sqrt{2\ln|x| + 1 - \ln 4}\,, \qquad x > \frac{2}{\sqrt{e}}\,.$$

5.3.1   Exercises

Solve the following initial values problems for homogeneous equations:

1. $\begin{cases} y'(x) = \dfrac{y^2(x)}{x^2 - x\,y(x)}\,, \\[4mm] y(1) = \dfrac{1}{3}\,, \end{cases}$

3. $\begin{cases} y'(x) = \dfrac{1}{x}\left(y(x) + x\,e^{\frac{y(x)}{x}}\right), \\[4mm] y(1) = -1\,, \end{cases}$

2. $\begin{cases} y'(x) = -\dfrac{1}{x^2}\left(x\,y(x) + y^2(x)\right), \\[4mm] y(1) = 1\,, \end{cases}$

4. $\begin{cases} y'(x) = \dfrac{2\,x - y(x)}{x + 3\,y(x)}\,, \\[4mm] y(1) = 1\,. \end{cases}$

Solutions:

1. $y(x) = \dfrac{\sqrt{3x^2 + 1} - 1}{3\,x}\,,$

3. $y(x) = -x\ln\left(e - \ln x\right),$

2. $y(x) = \dfrac{2\,x}{3\,x^2 - 1}\,,$

4. $y(x) = \dfrac{1}{3}\left(-x + \sqrt{7\,x^2 + 9}\right).$

5.4 Quasi homogeneous equations

It is possible to transform some differential equations, non–separable and non–homogeneous, into equivalent equations that are separable or homogeneous. Here, we deal with differential equations of the form:

$$y' = f\left(\frac{a\,x + b\,y + c}{\alpha\,x + \beta\,y + \gamma}\right), \tag{5.13}$$

where

$$\det\begin{bmatrix} a & b \\ \alpha & \beta \end{bmatrix} \neq 0\,. \tag{5.14}$$

In this situation, the linear system:

$$\begin{cases} a\,x + b\,y + c = 0 \\ \alpha\,x + \beta\,y + \gamma = 0 \end{cases} \tag{5.15}$$

has a unique solution, say, $(x\,,y) = (x_1\,,y_1)$. To obtain a homogeneous or a separable equation, it is possible to exploit the solution uniqueness, employing the change of variable:

$$\begin{cases} X = x - x_1\,, \\ Y = y - y_1\,, \end{cases} \qquad \Longleftrightarrow \qquad \begin{cases} x = X + x_1\,, \\ y = Y + y_1\,. \end{cases}$$

Example 5.14 illustrates the transformation procedure.

Example 5.14. Consider the equation:

$$\begin{cases} y' = \dfrac{3x+4}{y-1} \,, \\ y(0) = 2 \,. \end{cases}$$

The first step consists in solving the system:

$$\begin{cases} 3x + 4 = 0 \,, \\ y - 1 = 0 \,, \end{cases}$$

whose solution is $x_1 = -\dfrac{4}{3}$, $y_1 = 1$. In the second step, the change of variable is performed:

$$\begin{cases} X = x + \frac{4}{3} \,, \\ Y = y - 1 \,, \end{cases}$$

which leads to the separable equation:

$$\begin{cases} Y' = \dfrac{3X}{Y} \,, \\ Y\left(\dfrac{4}{3}\right) = 1 \,, \end{cases}$$

with solution:

$$3\left(X^2 - \frac{16}{9}\right) = Y^2 - 1.$$

Recovering the original variables:

$$3\left(\left(x + \frac{4}{3}\right)^2 - \frac{16}{9}\right) = (y-1)^2 - 1$$

and simplifying:

$$3\left(x^2 + \frac{8}{3}x\right) = y^2 - 2y$$

yields:

$$y = 1 \pm \sqrt{3x^2 + 8x + 1}\,.$$

Finally, recalling that $y(0) = 2$, the solution is given by:

$$y(x) = 1 + \sqrt{3x^2 + 8x + 1}\,, \qquad x > \frac{-4 + \sqrt{13}}{3}.$$

The worked–out Example 5.15 illustrates the procedure to be followed if, when considering equation (5.13), condition (5.14) is not fulfilled.

Example 5.15. Consider the equation:

$$\begin{cases} y' = -\dfrac{x+y+1}{2x+2y+1} \,, \\ y(1) = 2 \,. \end{cases}$$

Here, there is no solution to system:

$$\begin{cases} x+y+1 = 0 \,, \\ 2x+2y+1 = 0 \,. \end{cases}$$

In this situation, since the two equations in the system are proportional, the change of variable to be employed is:

$$\begin{cases} t = x \,, \\ z = x+y \,. \end{cases}$$

The given differential equation is, hence, transformed into the separable one:

$$\begin{cases} z' = \dfrac{z}{2z+1} \,, \\ z(1) = 3 \,. \end{cases}$$

Separating the variables leads to:

$$\int_3^z \frac{2w+1}{w}\,dw = \int_1^t ds \,.$$

Thus:

$$2z - 6 + \ln z - \ln 3 = t - 1 \implies x + 2y + \ln(x+y) = 5 + \ln 3 \,.$$

Observe that, in this example, it is not possible to express the dependent variable y in an elementary way, i.e., in terms of elementary functions.

5.5 Exact equations

Aim of this section is to provide full details on solving exact differential equations. To understand the idea behind the treatment of this kind of equation, we present Example 5.16, that will help in illustrating what an exact differential equation is, how its structure can be exploited, to arrive at a solution, and why the process works as it does.

Example 5.16. Consider the differential equation:

$$y' = \frac{3\,x^2 - 2\,x\,y}{2\,y + x^2 - 1}.$$ (5.16)

First, rewrite (5.16) as:

$$2\,x\,y - 3\,x^2 + (2\,y + x^2 - 1)\,y' = 0.$$ (5.16a)

Equation (5.16a) is solvable under the assumption that a suitable function $\Phi(x\,,y)$ can be found, that verifies:

$$\frac{\partial \Phi}{\partial x} = 2\,x\,y - 3\,x^2, \qquad \frac{\partial \Phi}{\partial y} = 2\,y + x^2 - 1.$$

Note that it is not always possible to determine such a $\Phi(x\,,y)$. In the current Example 5.16, though, we are able to define $\Phi(x\,,y) = y^2 + (x^2 - 1)\,y - x^3$. Therefore (5.16a) can be rewritten: as

$$\frac{\partial \Phi}{\partial x} + \frac{\partial \Phi}{\partial y}\,y' = 0.$$ (5.16b)

Invoking the multi–variable Chain Rule[1], we can write (5.16b) as:

$$\frac{d}{dx}\,\Phi(x\,,y(x)) = 0.$$ (5.16c)

Since, when the ordinary derivative of a function is zero, the function is constant, there must exist a real number c such that:

$$\Phi(x\,,y(x)) = y^2 + (x^2 - 1)\,y - x^3 = c.$$ (5.17)

Thus (5.17) is an implicit solution for the differential equation (5.16); if an initial condition is assigned, we can determine c.

It is not always possible to determine an explicit solution, expressed in terms of y. In the particular situation of Example 5.16, though, finding a solution in explicit form is feasible. For instance, setting $y(0) = 1$, we get $c = 0$ and:

$$y(x) = \frac{1}{2}\left(1 - x^2 + \sqrt{x^4 + 4\,x^3 - 2\,x^2 + 1}\right).$$

Let us, now, leave the particular case of Example 5.16, and return to the general situation, i.e., consider ordinary differential equation of the form:

$$M(x\,,y) + N(x\,,y)\,y' = 0.$$ (5.18)

We call *exact* the differential equation (5.18), if there exists a function $\Phi(x\,,y)$ such that:

$$\frac{\partial \Phi}{\partial x} = M(x\,,y), \qquad \frac{\partial \Phi}{\partial y} = N(x\,,y),$$ (5.19)

---

[1]See, for example, mathworld.wolfram.com/ChainRule.html

and the (implicit) solution to an exact differential equation is constant:

$$\Phi(x,y) = c.$$

In other words, finding $\Phi(x,y)$ constitutes the central task in determining whether a differential equation is exact and in computing its solution.

Establishing a necessary condition for (5.18) to be exact is easy. In fact, if we assume that (5.18) is exact and that $\Phi(x,y)$ satisfies the hypotheses of Theorem 3.4, then the equality holds:

$$\frac{\partial}{\partial x}\left(\frac{\partial \Phi}{\partial y}\right) = \frac{\partial}{\partial y}\left(\frac{\partial \Phi}{\partial x}\right).$$

Inserting (5.19), we obtain the necessary condition for an equation to be exact:

$$\frac{\partial}{\partial x} N(x,y) = \frac{\partial}{\partial y} M(x,y). \tag{5.20}$$

The result in Theorem 5.17 speeds up the search of solution for exact equations.

Theorem 5.17. Define $Q = \{(x,y) \in \mathbb{R}^2 \mid a < x < b, \quad c < y < d\}$ and let $M,N : Q \to \mathbb{R}$ be C^1, with $N(x,y) \neq 0$ for any $(x,y) \in Q$. Assume that M,N, verify the closure condition (5.20) for any $(x,y) \in Q$. Then, there exists a unique solution to the initial value problem:

$$\begin{cases} y' = -\dfrac{M(x,y)}{N(x,y)}, \\ y(x_0) = y_0, \qquad (x_0,y_0) \in Q. \end{cases} \tag{5.21}$$

Such a solution is implicitly defined by:

$$\int_{x_0}^{x} M(t,y_0)\, dt + \int_{y_0}^{y} N(x,s)\, ds = 0. \tag{5.22}$$

Example 5.18. Consider the differential equation:

$$\begin{cases} y' = -\dfrac{6x + y^2}{2xy + 1}, \\ y(1) = 1. \end{cases}$$

The closure condition (5.20) is fullfilled, since:

$$M(x,y) = 6x + y^2 \quad \Longrightarrow \quad \frac{\partial M(x,y)}{\partial y} = 2y,$$

$$N(x,y) = 2xy + 1 \quad \Longrightarrow \quad \frac{\partial N(x,y)}{\partial x} = 2y.$$

Formula (5.22) then yields:

$$\int_1^x M(t,1)\, dt = \int_1^x (6\, t + 1)\, dt = -4 + x + 3\, x^2\,,$$

$$\int_1^y N(x,s)\, ds = \int_1^y (2\, x\, s + 1)\, ds = -1 - x + y + x\, y^2\,.$$

Hence, the solution to the given initial value problem is implicitly defined by:

$$x\, y^2 + y + 3\, x^2 - 5 = 0$$

and, solving for y, two solutions are reached:

$$y = \frac{-1 \pm \sqrt{1 + 20\, x - 12\, x^3}}{2\, x}\,.$$

Recalling that $y(1) = 1$, we choose:

$$y = \frac{-1 + \sqrt{1 + 20\, x - 12\, x^3}}{2\, x}\,.$$

Example 5.19. Consider solving:

$$\begin{cases} y' = -\dfrac{3\, y\, e^{3\, x} - 2\, x}{e^{3\, x}}\,, \\[2mm] y(1) = 1\,. \end{cases}$$

Here, it holds:

$$M(x,y) = 3\, y\, e^{3\, x} - 2\, x \quad \Longrightarrow \quad \frac{\partial M(x,y)}{\partial y} = 3\, e^{3\, x}\,,$$

$$N(x,y) = e^{3\, x} \quad \Longrightarrow \quad \frac{\partial N(x,y)}{\partial x} = 3\, e^{3\, x}\,.$$

Using formula (5.22):

$$\int_1^x M(t,1)\, dt = \int_1^x (3\, e^{3\, t} - 2\, t)\, dt = -x^2 + e^{3\, x} - e^3 + 1\,,$$

$$\int_1^y N(x,s)\, ds = \int_1^y (e^{3\, x})\, ds = (y - 1)\, e^{3\, x}\,.$$

The solution to the given initial value problem is, therefore:

$$-x^2 + e^{3\, x} - e^3 + 1 + (y - 1)\, e^{3\, x} = 0 \quad \Longrightarrow \quad y = e^{-3\, x}\left(x^2 + e^3 - 1\right)\,.$$

5.5.1   Exercises

1. Solve the following initial value problems, for exact equations:

(a) $\begin{cases} y' = -\dfrac{2\,x^3 + 3\,y}{3\,x + y - 1}\,, \\ y(1) = 2\,, \end{cases}$

(d) $\begin{cases} y' = \dfrac{9\,x^2 - 2\,x\,y}{x^2 + 2\,y + 1}\,, \\ y(0) = -3\,, \end{cases}$

(b) $\begin{cases} y' = \dfrac{9\,x^2 - 2\,x\,y}{x^2 + 2\,y + 1}\,, \\ y(0) = -3\,, \end{cases}$

(e) $\begin{cases} y' = \dfrac{2\,x\,y^2 + 4}{2\,(3 - x^2\,y)}\,, \\ y(-1) = 8\,, \end{cases}$

(c) $\begin{cases} y' = \dfrac{2\,x\,y^2 + 4}{2\,(3 - x^2\,y)}\,, \\ y(-1) = -8\,, \end{cases}$

(f) $\begin{cases} y' = \dfrac{\dfrac{2\,x\,y}{1 + x^2} - 2\,x}{2 - \ln\,(1 + x^2)}\,, \\ y(0) = 1\,. \end{cases}$

2. Using the method for exact equations, described in this § 5.5, prove that the solution of the initial value problem:

$$\begin{cases} y' = \dfrac{1 - 3\,y^3\,e^{3\,x\,y}}{3\,x\,y^2\,e^{3\,x\,y} + 2\,y\,e^{3\,x\,y}} \\ y(0) = 1 \end{cases}$$

is implicitly defined by $y^2\,e^{3\,x\,y} - x = 1$, and verify this result using the Dini Implicit function Theorem 3.25.

5.6 Integrating factor for non–exact equations

In § 5.5, we faced differential equations of the form (5.18), with the closure condition (5.20), essential to detect the solution; we recall both formulæ, for convenience:

$$M(x, y) + N(x, y)\,y' = 0\,, \qquad \frac{\partial}{\partial x}\,N(x, y) = \frac{\partial}{\partial y}\,M(x, y)\,.$$

The case is more frequent, though, in which condition (5.19) is not satisfied, so that we are unable to express the solution of the given differential equation in terms of the known functions.

There is, however, a general method of solution which, at times, allows the solution of the general differential equation to be formulated using the known functions. In formula (5.18a) below, although it can hardly be considered an orthodox procedure, we split the derivative y' and, then, rewrite (5.18) in the so–called *Pfaffian*[2] form:

$$M(x\,, y)\,\mathrm{d}x + N(x\,, y)\,\mathrm{d}y = 0\,. \tag{5.18a}$$

---

[2] Johann Friedrich Pfaff (1765–1825), German mathematician.

We do not assume condition (5.20). In this situation, there exists a function $\mu(x,y)$ such that, multiplying both sides of (5.18a) by μ, an equivalent equation is obtained which is exact, namely:

$$\mu(x,y)\ M(x,y)\ \mathrm{d}x + \mu(x,y)\ N(x,y)\ \mathrm{d}y = 0. \qquad (5.18\text{b})$$

This represents a theoretical statement, in the sense that it is easy to formulate conditions that need to be satisfied by the *integrating factor* μ, namely:

$$\frac{\partial}{\partial x}\big(\mu(x,y)\ N(x,y)\big) = \frac{\partial}{\partial y}\big(\mu(x,y)\ M(x,y)\big). \qquad (5.23)$$

Evaluating the partial derivatives (and employing a simplified subscript notation for partial derivatives), the partial differential equation for μ is obtained:

$$M(x,y)\ \mu_y - N(x,y)\ \mu_x = \big(N_x(x,y) - M_y(x,y)\big)\ \mu. \qquad (5.23\text{a})$$

Notice that solving (5.23a) may turn out to be harder than solving the original differential equation (5.18a). However, depending on the particular structure of the functions $M(x,y)$ and $N(x,y)$, there exist favorable situations in which it is possibile to detect the integrating factor $\mu(x,y)$, provided that some restrictions are imposed on μ itself. In the following Theorems 5.20 and 5.23, we describe what happens when μ depends on one variable only.

Theorem 5.20. Equation (5.18a) admits an integrating factor μ depending on x only, if the quantity:

$$\rho(x) = \frac{M_y(x,y) - N_x(x,y)}{N(x,y)} \qquad (5.24)$$

also depends on x only. In this case, it is:

$$\mu(x) = e^{\displaystyle\int \rho(x)\ \mathrm{d}x} \qquad (5.25)$$

with $\rho(x)$ given by (5.24).

Proof. Assume that $\mu(x,y)$ is a function of one variable only, say, it is a function of x only, thus:

$$\mu(x,y) = \mu(x), \qquad \mu_x = \frac{\mathrm{d}\mu}{\mathrm{d}x} = \mu', \qquad \mu_y = 0.$$

In this situation, equation (5.23a) reduces to:

$$N(x,y)\ \mu' = \big(M_y(x,y) - N_x(x,y)\big)\ \mu, \qquad (5.23\text{b})$$

that is:

$$\frac{\mu'}{\mu} = \frac{M_y(x,y) - N_x(x,y)}{N(x,y)}. \qquad (5.23\text{c})$$

Now, if the left–hand side of (5.23c) depends on x only, then (5.23c) is separable: solving it leads to the integrating factor represented in thesis (5.25). $\square$

Example 5.21. Consider the initial value problem:

$$\begin{cases} y' = -\dfrac{3\,x\,y - y^2}{x\,(x-y)}\,, \\ y(1) = 3\,. \end{cases} \tag{5.26}$$

Equation (5.26) is not exact, nor separable. Let us rewrite it in Pfaffian form, temporarily ignoring the initial condition:

$$(3\,x\,y - y^2)\,\mathrm{d}x + x\,(x-y)\,\mathrm{d}y = 0\,. \tag{5.26a}$$

Setting $M(x,y) = 3\,x\,y - y^2$ and $N(x,y) = x\,(x-y)$ yields:

$$\frac{M_y(x,y) - N_x(x,y)}{N(x,y)} = \frac{3\,x - 2\,y - (2\,x - y)}{x\,(x-y)} = \frac{1}{x}\,,$$

which is a function of x only. The hypotheses of Theorem 5.20 are fulfilled, and the integrating factor comes from (5.25):

$$\mu(x) = e^{\displaystyle\int \frac{1}{x}\,\mathrm{d}x} = x\,.$$

Multiplying equation (5.26a) by the integrating factor x, we form an exact equation, namely:

$$(3\,x^2\,y - x\,y^2)\,\mathrm{d}x + x^2\,(x-y)\,\mathrm{d}y = 0\,. \tag{5.26b}$$

Now, we can define the modified functions that constitute (5.26b):

$$M_1(x,y) = x\,M(x,y) = 3\,x^2\,y - x\,y^2\,,$$

$$N_1(x,y) = x\,N(x,y) = x^2\,(x-y)\,,$$

and employ them in equation (5.22), which also incorporates the initial condition:

$$\int_1^x M_1(t,3)\,\mathrm{d}t + \int_3^y N_1(x,s)\,\mathrm{d}s = 0\,,$$

that is:

$$\int_1^x (9\,t^2 - 9\,t)\,\mathrm{d}t + \int_3^y (x^3 - x^2\,s)\,\mathrm{d}s = 0\,.$$

Evaluating the integrals:

$$x^3\,y - \frac{x^2\,y^2}{2} + \frac{3}{2} = 0\,.$$

Solving for y, and recalling the initial condition, leads to the solution of the initial value problem (5.26):

$$y = x + \frac{\sqrt{3 + x^4}}{x}\,.$$

Example 5.22. Consider the following initial value problem, in which the differential equation is not exact, nor separable:

$$\begin{cases} y' = -\dfrac{4\,x\,y + 3\,y^2 - x}{x\,(x + 2y)}\,, \\ y(1) = 1\,. \end{cases} \tag{5.27}$$

Here, $M(x\,,y) = 4\,x\,y + 3\,y^2 - x$, $N(x\,,y) = x\,(x + 2\,y)$, so that the quantity below turns out to be a function of x :

$$\frac{M_y(x\,,y) - N_x(x\,,y)}{N(x\,,y)} = \frac{2}{x}\,.$$

Since the hypotheses of Theorem 5.20 are fulfilled, the integrating factor is given by (5.25):

$$\mu(x) = e^{\displaystyle\int \frac{2}{x}\,dx} = x^2\,.$$

After defining the modified functions:

$$M_1(x\,,y) = x^2\,M(x\,,y) = 4\,x^3\,y + 3\,x^2\,y^2 - x^3\,,$$

$$N_1(x\,,y) = x^2\,N(x\,,y) = x^3\,(x + 2\,y)\,,$$

we can use them into equation (5.22), which also incorporates the initial condition, obtaining:

$$\int_1^x M_1(t\,,1)\,dt + \int_1^y N_1(x\,,s)\,ds = 0\,,$$

that is:

$$\int_1^x (3\,t^3 + 3\,t^2)\,dt + \int_1^y x^3\,(2\,s + x)\,ds = 0\,.$$

Evaluating the integrals yields:

$$x^4\,y - \frac{x^4}{4} + x^3\,y^2 - \frac{7}{4} = 0\,.$$

Solving for y and recalling the initial condition, we get the solution of the initial value problem (5.27):

$$y = \frac{\sqrt{x^5 + x^4 + 7}}{2\,x^{\frac{3}{2}}} - \frac{x}{2}\,.$$

We examine, now, the case in which the integrating factor μ is a function of y only. Given the analogy with Theorem 5.20, the proof of Theorem 5.23 is not provided here.

Theorem 5.23. Equation (5.18a) admits an integrating factor μ depending on y only, if the quantity:

$$\rho(y) = \frac{N_x(x,y) - M_y(x,y)}{M(x,y)} \tag{5.28}$$

also depends on y only. In this case, the integrating factor is:

$$\mu(y) = e^{\int \rho(y)\, dy} \tag{5.29}$$

with $\rho(y)$ given by (5.28).

Example 5.24. Consider the initial value problem, with non–separable and non–exact differential equation:

$$\begin{cases} y' = -\dfrac{y\,(x + y + 1)}{x\,(x + 3\,y + 2)}\,, \\ y(1) = 1\,. \end{cases} \tag{5.30}$$

Functions $M(x,y) = y\,(x + y + 1)$ and $N(x,y) = x\,(x + 3\,y + 2)$ are such that the following quantity is dependent on y only:

$$\frac{N_x(x,y) - M_y(x,y)}{M(x,y)} = \frac{1}{y}\,.$$

Formula (5.29) then leads to the integrating factor $\mu(y) = y$, which in turn leads to the following exact equation, written in Pfaffian form:

$$y^2\,(x + y + 1)\, dx + x\,y\,(x + 3\,y + 2)\, dy = 0\,.$$

Define the modified functions:

$$M_1(x,y) = x\,M(x,y) = y^2\,(x + y + 1)\,,$$

$$N_1(x,y) = x\,N(x,y) = x\,y\,(x + 3\,y + 2)\,,$$

and employ them into equation (5.22), which also incorporates the initial condition, obtaining:

$$\int_1^x M_1(t,1)\, dt + \int_1^y N_1(x,s)\, ds\,,$$

that is:

$$\int_1^x (2 + t)\, dt + \int_1^y s\,x\,(2 + 3\,s + x)\, ds = 0\,.$$

The solution to (5.30) can be thus expressed, in implicit form, as:

$$\frac{x^2\,y^2}{2} + x\,y^3 + x\,y^2 - \frac{5}{2} = 0\,.$$

To end this § 5.6, let us consider the situation of a family of differential equations for which an integrating factor μ is available.

Theorem 5.25. Let $Q = \{(x,y) \in \mathbb{R}^2 \mid 0 < a < x < b, \quad 0 < c < x < d\}$, and let f_1 and f_2 be $\mathcal{C}^1$ functions on Q, such that $f_1(x,y) - f_2(x,y) \neq 0$. Define the functions $M(x,y)$ and $N(x,y)$ as:

$$M(x,y) = y\, f_1(x,y), \qquad N(x,y) = x\, f_2(x,y).$$

Then:

$$\mu(x,y) = \frac{1}{x\,y\,(f_1(x,y) - f_2(x,y))}$$

is an integrating factor for:

$$y' = -\frac{y\, f_1(x,y)}{x\, f_2(x,y)}.$$

Proof. It suffices to insert the above expressions of μ, M and N into condition (5.23) and verify that it gets satisfied. $\qquad\square$

5.6.1   Exercises

1. Solve the following initial value problems, using a suitable integrating factor.

(a) $\begin{cases} y' = -\dfrac{y\,(x+y)}{x+2\,y-1}, \\ y(1) = 1, \end{cases}$

(b) $\begin{cases} y' = -\dfrac{y}{x\,(y^2 - \ln x)}, \\ y(1) = 1, \end{cases}$

(c) $\begin{cases} y' = \dfrac{y}{y - 3\,x - 3}, \\ y(0) = 0, \end{cases}$

(d) $\begin{cases} y' = x - y, \\ y(0) = 0, \end{cases}$

(e) $\begin{cases} y' = \dfrac{3\,x + 2\,y^2}{2\,x\,y}, \\ y(1) = 1, \end{cases}$

(f) $\begin{cases} y' = \dfrac{y - 2\,x^3}{x}, \\ y(1) = 1, \end{cases}$

(g) $\begin{cases} y' = \dfrac{y}{y^3 - 3\,x}, \\ y(0) = 1, \end{cases}$

(h) $\begin{cases} y' = -\dfrac{y^3 + 2\,y\,e^x}{e^x + 3\,y^2}, \\ y(0) = 0. \end{cases}$

5.7   Linear equations of first order

Consider the differential equation:

$$\begin{cases} y'(x) = a(x)\,y(x) + b(x), \\ y(x_0) = y_0. \end{cases} \tag{5.31}$$

Let functions $a(x)$ and $b(x)$ be continuous on the interval $I \subset \mathbb{R}$. The first–order differential equation (5.31) is called *linear*, since y is represented by a polynomial of degree 1. We can establish a formula for its integration, following a procedure that is similar to what we did for separable equations.

Theorem 5.26. The unique solution to (5.31) is:

$$y(x) = e^{\int_{x_0}^x a(t)\,dt} \left(y_0 + \int_{x_0}^x b(t)\, e^{-\int_{x_0}^t a(s)\,ds}\, dt \right)$$

i.e., in a more compact form:

$$y(x) = e^{A(x)} \left(y_0 + \int_{x_0}^x b(t)\, e^{-A(t)}\, dt \right) \tag{5.32}$$

where:

$$A(x) = \int_{x_0}^x a(s)\,ds\,. \tag{5.33}$$

Proof. To arrive at formula (5.32), we first examine the case $b(x) = 0$, for which (5.31) reduces to the separable (and linear) equation:

$$\begin{cases} y'(x) = a(x)\,y(x)\,, \\ y(x_0) = c\,, \end{cases} \tag{5.34}$$

having set $y_0 = c$. If $x_0 \in I$, the solution of (5.34) through point (x_0, c) is:

$$y(x) = c\,e^{A(x)}\,. \tag{5.35}$$

To find the solution to the more general differential equation (5.31), we use the method of the *variation of parameters* [3], due to Lagrange: we assume that c is a function of x and search for $c(x)$ such that the function:

$$y(x) = c(x)\,e^{A(x)} \tag{5.36}$$

becomes, indeed, a solution of (5.31). To this aim, differentiate (5.36):

$$y'(x) = c'(x)\,e^{A(x)} + c(x)\,a(x)\,e^{A(x)}$$

and impose that function (5.36) solves (5.31), that is:

$$c'(x)\,e^{A(x)} + c(x)\,a(x)\,e^{A(x)} = a(x)\,c(x)\,e^{A(x)} + b(x)\,,$$

from which:

$$c'(x) = b(x)\,e^{-A(x)}\,. \tag{5.37}$$

---

[3]See, for example, mathworld.wolfram.com/VariationofParameters.html

Integrating (5.37) between x_0 and x, we obtain:

$$c(x) = \int_{x_0}^{x} b(t) e^{-A(t)} \, dt + K \,,$$

with K constant. Finally, the solution to (5.31) is:

$$y(x) = e^{A(x)} \left(\int_{x_0}^{x} b(t) e^{-A(t)} \, dt + K \right).$$

Evaluating $y(x_0)$ and recalling the initial condition in (5.31), we see that $y_0 = K$. Thesis (5.32) thus follows. $\square$

Remark 5.27. An alternative proof to Theorem 5.26 can be provided, using Theorem 5.20 and the integrating factor procedure. In fact, if we assume:

$$M(x,y) = a(x)\, y(x) + b(x) \,, \qquad N(x,y) = -1 \,,$$

then:

$$\frac{M_y(x,y) - N_y(x,y)}{N(x,y)} = -a(x) \,,$$

which yields the integrating factor $\mu(x) = e^{-A(x)}$, with $A(x)$ defined as in (5.33). Considering the following exact equation, equivalent to (5.31):

$$y'(x) = -\frac{e^{-A(x)} \left(a(x)\, y + b(x) \right)}{-e^{-A(x)}} \,,$$

and employing relation (5.22), we obtain:

$$\int_{x_0}^{x} \left(a(t)\, y_0 + b(t) \right) e^{-A(t)} \, dt - \int_{y_0}^{y} e^{-A(x)} \, ds = 0 \,,$$

which, after some straightforward computations, yields formula (5.32).

Remark 5.28. The general solution of the linear differential equation (5.31) can be described when a particular solution of it is known, together with the general solution of the linear and separable equation (5.34).

If y_1 and y_2 are both solutions of (5.31), in fact, there exist $v_1, v_2 \in \mathbb{R}$ such that:

$$y_1(x) = e^{A(x)} \left(v_1 + \int_{x_0}^{x} b(t) e^{-A(t)} \, dt \right),$$

$$y_2(x) = e^{A(x)} \left(v_2 + \int_{x_0}^{x} b(t) e^{-A(t)} \, dt \right).$$

Subtracting, we obtain:

$$y_1(x) - y_2(x) = \left(v_1 - v_2 \right) e^{A(x)} \,,$$

which means that $y_1 - y_2$ has the form (5.35) and, therefore, solves (5.34). Now, using the fact that y_1 is a solution of (5.31), the general solution to (5.31) can be written as:

$$y(x) = c\,e^{A(x)} + y_1(x), \qquad c \in \mathbb{R},$$

and all this is equivalent to saying that the general solution $y = y(x)$ of (5.31) can be written in the form:

$$\frac{y - y_1}{y_2 - y_1} = c, \qquad c \in \mathbb{R}.$$

Example 5.29. Consider the equation:

$$\begin{cases} y'(x) = 3\,x^2\,y(x) + x\,e^{x^3}, \\[2mm] y(0) = 1. \end{cases}$$

Here, $a(x) = 3\,x^2$ and $b(x) = x\,e^{x^3}$. Using (5.32)–(5.33), we get:

$$A(x) = \int_{x_0}^{x} a(s)\,ds = \int_{0}^{x} 3\,s^2\,ds = x^3$$

and

$$\int_{x_0}^{x} b(t)\,e^{-A(t)}\,dt = \int_{0}^{x} t\,e^{t^3}\,e^{-t^3}\,dt = \frac{x^2}{2},$$

so that:

$$y(x) = e^{x^3}\left(1 + \frac{x^2}{2}\right).$$

Example 5.30. Consider the equation:

$$\begin{cases} y'(x) = 2\,x\,y(x) + x, \\ y(0) = 2. \end{cases}$$

Note that $a(x) = 2\,x$, $b(x) = x$. Then:

$$A(x) = \int_{0}^{x} a(s)\,ds = \int_{0}^{x} 2\,s\,ds = x^2$$

and

$$\int_{0}^{x} b(t)\,e^{-A(t)} = \int_{0}^{x} t\,e^{-t^2}\,dt = \left[-\frac{e^{-t^2}}{2}\right]_{0}^{x} = \frac{1}{2} - \frac{e^{-x^2}}{2}.$$

Therefore, the solution is:

$$y(x) = e^{x^2}\left(2 + \frac{1}{2} - \frac{e^{-x^2}}{2}\right) = \frac{5}{2}\,e^{x^2} - \frac{1}{2}.$$

Remark 5.31. When, in equation (5.32), functions $a(x)$ and $b(x)$ are constant, we obtain:

$$\begin{cases} y'(x) = a\, y(x) + b\,, \\ y(x_0) = y_0\,, \end{cases} \tag{5.32a}$$

and the solution is given by:

$$y(x) = \left(y_0 + \frac{b}{a} \right) e^{a\,(x - t_0)} - \frac{b}{a}\,.$$

5.7.1   Exercises

Solve the following initial value problems for linear equations.

1. $\begin{cases} y'(x) = -\dfrac{1}{1+x^2}\, y(x) + \dfrac{1}{1+x^2}\,, \\ y(0) = 0\,, \end{cases}$

4. $\begin{cases} y'(x) = \dfrac{1}{x}\, y(x) + x^2\,, \\ y(1) = 0\,, \end{cases}$

2. $\begin{cases} y'(x) = -\sin(x)\, y(x) + \sin(x)\,, \\ y(0) = 0\,, \end{cases}$

5. $\begin{cases} y'(x) = 3\,x^2\, y(x) + x\, e^{x^3}\,, \\ y(0) = 1\,, \end{cases}$

3. $\begin{cases} y'(x) = \dfrac{x}{1+x^2}\, y(x) + 1\,, \\ y(0) = 0\,, \end{cases}$

6. $\begin{cases} y'(x) = x - \dfrac{1}{3\,x}\, y(x)\,, \\ y(1) = 1\,. \end{cases}$

5.8   Bernoulli equation

A *Bernoulli*[4] differential equation, with *exponent* α, has the form:

$$y'(x) = a(x)\, y(x) + b(x)\, y^\alpha(x)\,. \tag{5.38}$$

Let us assume $\alpha \neq 0, 1$, so that (5.38) is non–linear. The change of variable:

$$v(x) = y^{1-\alpha}(x)$$

transforms (5.38) into a linear equation:

$$v'(x) = (1 - \alpha)\, a(x)\, v(x) + (1 - \alpha)\, b(x)\,. \tag{5.39}$$

---

[4] Jacob Bernoulli (1654–1705), Swiss mathematician.

Example 5.32. Consider the differential equation:

$$\begin{cases} y' = -\dfrac{1}{x}\, y - \dfrac{1}{x}\, y^4 \,, \\ y(2) = 1 \,. \end{cases}$$

Here $\alpha = 4$, and the change of variable is $v(x) = y^{-3}(x)$, i.e., $y = v^{-\frac{1}{3}}$, leading to:

$$-\frac{1}{3}\, v^{-\frac{4}{3}}\, v' = -\frac{1}{x}\, v^{-\frac{1}{3}} - \frac{1}{x}\, v^{-\frac{4}{3}} \,.$$

Multiplication by $v^{\frac{4}{3}}$ yields a linear differential equation in v :

$$-\frac{1}{3}\, v' = -\frac{1}{x}\, v - \frac{1}{x} \,.$$

Simplifying and recalling the initial condition, we obtain a linear initial value problem:

$$\begin{cases} v' = \dfrac{3}{x}\, v + \dfrac{3}{x} \,, \\ v(2) = 1 \,, \end{cases}$$

solved by $v(x) = \dfrac{1}{4}\, x^3 - 1$. Hence, the solution to the original problem is:

$$y(x) = \sqrt[3]{\frac{4}{x^3 - 4}} \,.$$

Example 5.33. Given $x > 0$, solve the initial value problem:

$$\begin{cases} y'(x) = \dfrac{1}{2\,x}\, y(x) + 5\,x^2\, y^3(x) \,, \\ y(1) = 1 \,. \end{cases}$$

This is a Bernoulli equation with exponent $\alpha = 3$. Consider the change of variable:

$$y(x) = \left(v(x)\right)^{\frac{1}{1-3}} = \left(v(x)\right)^{-\frac{1}{2}} \,.$$

The associated linear equation in $v(x)$ is:

$$v'(x) = -\frac{1}{x}\, v(x) - 10\,x^2 \,,$$

which is solved by:

$$v(x) = \frac{7}{2\,x} - \frac{5}{2}\, x^3 \,.$$

To recover $y(x)$, we have to assume:

$$\frac{7}{2\,x} - \frac{5}{2}\, x^3 > 0 \qquad \Longleftrightarrow \qquad 0 < x < \sqrt[4]{\frac{7}{5}} \,.$$

Finally:

$$y(x) = \frac{1}{\sqrt{\dfrac{7}{2x} - \dfrac{5}{2}\, x^3}}, \qquad \text{with} \quad 0 < x < \sqrt[4]{\dfrac{7}{5}}.$$

Example 5.34. Solve the initial value problem:

$$\begin{cases} y'(x) = -\dfrac{1}{4x}\, y(x) - \dfrac{1}{4}\, x^2\, y^5(x), \\ y(1) = 1. \end{cases}$$

We have to solve a Bernoulli equation, with exponent $\alpha = 5$. The change of variable is:

$$y(x) = \left(v(x)\right)^{\frac{1}{1-5}} = \left(v(x)\right)^{-\frac{1}{4}},$$

which leads to the transformed linear equation:

$$\begin{cases} v'(x) = \dfrac{1}{x}\, v(x) + x^2, \\ v(1) = 1, \end{cases}$$

solved by:

$$v(x) = \frac{x + x^3}{2}.$$

Recovering $y(x)$, we find:

$$y(x) = \sqrt[4]{\frac{2}{x + x^3}},$$

that is defined for $x > 0$.

Remark 5.35. Bernoulli equation (5.38) can also be solved using the same approach used for linear equations, that is, imposing a solution of the form:

$$y(x) = c(x)\, e^{A(x)}$$

and obtaining the separable equation for $c(x)$:

$$c'(x) = b(x)\, e^{(\alpha-1)\, A(x)}\, c^{\alpha}(x),$$

so that:

$$c(x) = \left((1-\alpha)\, F(x) + c_0^{1-\alpha}\right)^{\frac{1}{1-\alpha}},$$

where:

$$F(x) = \int_{x_0}^{x} b(z)\, e^{(\alpha-1)\, A(z)}\, dz.$$

5.8.1   Exercises

Solve the following initial value problems for Bernoulli equations.

1. $\begin{cases} y' = -\dfrac{1}{x} - \dfrac{1}{\ln x}\, y^2\,, \\ y(2) = 1\,, \end{cases}$

3. $\begin{cases} y' = \dfrac{2\,x}{1+x}\, y + 2\sqrt{y}\,, \\ y(0) = 1\,, \end{cases}$

2. $\begin{cases} y' = x^2\, y + \left(x^5 + x^2\right) y^{\frac{2}{3}}\,, \\ y(0) = 1\,, \end{cases}$

4. $\begin{cases} y' = y + x\, y^2\,, \\ y(0) = 1\,, \end{cases}$

5. $\begin{cases} y'(x) - y(x) + (\cos x)\, y(x)^2 = 0\,, \\ y(0) = 1\,. \end{cases}$

5.9   Riccati equation

A *Riccati differential equation* has the following form:

$$y' = a(x) + b(x)\, y + c(x)\, y^2\,. \tag{5.40}$$

The solving strategy is based on knowing one particular solution $y_1(x)$ of (5.40). Then, it is assumed that the other solutions of (5.40) have the form $y(x) = y_1(x) + u(x)$, where $u(x)$ is an unknown function, to be found, and that solves the associated Bernoulli equation:

$$u'(x) = \big(\, b(x) + 2\, c(x)\, y_1(x)\,\big)\, u(x) + c(x)\, u^2(x)\,. \tag{5.40a}$$

Indeed (temporarily discarding the dependence on x), from $y = y_1 + u$ it follows:

$$(y_1 + u)' = a + b\,(y_1 + u) + c\,(y_1 + u)^2\,,$$
$$y_1' + u' = a + b\, y_1 + b\, u + c\, y_1^2 + 2\, c\, y_1\, u + c\, u^2\,,$$

where y_1' (in the left–hand side) and $a + b\, y_1 + c\, y_1^2$ (in the right–hand side) cancel out.

Another way to form (5.40a) is via the substitution:

$$y(x) = y_1(x) + \frac{1}{u(x)}\,,$$

which transforms (5.40) directly into the linear equation:

$$u'(x) = -c(x) - \big(\, b(x) + 2\, c(x)\, y_1(x)\,\big)\, u(x)\,. \tag{5.41}$$

Notice that, in the latter way, we combine together two substitutions: the first one maps the Riccati[5] equation into a Bernoulli equation; the second one linearizes the Bernoulli equation.

Example 5.36. Knowing that $y_1(x) = 1$ solves the Riccati equation:

$$y' = -\frac{1+x}{x} + y + \frac{1}{x}\, y^2\,, \tag{5.42}$$

---

[5] Jacopo Francesco Riccati (1676–1754), Italian mathematician and jurist.

we want to show that the general solution to equation (5.42) is:

$$y(x) = 1 + \frac{x^2 e^x}{c + (1 - x) e^x}.$$

Let us use the change of variable:

$$y(x) = 1 + \frac{1}{u(x)},$$

to obtain the linear equation:

$$u'(x) = -\frac{1}{x} - \left(1 + \frac{2}{x}\right) u(x). \qquad (5.42a)$$

To solve it, we proceed as we learned. First, compute $A(x)$:

$$A(x) = -\int \left(1 + \frac{2}{x}\right) dx = -x - 2\ln x \qquad \Longrightarrow \qquad e^{A(x)} = \frac{e^{-x}}{x^2}.$$

Then, form:

$$\int b(x)\, e^{-A(x)}\, dx = -\int \frac{1}{x} x^2\, e^x\, dx = -\int x\, e^x\, dx = (1 - x)\, e^x.$$

The solution to (5.42a) is, therefore:

$$u(x) = \frac{e^{-x}}{x^2}\left(c + (1 - x)\, e^x\right) = \frac{c e^{-x} + 1 - x}{x^2} = \frac{c + (1 - x)\, e^x}{x^2\, e^x}.$$

Finally, the solution to (5.42) is:

$$y(x) = 1 + \frac{1}{u(x)} = 1 + \frac{x^2\, e^x}{c + (1 - x)\, e^x}.$$

Example 5.37. Using the fact that $y_1(x) = x$ solves:

$$y'(x) = -x^5 + \frac{y(x)}{x} + x^3\, y^2(x), \qquad (5.43)$$

find the general solution of (5.43).

The substitution $y = x + \frac{1}{v}$ leads to the linear differential equation:

$$1 - \frac{v'}{v^2} = -x^5 + \frac{x + \frac{1}{v}}{x} + x^3\left(x + \frac{1}{v}\right)^2 \qquad \Longrightarrow \qquad v' = -\frac{2x^5 + 1}{x}\, v - x^3,$$

whose solution is:

$$v(x) = \frac{c\, e^{-\frac{2x^5}{5}}}{x} - \frac{1}{2x},$$

where c is an integration constant. The general solution of (5.43) is, therefore:

$$y = \frac{2x}{2\, c\, e^{-\frac{2x^5}{5}} - 1} + x.$$

Remark 5.38. In applications, it may be useful to state conditions on the co-efficient functions $a(x), b(x)$ and $c(x)$, with the aim that the relevant Riccati equation (5.40) is solved by some particular function having simple form. The following list summarises such conditions and, for each one, the correspondent simple–form solution y_1.

1. Monomial solution:
   if $a(x) + x^{n-1} \left(x\, b(x) + c(x)\, x^{n+1} - n \right) = 0$, then $y_1(x) = x^n$.

2. Exponential solution:
   if $a(x) + e^{n\,x} \left(b(x) + c(x)\, e^{n\,x} - n \right) = 0$, then $y_1(x) = e^{n\,x}$.

3. Exponential monomial solution:
   if $a(x) + e^{n\,x} \left(x\, b(x) + x^2\, c(x)\, e^{n\,x} - n\,x - 1 \right) = 0$, then $y_1(x) = x\, e^{n\,x}$.

4. Sine solution: if $a(x) + b(x)\, \sin(n\,x) + c(x)\, \sin^2(n\,x) - n\, \cos(n\,x) = 0$,
   then $y_1(x) = \sin(n\,x)$.

5. Cosine solution: if $a(x) + b(x)\, \cos(n\,x) + c(x)\, \cos^2(n\,x) + n\, \sin(n\,x) = 0$,
   then $y_1(x) = \cos(n\,x)$.

5.9.1   Cross–Ratio property

Solutions of Riccati equations posses some peculiar properties, due to the connection with linear equations, as explained in the following Theorem 5.39.

Theorem 5.39. Given any three functions y_1, y_2, y_3, which satisfy (5.40), then the general solution y of (5.40) can be expressed in the form:

$$\frac{y - y_2}{y - y_1} = c\, \frac{y_3 - y_2}{y_3 - y_1}. \tag{5.44}$$

Proof. We saw in § 5.9 that, if y_1 is a solution of (5.40), then solutions y_2 and y_3 will be determined by two particular choices of u in the substitution:

$$y = y_1 + \frac{1}{u}.$$

Let us denote such functions with u_2 and u_3, respectively:

$$y_2 = y_1 + \frac{1}{u_2}, \qquad y_3 = y_1 + \frac{1}{u_3}.$$

Recalling that u_2 and u_3 are solutions to the linear equation (5.41), we know that the general solution of (5.41) can be written as shown in Remark 5.28:

$$\frac{u - u_2}{u_3 - u_2} = c.$$

At this point, employing the reverse substitution, and following [32] (page 23):

$$u = \frac{1}{y - y_1}, \quad u_2 = \frac{1}{y_2 - y_1}, \quad u_3 = \frac{1}{y_3 - y_1},$$

we arrive at formula (5.44), representing the general solution of (5.40). $\square$

A consequence of Theorem 5.39 is the so-called *Cross–Ratio* property of the Riccati equation, illustrated in the following Corollary 5.40.

Corollary 5.40. Given any four solutions $y_1, \ldots, y_4$ of the Riccati equation (5.40), their Cross–Ratio is constant and is given by the quantity:

$$\frac{y_4 - y_2}{y_4 - y_1} \frac{y_3 - y_1}{y_3 - y_2}. \tag{5.45}$$

Proof. Relation (5.44) implies that, if y_4 is a fourth solution of (5.40), then:

$$\frac{y_4 - y_2}{y_4 - y_1} = c \, \frac{y_3 - y_2}{y_3 - y_1},$$

which, since c is constant, demonstrates thesis (5.45). $\square$

5.9.2 Reduced form of the Riccati equation

The particular differential equation:

$$u' = A_0(x) + A_1(x) \, u^2$$

is known as *reduced form* of the Riccati equation (5.40). Functions $A_0(x)$ and $A_1(x)$ are related to functions $a(x), b(x)$ and $c(x)$ appearing in (5.40). In fact, if $B(x)$ is a primitive of $b(x)$, i.e., $B'(x) = b(x)$, the change of variable:

$$u = e^{-B(x)} y \tag{5.46}$$

trasforms (5.40) into the *reduced Riccati* equation:

$$u' = a(x) \, e^{-B(x)} + c(x) \, e^{B(x)} \, u^2. \tag{5.40b}$$

This can be seen by computing $u' = e^{-B(x)} \left(y' - y \, B'(x) \right)$ from (5.46) and then substituting, in the factor $y' - y \, B'(x)$, the equalities $B'(x) = b(x)$ and $y' = a(x) + b(x) \, y + c(x) \, y^2$, and finally $y = e^{B(x)} \, u$.

Sometimes, given a Riccati equation, its solution can be obtained by simply transforming it to its reduced form. Example 5.41. illustrates this fact.

Example 5.41. Consider the initial value problem for the Riccati equation:

$$\begin{cases} y' = 1 - \dfrac{1}{2\,x} \, y + x \, y^2, \\ y(1) = 0. \end{cases}$$

To obtain its reduced form, define:

$$B(x) = \int -\frac{1}{2\,x}\mathrm{d}x = -\frac{1}{2}\ln x\,,$$

and employ the change of variable:

$$y = e^{B(x)}\,u = e^{-\frac{1}{2}\ln x}\,u = x^{-\frac{1}{2}}\,u\,.$$

The reduced (separable) Riccati equation is, then:

$$\begin{cases} u' = x^{\frac{1}{2}}\left(1 + u^2\right)\,, \\ u(1) = 0\,, \end{cases}$$

whose solution is:

$$u(x) = \tan\left(\frac{1}{3}\left(2\,x^{\frac{3}{2}} - 2\right)\right)\,.$$

Remark 5.42. The reduced Riccati equation is also separable if and only if there exists a real number λ such that:

$$\lambda\,a(x)\,e^{-B(x)} = c(x)\,e^{B(x)}\,.$$

In other words, to have separability of the reduced equation, the function:

$$\frac{c(x)}{a(x)}\,e^{2\,B(x)}$$

has to be constant and equal to a certain real number λ. The topic of separability of the Riccati equation is presented in [2, 48, 49, 57, 61].

5.9.3 Connection with the linear equation of second order

Second–order differential equations will be discussed in Chapter 6, but the study of a particular second–order differential equation is anticipated here, since it is related to the Riccati equation.

A linear differential equation of second order has the form:

$$y'' + P(x)\,y' + Q(x)\,y = 0\,, \tag{5.47}$$

where $P(x)$ and $Q(x)$ are given continuous functions, defined on an interval $I \subset \mathbb{R}$. The term *linear* indicates that the unknown function $y = y(x)$ and its derivatives appear in polynomial form of degree one.

The second–order linear differential equation (5.47) is equivalent to a particular Riccati equation. We follow the fair exposition given in Chapter 15 of [46]. Let us introduce a new variable $u = u(x)$, setting:

$$y = e^{-U(x)}\,, \qquad \text{with}\quad U(x) = -\int u(x)\,\mathrm{d}x\,. \tag{5.48}$$

Compute the first and second derivatives of y, respectively:

$$y' = -u\,e^{-U(x)}, \qquad\qquad y'' = e^{-U(x)}\left(u^2 - u'\right).$$

Equation (5.47) gets then transformed into:

$$e^{-U(x)}\left(u^2 - u'\right) - P(x)\,u\,e^{-U(x)} + Q(x)\,e^{-U(x)} = 0,$$

simplifying which leads to the non–linear Riccati differential equation of first order:

$$u' = Q(x) - P(x)\,u + u^2. \tag{5.47a}$$

Vice versa, to find a linear differential equation, of second order, that is equivalent to the first–order Riccati equation (5.40), let us proceed as follows. Consider the transformation:

$$y = -\frac{w'}{c(x)\,w}, \tag{5.49}$$

with first derivative:

$$y' = \frac{w\,c'(x)\,w' - c(x)\,w\,w'' + c(x)\,(w')^2}{c^2(x)\,w^2}.$$

Now, apply transformation (5.49) to the right–hand side of (5.40), i.e., to $a(x) + b(x)\,y + c(x)\,y^2$:

$$a(x) - \frac{b(x)\,w'}{c(x)\,w} + \frac{(w')^2}{c(x)\,w^2}.$$

By comparison, and after some algebra, we arrive at:

$$\frac{-a(x)\,c^2 w + b(x)\,c(x)\,w' + c'(x)\,w' - c(x)\,w''}{c^2\,w} = 0,$$

that is a linear differential equation of second order:

$$c(x)\,w'' - (b(x)\,c(x) + c'(x))\,w' + a(x)\,c^2(x)\,w = 0 \tag{5.47b}$$

equivalent to the Riccati equation (5.40).

Example 5.43. Consider the linear differential equation of order 2 :

$$y'' - \frac{x}{1 - x^2}\,y' + \frac{1}{1 - x^2}\,y = 0, \qquad \text{with} \quad -1 < x < 1. \tag{5.50}$$

Following the notations in (5.47):

$$P(x) = -\frac{x}{1 - x^2}, \qquad Q(x) = \frac{1}{1 - x^2},$$

and using the transformation (5.48), we arrive at the Riccati equation:

$$u' = \frac{1}{1-x^2} + \frac{x}{1-x^2}\,u + u^2\,. \tag{5.50a}$$

To obtain the reduced form of (5.50a), we employ the transformation (5.46), observing that, here, such a transformation works in the following way:

$$b(x) = \frac{x}{1-x^2} \quad\Longrightarrow\quad B(x) = \int b(x)\,dx = -\frac{1}{2}\ln(1-x^2)$$

$$\Longrightarrow\quad v = e^{-B(x)}\,u = \sqrt{1-x^2}\,u\,.$$

The reduced Riccati (separable) differential equation is, then:

$$v' = \frac{1}{\sqrt{1-x^2}} + \frac{1}{\sqrt{1-x^2}}\,v^2\,, \tag{5.51}$$

whose general solution is:

$$v(x) = \tan\big(\arcsin(x)+c\big)\,.$$

Recovering the u variable, we get the solution to equation (5.50a):

$$u(x) = \frac{\tan\big(\arcsin(x)+c\big)}{\sqrt{1-x^2}}\,.$$

To get the solution of the linear equation (5.50), we reuse relation (5.48). First, a primitive $U(x)$ of $u(x)$ must be found:

$$U(x) = \int u(x)\,dx = \int \frac{\tan\big(\arcsin(x)+c\big)}{\sqrt{1-x^2}}\,dx = -\ln\big(\cos(\arcsin(x)+c)\big) - K\,.$$

where $\pm K$ is a constant whose sign can be made positive. Then, using (5.48), we can conclude that the general solution to (5.50) is:

$$y(x) = e^{\ln\big(\cos\big(\arcsin(x)+c\big)\big)+K} = e^K\,\cos\big(\arcsin(x)+c\big)\,.$$

5.9.4   Exercises

1. Knowing that $y_1(x) = 1$ is a solution of:

$$y'(x) = -1 - e^x + e^x\,y(x) + y^2(x)\,, \tag{5.52}$$

show that the general solution to (5.52) is:

$$y(x) = 1 + \frac{e^{2x+e^x}}{c - e^{e^x}\,(e^x - 1)}\,.$$

Hint: $\displaystyle\int e^{2x+e^x}\,dx = e^{e^x}\,(e^x - 1)\,.$

2. Find the general solution of the equation:

$$y'(x) = -\frac{2}{x^2} + y^2(x) \,,$$

knowing that $y_1(x) = \dfrac{1}{x}$ is a particular solution.

3. Find the general solution of the equation:

$$y'(x) = -1 + \frac{1}{x} y + \frac{1}{x^2} y^2(x) \,,$$

knowing that $y_1(x) = x$ is a particular solution.

4. Solve the linear differential equation of second order:

$$x^2 y'' + 3 x y' + y = 0 \,.$$

Hint: Transform the equation into a Riccati equation, in reduced form, and then use the fact that $v_1(x) = x^2$ is a particular solution.

5. Solve the linear differential equation of order 2 :

$$(1 + x^2) y'' - 2 x y' + 2 y = 0 \,.$$

Hint: Transform the equation into a Riccati equation, then solve it, using the fact that $u_1(x) = -\dfrac{1}{x}$ is a particular solution.

5.10   Change of variable

A differential equation can be solved, at times, using an appropriate change of variable. Aim of this § 5.10 is to provide a short account on how to apply any change of variable, to a given differential equation, in a correct way. Details on how to find a change of variable, capable of transforming a given differential equation into a simpler (possibly, the simplest) form, can be found in [31, 4].

Consider the differential equation:

$$y'_x = f(x \,, y(x)) \,, \tag{5.53}$$

where, the subscript x emphasizes that we are considering the x-derivative, in contrast with the fact that, below, we are also going to form the derivative with respect to a new variable.

Consider, in fact, the mapping:

$$(x, y) \mapsto (X, Y) \,,$$

where $X = X(x, y)$ and $Y = Y(x, y)$ are C^1 functions, with:

$$\det \begin{bmatrix} X_x(x, y) & X_y(x, y) \\ Y_x(x, y) & Y_y(x, y) \end{bmatrix} \neq 0.$$

This last condition ensures uniqueness for the (x, y)–solution of the system:

$$\begin{cases} X = X(x, y), \\ Y = Y(x, y). \end{cases}$$

Equation (5.53) gets, thus, changed into:

$$\frac{D_x Y(x, y)}{D_x X(x, y)} = \frac{Y_x + Y_y \, y'_x}{X_x + X_y \, y'_x}$$

i.e.,

$$Y'_X = \frac{Y_x + f(x, y) \, Y_y}{X_x + f(x, y) \, X_y}. \tag{5.53a}$$

The right–hand side of (5.53a) contains (x, y). To complete the coordinate change, we have to reverse the mapping, solving the system:

$$\begin{cases} X = X(x, y), \\ Y = Y(x, y), \end{cases} \implies \begin{cases} x = \hat{x}(X, Y), \\ y = \hat{y}(X, Y), \end{cases}$$

and substituting the found expressions for x and y into (5.53a).

Example 5.44. Consider the Riccati equation presented in [31]:

$$y'_x = x \, y^2 - \frac{2 \, y}{x} - \frac{1}{x^3}. \tag{5.54}$$

Use the change of variables $(X, Y) = (x^2 \, y, \ln x)$, so that:

$$Y_x = D_x Y(x, y) = D_x(\ln x) = \frac{1}{x}, \qquad Y_y = D_y Y(x, y) = D_y(\ln x) = 0,$$

$$X_x = D_x X(x, y) = D_x(x^2 \, y) = 2 \, x \, y, \qquad X_y = D_y X(x, y) = D_y(x^2 \, y) = x^2,$$

thus:

$$Y'_X = \frac{\frac{1}{x} + \left(x \, y^2 - \frac{2 \, y}{x} - \frac{1}{x^3}\right) 0}{2 \, x \, y + \left(x \, y^2 - \frac{2 \, y}{x} - \frac{1}{x^3}\right) x^2} = \frac{1}{x^4 \, y^2 - 1}. \tag{5.54a}$$

Coordinates (x, y) can be expressed in terms of the new coordinates (X, Y) as:

$$\begin{cases} X = x^2 \, y, \\ Y = \ln x, \end{cases} \implies \begin{cases} x = e^Y, \\ y = X \, e^{-2Y}. \end{cases}$$

Substitution into (5.54a) leads to:

$$Y'_X = \frac{1}{e^{4Y} X^2 e^{-4Y} - 1} = \frac{1}{X^2 - 1} = \frac{1}{2}\left(\frac{1}{X-1} - \frac{1}{X+1}\right). \qquad (5.54b)$$

Integrating (5.54b):

$$Y = \ln c + \frac{1}{2} \ln \frac{X+1}{X-1},$$

where $\ln c$ is an integration constant. Recovering the original variables:

$$\ln x = \ln c + \frac{1}{2} \ln \frac{x^2 y + 1}{x^2 y - 1} = \ln \left(c \sqrt{\frac{x^2 y + 1}{x^2 y - 1}}\right).$$

The solution of (5.54) is, therefore:

$$x = c \sqrt{\frac{x^2 y + 1}{x^2 y - 1}} \qquad \Longrightarrow \qquad y = \frac{c^2 + x^2}{x^2 (c^2 - x^2)}.$$

5.10.1   Exercises

1. Prove that the differential equation:

$$y'_x = \frac{y - 4xy^2 - 16x^3}{y^3 + 4x^2 y + x}$$

is transformed, by the change of variables $(x,y) \mapsto (X,Y)$, into:

$$Y'_X = -2X .$$

where $X(x,y) = \sqrt{4x^2 + y^2}$ and $Y(x,y) = \arctan\dfrac{y}{2x}$.

Then, use this fact to integrate the original differential equation.

2. Prove that the differential equation:

$$y'_x = \frac{y^3 + x^2 y - y - x}{x y^2 + x^3 + y - x}$$

is transformed, by the change of variables $(x,y) \mapsto (X,Y)$, into:

$$Y'_X = Y (1 - Y^2) ,$$

where $X(x,y) = \arctan\dfrac{y}{x}$ and $Y(x,y) = \sqrt{x^2 + y^2}$.

Then, use this fact to integrate the original differential equation.

3. Given the differential equation, in the unknown $y = y(x)$:

$$y' = \frac{y^3 \, e^y}{y^3 \, e^y - e^x} \, ,$$ (5.55)

transform it into an equation for $Y = Y(X)$, using the change of variables:

$$\begin{cases} X(x,y) = -\dfrac{1}{y} \, , \\ Y(x,y) = e^{x-y} \, . \end{cases}$$

Then, express the solution to (5.55) in implicit form.

4. Given the differential equation, in the unknown $y = y(x)$:

$$\begin{cases} y' = \dfrac{y^2}{x^3} + \dfrac{y+1}{x} \, , \\ y(1) = 0 \, , \end{cases}$$ (5.56)

transform it into an equation for $Y = Y(X)$, using the change of variables:

$$\begin{cases} X(x,y) = \dfrac{y}{x} \, , \\ Y(x,y) = -\dfrac{1}{x} \, , \end{cases}$$

and then give the solution to (5.56).

Chapter 6

Linear differential equations of second order

The general form of a differential equation of order $n \in \mathbb{N}$ was briefly introduced in equation (4.10) of Chapter 4. The current Chapter 6 is devoted to the particular situation of linear equations of second order:

$$a(x)\, y'' + b(x)\, y' + c(x)\, y = d(x)\,, \tag{6.1}$$

where $a\,, b\,, c$ and d are continuous real functions of the real variable $x \in I$, being I an interval in $\mathbb{R}$ and $a(x) \neq 0$. Equation (6.1) may be represented, at times, in operational notation:

$$M\, y = d(x)\,,$$

where $M : \mathcal{C}^2(I) \to \mathcal{C}(I)$ is a differential operator that acts on the function $y \in \mathcal{C}^2(I)$:

$$M\, y = a(x)\, y'' + b(x)\, y' + c(x)\, y\,. \tag{6.2}$$

In this situation, existence and uniqueness of solutions are verified, for any initial value problem associated to (6.1).

Before dealing with the simplest case, in which the coefficient functions $a(x)\,, b(x)\,, c(x)$ are constant, we examine general properties, that hold in any situation. We will study some variable–coefficient equations, that are meaningful in applications. Our treatment can be easily extended to equations of any order; for details, refer to Chapter 5 of [47] or Chapter 6 of [3].

6.1 Homogeneous equations

Assume that $a(x) \neq 0$ on a certain interval I. Hence, we can set:

$$p(x) = \frac{b(x)}{a(x)}\,, \qquad q(x) = \frac{c(x)}{a(x)}\,, \qquad r(x) = \frac{d(x)}{a(x)}\,, \tag{6.3}$$

and represent the differential equation (6.1) in the explicit form:

$$L\, y(x) = r(x)\,, \tag{6.4}$$

where L is the differential operator:

$$L\,y = y'' + p(x)\,y' + q(x)\,y\,,\qquad(6.5)$$

The *homogeneous equation* associated to (6.4) is:

$$L\,y = 0\,.\qquad(6.4a)$$

The first step in studying (6.4) consists in the change of variable:

$$y(x) = f(x)\,u(x)\,,$$

where $u(x)$ is the new dependent variable, while $f(x)$ is a function to be specified, in order to simplify computations. It is, therefore:

$$L(f\,u) = f\,u'' + (2\,f' + p\,f)\,u' + (f'' + p\,f' + q\,f)\,u = r\,.\qquad(6.6)$$

In (6.6), we can choose f so that the coefficient of u vanishes, that is:

$$f'' + p\,f' + q\,f = 0\,.\qquad(6.7)$$

In this way, equation (6.6) becomes easily solvable, since it reduces to a first–order linear equation in the unknown $v = u'$:

$$f\,v' + (2\,f' + p\,f)\,v = r\,.\qquad(6.6a)$$

At this point, if any particular solution to the homogeneous equation (6.4a) is available, the solution of the non–homogeneous equation (6.4) can be obtained.

The set of solutions to a homogeneous equation forms a two–dimensional vector space, as illustrated in the following Theorem 6.1. The first, and easy, step is to recognise, that given two solutions y_1 and y_2 of (6.4a), their linear combination:

$$y = \alpha_1\,y_1 + \alpha_2\,y_2\,,\qquad \alpha_1,\alpha_2 \in \mathbb{R}\,,$$

is also a solution to (6.4a). In particular, if y_1 and y_2 are solution to (6.4a), then:

$$(L\,y_1)(x) = y_1'' + p(x)\,y_1' + q(x)\,y_1 = 0\,,\qquad(6.8)$$
$$(L\,y_2)(x) = y_2'' + p(x)\,y_2' + q(x)\,y_2 = 0\,.\qquad(6.9)$$

To form their linear combination, we multiply both sides of (6.8) and (6.9) by α_1 and α_2, respectively, and add up the results, obtaining:

$$(\alpha_1\,y_1'' + \alpha_2\,y_2'') + p(x)\,(\alpha_1\,y_1' + \alpha_2\,y_2') + q(x)\,(\alpha_1\,y_1 + \alpha_2\,y_2) = 0\,.$$

Using the elementary properties of differentiation, we see that:

$$\alpha_1\,y_1'' + \alpha_2\,y_2'' = (\alpha_1\,y_1 + \alpha_2\,y_2)''$$

and

$$\alpha_1\,y_1' + \alpha_2\,y_2' = (\alpha_1\,y_1 + \alpha_2\,y_2)'\,,$$

which shows that $\alpha_1\,y_1 + \alpha_2\,y_2$ is indeed a solution to (6.4a). This also demonstrates that, when $\alpha_2 = 0$, any multiple of one solution of (6.4a) solves (6.4a) too. By iteration, it holds that any linear combination of solutions of (6.4a) solves (6.4a) too.

6.1.1 Operator notation

The operator notation (6.4) comes out handy in understanding, in details, the structure of the solution set of a linear homogeneous differential equation.

The L introduced in (6.2) represents a linear operator between the (infinite dimensional) vector space $C^2(I)$, formed by all the functions f whose first and second derivatives, f', f'', exist and are continuous on I, and the (infinite dimensional) space $C(I)$ of the continuous functions on I:

$$L : C^2(I) \to C(I). \tag{6.10}$$

The task of solving the linear homogeneous differential equation (6.4a) becomes, thus, equivalent to describing the *kernel*, denoted by $\ker(L)$, of the linear operator L, that is the space of the solutions of the linear homogeneous equation (6.4a):

$$\ker(L) = \{\, y \in C^2(I) \mid Ly = 0 \,\}. \tag{6.11}$$

Even if $C^2(I)$ is a infinite–dimensional vector space, $\ker(L)$ is a subspace of dimension 2, as stated in Theorem 6.1.

Theorem 6.1. Consider the linear differential operator $L : C^2(I) \to C(I)$, defined by (6.5). Then, the kernel of L has dimension 2.

Proof. Fix $x_0 \in I$ and define the linear operator $T : \ker(L) \to \mathbb{R}^2$, which maps each function $y \in \ker(L)$ onto its initial value, evaluated at x_0, i.e.:

$$T y = (\, y(x_0),\, y'(x_0)\,).$$

The existence and uniqueness Theorem 4.17 means that $T y = (0,0)$ implies $y = 0$. Hence, by the theory of linear operators, T is a one–to–one operator and it holds:

$$\dim \ker(L) = \dim \mathbb{R}^2 = 2.$$

$\square$

6.1.2 Wronskian determinant

To study the vector space structure of the set of solutions to (6.4a), it is useful to examine some properties, related to the linear independence of functions. Given n real functions $f_1, \ldots, f_n$ of a real variable, all defined on the same interval I, we say that $f_1, \ldots, f_n$ are *linearly dependent* if there exist n numbers $\alpha_1, \ldots, \alpha_n$, not all zero and such that, for any $x \in I$:

$$\sum_{k=1}^{n} \alpha_k\, f_k(x) = 0. \tag{6.12}$$

If condition (6.12) holds only when all the α_k are zero (i.e., $\alpha_k = 0$ for all $k = 1, \ldots, n$), then functions $f_1, \ldots, f_n$ are *linearly independent*.

We now provide a sufficient condition for linear independence of a set of functions. Let us assume that $f_1, \ldots, f_n$ are n–times differentiable. Then, from equation (6.12), applying successive differentiations, we can form a system of n linear equations in the variables $\alpha_1, \ldots, \alpha_n$:

$$
\begin{aligned}
\alpha_1 f_1 &+ \alpha_2 f_2 &+\ldots+ \alpha_n f_n &= 0, \\
\alpha_1 f_1' &+ \alpha_2 f_2' &+\ldots+ \alpha_n f_n' &= 0, \\
\alpha_1 f_1'' &+ \alpha_2 f_2'' &+\ldots+ \alpha_n f_n'' &= 0, \\
&\vdots & &\vdots \\
\alpha_1 f_1^{(n-1)} &+\alpha_2 f_2^{(n-1)} &+\cdots+ \alpha_n f_n^{(n-1)} &= 0.
\end{aligned}
\tag{6.13}
$$

Functions $f_1, \ldots, f_n$ are linearly independent, if it holds:

$$
\det
\begin{bmatrix}
f_1 & f_2 & \cdots & f_n \\
f_1' & f_2' & \cdots & f_n' \\
\vdots & \vdots & & \vdots \\
f_1^{(n-1)} & f_2^{(n-1)} & \cdots & f_n^{(n-1)}
\end{bmatrix}
\neq 0.
\tag{6.14}
$$

The determinant in (6.14) is called the *Wronskian*[1] of functions $f_1, \ldots, f_n$, denoted as:

$$
W(f_1, \ldots, f_n)(x) = \det
\begin{bmatrix}
f_1(x) & f_2(x) & \cdots & f_n(x) \\
f_1'(x) & f_2'(x) & \cdots & f_n'(x) \\
\vdots & \vdots & & \vdots \\
f_1^{(n-1)}(x) & f_2^{(n-1)}(x) & \cdots & f_n^{(n-1)}(x)
\end{bmatrix}.
$$

For example, functions $f_1(x) = \sin^2 x$, $f_2(x) = \cos^2 x$, $f_3(x) = \sin(2x)$, are linearly independent on $I = \mathbb{R}$, since their Wronskian is non–zero:

$$
W(f_1, f_2, f_3)(x) = \det
\begin{bmatrix}
f_1(x) & f_2(x) & f_3(x) \\
f_1'(x) & f_2'(x) & f_3'(x) \\
f_1''(x) & f_2''(x) & f_3''(x)
\end{bmatrix}
$$

$$
= \det
\begin{bmatrix}
\sin^2 x & \cos^2 x & \sin(2x) \\
2\cos x \sin x & -2\cos x \sin x & 2\cos(2x) \\
2\left(\cos^2 x - \sin^2 x\right) & 2\left(\sin^2 x - \cos^2 x\right) & -4\sin(2x)
\end{bmatrix}
= 4.
$$

A non–vanishing Wronskian represents a sufficient condition for linear independence of functions. It is worth noting that, in general, the Wronskian of

---

[1] Josef–Maria Hoëne de Wronski (1778–1853), Polish philosopher, mathematician, physicist, lawyer and economist.

a set of linearly independent functions may vanish, but this situation cannot occurr when the functions are solutions to a linear differential equation. There exists, in fact, the important result, due to Abel, stated in the following Theorem 6.2.

Theorem 6.2. Let functions $y_1(x)$ and $y_2(x)$, defined on the interval I, be solutions to the linear differential equation (6.4a). Then, a necessary and sufficient condition, for y_1 and y_2 to be linearly independent, is provided by their Wronskian being non–zero on I.

Proof. The Wronskian of $y_1(x)$ and $y_2(x)$ is a function $W : I \to \mathbb{R}$ defined as:

$$W(x) = W(y_1, y_2)(x) = \det \begin{bmatrix} y_1(x) & y_2(x) \\ y_1'(x) & y_2'(x) \end{bmatrix} = \Big(y_1(x)\, y_2'(x) - y_2(x)\, y_1'(x) \Big).$$

Differentiating, we obtain:

$$W'(x) = \frac{\mathrm{d}}{\mathrm{d}x} W(y_1, y_2)(x) = y_1(x)\, y_2''(x) - y_2(x)\, y_1''(x).$$

Since y_1 and y_2 are solution to (6.4a), recalling that we assume $a(x) \neq 0$ in (6.1), it holds:

$$y_1''(x) = -p(x)\, y_1'(x) - q(x)\, y_1(x),$$
$$y_2''(x) = -p(x)\, y_2'(x) - q(x)\, y_2(x),$$

where $p(x), q(x)$ are as in (6.3). Then:

$$W'(x) = -p(x) \Big(y_1(x)\, y_2'(x) - y_2(x)\, y_1'(x) \Big) = -p(x)\, W(x).$$

In other words, the Wronskian solves the separable differential equation:

$$W' = -p(x)\, W. \tag{6.15}$$

Solving (6.15) yields:

$$W(x) = W(x_0)\, e^{-\int_{x_0}^{x} p(s)\,\mathrm{d}s}, \tag{6.16}$$

with $p(s)$ as in (6.3).

Equation (6.16) implies that, if the Wronskian $W(x)$ vanishes at $x_0 \in I$, then $W(x)$ is the zero function; vice versa, if there exists $x_0 \in I$ such that $W(x_0) \neq 0$, then $W(x) \neq 0$ for each $x \in I$. Hence, to prove the thesis of Theorem 6.2, we need to prove that there exists $x_0 \in I$ such that $W(x_0) \neq 0$. The demonstration is by contradiction. Let us negate the assumption:

$$\exists\, x_0 \in I \quad \text{such that} \quad W(x_0) \neq 0,$$

which means that the following holds true:

$$W(x_0) = 0 \qquad \forall\, x_0 \in I.$$

Construct the 2×2 linear system of algebraic equations in the unknowns α_1, α_2 :

$$\begin{bmatrix} y_1(x_0) & y_2(x_0) \\ y_1'(x_0) & y_2'(x_0) \end{bmatrix} \begin{bmatrix} \alpha_1 \\ \alpha_2 \end{bmatrix} = \begin{bmatrix} 0 \\ 0 \end{bmatrix}. \tag{6.17}$$

By assumption, the determinant of this homogeneous system is zero, hence the system admits a non–trivial solution $(\alpha_1, \alpha_2)^T$, with α_1, α_2 not simultaneously null. Now, define the function:

$$y(x) = \alpha_1 y_1(x) + \alpha_2 y_2(x).$$

Since $y(x)$ is a linear combination of solutions $y_1(x)$ and $y_2(x)$ of (6.4a), then $y(x)$ is also a solution to (6.4a). And since, by construction, (α_1, α_2) solves (6.17), then it also holds that:

$$y(x_0) = \alpha_1 y_1(x_0) + \alpha_2 y_2(x_0) = 0 ,$$
$$y'(x_0) = \alpha_1 y_1'(x_0) + \alpha_2 y_2'(x_0) = 0 .$$

At this point, from the existence and uniqueness of the solutions of the initial value problem:

$$\begin{cases} Ly = 0 , \\ y(x_0) = y'(x_0) = 0 , \end{cases}$$

it turns out that $y(x) = 0$ identically, implying that $\alpha_1 = \alpha_2 = 0$. So, we arrive at a contradiction. The theorem is proved.   □

Putting together Theorems 6.1 and 6.2, it is possible to establish if a pair of solutions to (6.4a) is a basis for the set of solutions to the equation (6.4a), as illustrated in Theorem 6.3.

Theorem 6.3. Consider the linear differential operator $L : \mathcal{C}^2(I) \to \mathcal{C}(I)$ defined in (6.4a). If y_1 and y_2 are two independent elements of $\ker(L)$, then any other element of $\ker(L)$ can be expressed as a linear combination of y_1 and y_2 :

$$y(x) = c_1 y_1(x) + c_2 y_2(x)$$

for suitable constants $c_1, c_2 \in \mathbb{R}$.

6.1.3   Order reduction

When an integral of the homogeneous equation (6.4a) is known, possibly by inspection or by an educated guess, a second independent solution to (6.4a) can be obtained, with a procedure illustrated in Example 6.4.

Example 6.4. Knowing that $y_1(x) = x$ solves the differential equation:

$$x^2 (1 + x) y'' - x (2 + 4x + x^2) y' + (2 + 4x + x^2) y = 0, \tag{6.18}$$

a second independent solution to (6.18) can be found, by seeking a solution of the form:

$$y_2(x) = y_1(x)\, u(x) = x\, u(x).$$

Let us evaluate the first and the second derivative of y_2 :

$$y_2'(x) = u(x) + x\, u'(x), \qquad\qquad y_2''(x) = 2\, u'(x) + x\, u''(x),$$

and substitute the derivatives above into (6.18):

$$x^3 \left((x+1)\, u''(x) - (x+2)\, u'(x) \right) = 0.$$

Now, introducing $v(x) = u'(x)$, we see that v has to satisfy the first–order linear separable differential equation:

$$v' = \frac{x+2}{x+1}\, v,$$

which is solved by:

$$v(x) = c\,(1+x)\, e^x.$$

We can assume $c = 1$, since we are only interested in finding one particular solution of (6.18). Function $u(x)$ is then found by integration:

$$u(x) = \int (1+x)\, e^x \, \mathrm{d}x = x\, e^x.$$

Therefore, a second solution to (6.18) is $y_2(x) = x^2\, e^x$, where, again, we do not worry about the integration constant. Functions y_1, y_2 form an independent set of solutions to (6.18) if their Wronskian:

$$\det \begin{bmatrix} y_1(x) & y_2(x) \\ y_1'(x) & y_2'(x) \end{bmatrix} = \det \begin{bmatrix} x & x^2\, e^x \\ 1 & x\,(x+2)\, e^x \end{bmatrix} = x^2\,(x+1)\, e^x.$$

is different from zero. Now, observe that the differential equation (6.18) has to be considered in one of the intervals $(-\infty, -1), (-1, 0), (0, +\infty)$, where the leading coefficient $x^2\,(1+x)$ of (6.18) does not vanish. On such intervals, the Wronskian does not vanish as well, thus f_1, f_2 are linearly independent. In conclusion, the general solution to (6.18) is:

$$y(x) = c_1\, x + c_2\, x^2\, e^x, \qquad c_1, c_2 \in \mathbb{R}. \tag{6.19}$$

The procedure illustrated in Example 6.4 can be repeated in the general case. For convenience, we recall (6.4a) written in explicit form:

$$y'' + p(x)\, y' + q(x)\, y = 0. \tag{6.20}$$

If a solution $y_1(x)$ of (6.20) is known, we look for a second solution of the form:

$$y_2(x) = u(x)\, y_1(x),$$

where u is a function to be determined. Computing the first and second derivatives of y_2 :

$$y'_2 = y_1 \, u' + y'_1 u , \qquad y''_2 = y_1 \, u'' + 2 \, y'_1 \, u' + y''_1 \, u ,$$

and inserting them into (6.20), yields, after some computations:

$$y_1 \, u'' + (2 \, y'_1 + p \, y_1) \, u' + (y''_1 + p \, y'_1 + q \, y_1) \, u = 0 .$$

Now, since, y_1 is a solution to (6.20), the previous equation reduces to:

$$y_1 \, u'' + (2 \, y'_1 + p \, y_1) \, u' = 0 . \tag{6.21}$$

Equation (6.21) is a first–order separable equation in the unknown u', exactly in the same way as in the Example 6.4, and it can be integrated to obtain the second solution to (6.20).

The search for a second independent solution to (6.20) can also be pursued using the Wronskian equation (6.16), without explicitly computing two solutions of (6.20). This is stated in the following Theorem 6.5.

Theorem 6.5. If $y_1(x)$ is a non–vanishing solution of the second–order equation (6.20), then a second independent solution is given by:

$$y_2(x) = y_1(x) \int_{x_0}^{x} \frac{e^{\int_{x_0}^{t} -p(s) \, ds}}{y_1^2(t)} \, dt , \tag{6.22}$$

with $p(s)$ as in (6.3).

Proof. Given the assumption that y_1 is a non–vanishing function, rewrite the Wronskian as:

$$W(y_1 , y_2) = \det \begin{bmatrix} y_1 & y_2 \\ y'_1 & y'_2 \end{bmatrix} = y_1 \, y'_2 - y_2 \, y'_1 = y_1^2 \left(\frac{y_1 \, y'_2 - y_2 \, y'_1}{y_1^2} \right) ,$$

and observe that:

$$\frac{y_1 \, y'_2 - y_2 \, y'_1}{y_1^2} = \frac{d}{dx} \left(\frac{y_2}{y_1} \right) .$$

In other words:

$$\frac{d}{dx} \left(\frac{y_2}{y_1} \right) = \frac{W(y_1 , y_2)}{y_1^2} , \tag{6.23}$$

integrating which leads to:

$$y_2(x) = y_1(x) \int_{x_0}^{x} \frac{W(y_1 , y_2)(s)}{y_1^2(s)} \, ds , \tag{6.24}$$

setting to zero the constant of integration. At this point, thesis (6.22) follows from inserting equation (6.16) into (6.24). □

Example 6.6. Consider again Example 6.4. The solution $y_1(x) = x$ of (6.18) can be used in formula (6.22), to detect a second solution to such equation. Observe that, in this case:

$$p(x) = \frac{-x\,(2 + 4\,x + x^2)}{x^2\,(1 + x)} = -\frac{2 + 4\,x + x^2}{x\,(1 + x)},$$

so that:

$$\int -p(x)\ dx = x + 2\,\ln x + \ln(1 + x),$$

and:

$$e^{\int -p(s)\ ds} = x^2\,(1 + x)\,e^x.$$

The second solution is, hence:

$$y_2(x) = x \int (1 + x)\,e^x dx = x^2\,e^x,$$

in accordance with the solution y_2 found using the order reduction method in Example 6.4.

Example 6.7. Find the general solution of the homogeneous linear differential equation of second order:

$$y'' + \frac{1}{x}\,y' + \left(1 - \frac{1}{4\,x^2}\right) y = 0.$$

We seek a solution of the form $y_1 = x^m \sin x$. For such a solution, the first and second derivatives are:

$$\begin{cases} y_1' &= x^{m-1}\,(x\cos x + m\sin x),\\ y_1'' &= x^{m-2}\,(2\,m\,x\cos x + m\,(m-1)\sin x - x^2\,\sin x). \end{cases}$$

Imposing that y_1 solves the considered differential equation, we obtain:

$$x^{m-2}\,\left((2\,m + 1)\,x\,\cos x + \left(m^2 - \frac{1}{4}\right) \sin x\right) = 0,$$

which implies, in particular, $m = -\frac{1}{2}$. In this way, we have proved that:

$$y_1 = \frac{\sin x}{\sqrt{x}}$$

solves the given differential equation.
To obtain a second independent solution, we employ (6.22); in our case, it is $p(x) = \frac{1}{x}$, which gives:

$$y_2 = \frac{\sin x}{\sqrt{x}} \int \frac{dx}{\sin^2 x} = -\frac{\cos x}{\sqrt{x}}.$$

Remark 6.8. To facilitate the search for some particular solution of a linear differential equation, conditions on the coefficients $p(x)$ and $q(x)$, defined in (6.3), are provided below, each leading to a function y_1 that solves (6.20).

1. Monomial solution: if $n^2 - n + n\, x\, p(x) + x^2\, q(x) = 0$, then $y_1(x) = x^n$.

2. Exponential solution: if $n^2 + n\, p(x) + q(x) = 0$, then $y_1(x) = e^{n\, x}$.

3. Exponential monomial solution: if $n^2 x + 2n + (1 + n\, x)\, p(x) + x\, q(x) = 0$, then $y_1(x) = x\, e^{n\, x}$.

4. Exponential Gaussian solution: if $2\, m + 4\, m^2\, x^2 + 2\, m\, x\, p(x) + q(x) = 0$, then $y_1(x) = e^{m\, x^2}$.

5. Sine solution: if $n\, p(x)\, \cos(n\, x) - n^2\, \sin(n\, x) + q(x)\, \sin(n\, x) = 0$, then $y_1(x) = \sin(n\, x)$.

6. Cosine solution: if $q(x)\, \cos(n\, x) - n^2\, \cos(n\, x) - n\, p(x)\, \sin(n\, x) = 0$, then $y_1(x) = \cos(n\, x)$.

6.1.4   Constant–coefficient equations

In equation (6.1), the easiest case occurs when the coefficients functions $a(x)$, $b(x)$, $c(x)$ are constant. Suppose that $a, b, c \in \mathbb{R}$, with $a \neq 0$, and let u be a continuous function. A *constant–coefficient*, homogeneous differential equation has the form:

$$M\, y = a\, y'' + b\, y' + c\, y = 0. \tag{6.25}$$

We seek for solutions of (6.25) in the exponential form

$$y(x) = e^{\lambda x},$$

where λ is a constant to be determined. Computing the first two derivatives:

$$y'(x) = \lambda\, y(x), \qquad y''(x) = \lambda^2\, y(x),$$

and imposing that $y(x)$ solves (6.25), leads to the algebraic equation, called *characteristic equation* of (6.25):

$$a\, \lambda^2 + b\, \lambda + c = 0. \tag{6.26}$$

The roots of (6.26) determine solutions of (6.25).
If the *discriminant* $\Delta = b^2 - 4\, a\, c$ is positive, so that equation (6.26) admits two distinct real roots λ_1 and λ_2, then the general solution to (6.25) is:

$$y = c_1\, e^{\lambda_1 x} + c_2\, e^{\lambda_2 x}. \tag{6.27}$$

The independence of solutions $y_1 = c_1 e^{\lambda_1 x}$ and $y_2 = c_2 e^{\lambda_2 x}$ follows from the analysis of their Wronskian, which is non–vanishing for any $x \in \mathbb{R}$:

$$W(y_1, y_2)(x) = \det \begin{pmatrix} y_1 & y_2 \\ y_1' & y_2' \end{pmatrix} = (\lambda_2 - \lambda_1) e^{(\lambda_1 + \lambda_2) x} = \frac{\sqrt{\Delta}}{a} e^{-\frac{b}{a} x} \neq 0.$$

When $\Delta < 0$, equation (6.26) admits two distinct complex conjugate roots $\lambda_1 = \alpha + i\beta$ and $\lambda_2 = \alpha - i\beta$, and the general solution to (6.25) is:

$$y = e^{\alpha x} \left(c_1 \cos(\beta x) + c_2 \sin(\beta x) \right). \tag{6.28}$$

Forming the complex exponential of λ_1 and that of λ_2, two complex–valued functions z_1, z_2 are obtained:

$$z_1 = e^{\lambda_1 x} = e^{(\alpha + i\beta) x} = e^{\alpha x} e^{i\beta x} = e^{\alpha x} \left(\cos(\beta x) + i \sin(\beta x) \right),$$

$$z_2 = e^{\lambda_2 x} = e^{(\alpha - i\beta) x} = e^{\alpha x} e^{-i\beta x} = e^{\alpha x} \left(\cos(\beta x) - i \sin(\beta x) \right),$$

that have the same real and imaginary parts:

$$\Re(z_1) = \Re(z_2) = e^{\alpha x} \cos(\beta x), \qquad \Im(z_1) = \Im(z_2) = e^{\alpha x} \sin(\beta x).$$

Set, for example, $y_1 = e^{\alpha x} \cos(\beta x)$ and $y_2 = e^{\alpha x} \sin(\beta x)$. Then, the real solution presented in (6.28) is a linear combination of the real functions y_1 and y_2, which are are independent, since their Wronskian is non–vanishing:

$$W(y_1, y_2)(x) = \det \begin{pmatrix} y_1 & y_2 \\ y_1' & y_2' \end{pmatrix} = \beta e^{2\alpha x} \neq 0.$$

When $\Delta = 0$, equation (6.26) has one real root with multiplicity 2, and the correspondent solution to (6.25) is:

$$y_1 = e^{-\frac{b}{2a} x}.$$

In this situation, we need a second independent solution, that is obtained from formula (6.22) of Theorem 6.5, using the just found y_1 and with $p(s)$ built for equation (6.26), thus:

$$y_2 = e^{-\frac{b}{2a} x} \int \frac{e^{\int -\frac{b}{a} dx}}{e^{-\frac{b}{a} x}} dx = x \, e^{-\frac{b}{2a} x}.$$

In other words, when $\Delta = 0$ the general solution of (6.25) is:

$$y = e^{-\frac{b}{2a} x} \left(c_1 + c_2 x \right). \tag{6.29}$$

Observe that the Wronskian is:

$$W(y_1, y_2)(x) = \det \begin{pmatrix} y_1 & y_2 \\ y_1' & y_2' \end{pmatrix} = e^{-\frac{b}{a}x} \neq 0.$$

Note that the knowledge of the Wronskian expression is useful in the study of non–homogeneous differential equations too, as it will be shown in § 6.2.

Example 6.9. Consider the initial value problem:

$$\begin{cases} y'' - 2y' + 6y = 0, \\ y(0) = 0, \quad y'(0) = 1. \end{cases}$$

The characteristic equation is $\lambda^2 - 2\lambda + 6 = 0$, with roots $\lambda = 1 \pm i\sqrt{5}$. Hence, two independent solutions are:

$$y_1(x) = e^x \cos\left(\sqrt{5}\,x\right), \qquad y_2(x) = e^x \sin\left(\sqrt{5}\,x\right),$$

and the general solution can be expressed as $y(x) = c_1 y_1(x) + c_2 y_2(x)$. Now, forming the initial conditions:

$$\begin{cases} y(0) = c_1 y_1(0) + c_2 y_2(0) = c_1, \\ y'(0) = c_1 y_1'(0) + c_2 y_2'(0) = c_1 + c_2 \sqrt{5}, \end{cases}$$

we see that constants c_1 and c_2 must verify:

$$\begin{cases} c_1 = 0, \\ c_1 + c_2 \sqrt{5} = 1, \end{cases} \qquad \Longrightarrow \qquad \begin{cases} c_1 = 0, \\ c_2 = \dfrac{1}{\sqrt{5}}. \end{cases}$$

In conclusion, the considered initial value problem is solved by:

$$y(x) = \frac{e^x}{\sqrt{5}} \sin\left(\sqrt{5}\,x\right).$$

6.1.5   Cauchy–Euler equations

A Cauchy–Euler differential equation is a particular second–order linear equation, with variable coefficients, of the form:

$$a\,x^2\,y'' + b\,x\,y' + c\,y = 0, \tag{6.30}$$

where $a, b, c \in \mathbb{R}$, and with $x > 0$. We seek solutions of (6.30) in power form, that is, $y = x^m$, being m a constant to be determined and that must satisfy the algebraic equation:

$$a\,m\,(m-1) + b\,m + c = 0, \tag{6.31}$$

i.e., $a\,m^2 + (b-a)\,m + c = 0$. Let m_1 and m_2 be the roots of equation (6.31) and denote its discriminat with $\Delta = (a-b)^2 - 4\,a\,c$. According to the sign of Δ, the differential equation (6.30) is solved by a power–form function y defined as follows, with $c_1, c_2 \in \mathbb{R}$ in all cases:

(i) if $\Delta > 0$ then m_1, m_2 are real and distinct, hence $y = c_1\,x^{m_1} + c_2\,x^{m_2}$;

(ii) if $\Delta < 0$ then m_1, m_2 are complex and conjugate, say $m_{1,2} = \alpha \pm i\beta$, thus $y = x^\alpha \left(c_1\,\cos(\beta\,\ln x) + c_2\,\sin(\beta\,\ln x) \right)$;

(iii) if $\Delta = 0$ then $m_1 = m_2 = m$, and the solution is $y = c_1\,x^m + c_2\,x^m\,\ln x$.

Remark 6.10. The solution y of the Cauchy–Euler equation, illustrated in each of the three cases above, has components y_1, y_2, that are linearly independent. This statement can be verified using the Wronskian.

When $\Delta > 0$, equation (6.31) has two distinct real roots $m_1 \neq m_2$, which implies, defining $y_1 = x^{m_1}$ and $y_2 = x^{m_2}$, that the Wronskian is non–null:

$$W(y_1, y_2)(x) = \det \begin{pmatrix} y_1 & y_2 \\ y_1' & y_2' \end{pmatrix} = (m_2 - m_1)\,x^{m_1 + m_2 - 1} \neq 0, \qquad \text{for } x > 0.$$

When $\Delta < 0$, there are two complex conjugate roots $m_{1,2} = \alpha \pm i\beta$ of (6.31); then, setting for example $y_1 = x^\alpha\,\cos(\beta\,\ln x)$ and $y_2 = x^\alpha\,\sin(\beta\,\ln x)$, the Wronskian does not vanish:

$$W(y_1, y_2)(x) = \det \begin{pmatrix} y_1 & y_2 \\ y_1' & y_2' \end{pmatrix} = \beta\,x^{2\alpha - 1} \neq 0, \qquad \text{for } x > 0.$$

When $\Delta = 0$, equation (6.31) has one real root m of multiplicity 2; in this case $y_1 = x^m$ and $y_2 = y_1\,\ln x$; again, the Wronskian does not vanish:

$$W(y_1, y_2)(x) = \det \begin{pmatrix} y_1 & y_2 \\ y_1' & y_2' \end{pmatrix} = x^{2m - 1} \neq 0, \qquad \text{for } x > 0.$$

Notice, again, that knowing the Wronskian turns out useful, also, when studying the non–homogeneous differential equations case (refer to § 6.2).

Example 6.11. Consider the initial value problem:

$$\begin{cases} x^2\,y'' - 2\,x\,y' + 2\,y = 0, \\ y(1) = 1, \quad y'(1) = 0. \end{cases}$$

Here equation (6.30) assumes the form:

$$m\,(m-1) - 2\,m + 2 = 0,$$

that is $m = 1, m = 2$. Therefore, the general solution of the given equation is $y = c_1 x + c_2 x^2$; imposing the initial conditions yields the system:

$$\begin{cases} c_1 + c_2 = 1, \\ c_1 + 2 c_2 = 0, \end{cases} \qquad \Longrightarrow \qquad \begin{cases} c_1 = 2, \\ c_2 = 1. \end{cases}$$

In conclusion, the solution of the initial vale problem is $y = 2x - x^2$.

Example 6.12. Consider the initial value problem:

$$\begin{cases} x^2 y'' - x y' + 5 y = 0, \\ y(1) = 1, \quad y'(1) = 0. \end{cases}$$

Here, equation (6.30) assumes the form:

$$m(m - 1) - m + 5 = 0,$$

that is $m = 1 \pm 2 i$. Hence, the general solution of the given equation is $y = x\ (c_1 \cos(2 \ln x) + c_2 \sin(2 \ln x))$; again, imposing the initial conditions, we obtain the system:

$$\begin{cases} c_1 = 1, \\ c_1 + 2 c_2 = 0, \end{cases}$$

leading to the solution:

$$y = x\ \left(\cos(2 \ln x) - \frac{1}{2} \sin(2 \ln x) \right).$$

6.1.6 Invariant and normal form

To conclude § 6.1, we introduce the fundamental notion of *invariant* of a homogeneous linear differential equation of second order (6.4a) The basic fact is that any such equation can be transformed into a differential equation without the term containing the first derivative, namely:

$$u'' + I(x)\ u = 0, \tag{6.32}$$

with:

$$I(x) = q(x) - \frac{1}{2}\ p'(x) - \frac{1}{4}\ p^2(x). \tag{6.33}$$

In fact, assuming the knowledge of a solution to (6.4a) of the form $y = f\ u$ leads to, in the same way followed to obtain equation (6.6):

$$L(f\ u) = f\ u'' + (2 f' + p\ f)\ u' + (f'' + p\ f' + q\ f)\ u = 0, \tag{6.34}$$

in which $f(x)$ can be chosen so that the coefficient of u' vanishes, namely:

$$2 f' + p\ f = 0 \qquad \Longrightarrow \qquad f(x) = e^{-\frac{1}{2} \int p(x)\, dx}.$$

Function $f(x)$ does not vanish, and has first and second derivatives given by:

$$f' = -\frac{fp}{2}\,, \qquad\qquad f'' = \frac{f\,(p^2 - 2p')}{4}\,.$$

Hence, f can be simplified out in (6.34), yielding the reduced form:

$$u'' + \left(q - \frac{1}{2}p' - \frac{1}{4}p^2\right)u = 0\,, \tag{6.35}$$

which is stated in (6.32)–(6.33). Equation (6.32) is called the *normal* form of equation (6.4a). Function $I(x)$, introduced in (6.33), is called *invariant* of the homogeneous differential equation (6.4a) and represents a mathematical invariant, in the sense expressed by the following Theorem 6.13.

Theorem 6.13. If the equation:

$$L_1\,y = y'' + p_1\,y' + q_1\,y = 0 \tag{6.36}$$

can be transformed into the equation:

$$L_2\,y = y'' + p_2\,y' + q_2\,y = 0\,, \tag{6.37}$$

by the change of dependent variable $y = f\,u$, then the invariants of (6.36)–(6.37) coincide:

$$I_1 = q_1 - \frac{1}{2}p_1' - \frac{1}{4}p_1^2 = q_2 - \frac{1}{2}p_2' - \frac{1}{4}p_2^2 = I_2\,.$$

Vice versa, when equations (6.36) and (6.37) admit the same invariant, each equation can be transformed into the other, by:

$$y(x) = u(x)\,e^{-\frac{1}{2}\int\,(p_1(x) - p_2(x))\,\mathrm{d}x}\,.$$

Remark 6.14. We can transform any second–order linear differential equation into its normal form. Moreover, if we are able to solve the equation in normal form, then we can obtain, easily, the general solution to the original equation. The next Example 6.15 clarifies this idea.

Example 6.15. Consider the homogeneous differential equation, depending on the real positive parameter a :

$$y'' - \frac{2}{x}y' + \left(a^2 + \frac{2}{a^2}\right)y = 0\,. \tag{6.38}$$

Hence:

$$p(x) = -\frac{2}{x}\,, \qquad q(x) = a^2 + \frac{2}{a^2} \quad\Longrightarrow\quad I(x) = q(x) - \frac{1}{2}p'(x) - \frac{1}{4}p^2(x) = a^2\,.$$

In this example, the invariant is not dependent of x, and the normal form is:

$$u'' + a^2 u = 0. \tag{6.39}$$

The general solution to (6.39) is $u(x) = c_1 \cos(a x) + c_2 \sin(a x)$, where c_1, c_2 are real parameter, and the solution to the original equation (6.38) is:

$$y(x) = u(x)\, e^{-\frac{1}{2}\int p(x)\,dx} = c_1\, x\, \cos(a x) + c_2\, x\, \sin(a x).$$

Example 6.16. Find the general solution of the homogeneous linear differential equation of second order:

$$y'' - 2\tan(x)\, y' + y = 0.$$

The first step consists in transforming the given equation into normal form, with the change of variable:

$$y = u\, e^{-\frac{1}{2}\int p(x)\,dx} = u\, e^{\int \tan x\,dx} = \frac{u}{\cos x}.$$

The normal form is:

$$u'' + 2\,u = 0,$$

which is a constant–coefficient equation, solved by:

$$u = c_1 \cos(\sqrt{2}\, x) + c_2 \sin(\sqrt{2}\, x).$$

The solution to the given differential equation is:

$$y = c_1\, \frac{\cos(\sqrt{2}\, x)}{\cos x} + c_2\, \frac{\sin(\sqrt{2}\, x)}{\cos x}.$$

The normal form clarifies the structure of the solutions to a constant–coefficient, linear equation of second–order.

Remark 6.17. Consider the constant–coefficient equation (6.25). The change of variable:

$$y = u\, e^{-\frac{b}{2a}\, x}$$

allows to transform (6.25) into normal form:

$$u'' - \frac{b^2 - 4\,a\,c}{4\,a^2}\, u = 0, \tag{6.40}$$

since, in this case:

$$p = -\frac{b}{a}, \qquad q = -\frac{c}{a} \implies I = q - \frac{1}{2}\, p' - \frac{1}{4}\, p^2 = \frac{-b^2 + 4\,a\,c}{4\,a^2}.$$

The normal form (6.40) explains the nature of the following formulæ (6.41), (6.42) and (6.43), namely describing the structure of the solution to a constant–coefficient, homogeneous linear differential equation of second order. In the following, the discriminant is $\Delta = b^2 - 4\,a\,c$ and c_1, c_2 are constant:

(i) if $\Delta > 0$,

$$y(x) = e^{-\frac{b}{2a}x} \left(c_1 \cosh\left(\frac{\sqrt{\Delta}}{2a} x\right) + c_2 \sinh\left(\frac{\sqrt{\Delta}}{2a} x\right) \right); \qquad (6.41)$$

(ii) if $\Delta < 0$,

$$y(x) = e^{-\frac{b}{2a}x} \left(c_1 \cos\left(\frac{\sqrt{-\Delta}}{2a} x\right) + c_2 \sin\left(\frac{\sqrt{-\Delta}}{2a} x\right) \right); \qquad (6.42)$$

(iii) if $\Delta = 0$,

$$y(x) = e^{-\frac{b}{2a}x} (c_1 x + c_2). \qquad (6.43)$$

6.2 Non–homogeneous equation

We finally deal with the non–homogeneous equation (6.4), which we recall and label again, for convenience:

$$L\, y = y'' + p(x)\, y' + q(x)\, y = r(x), \qquad (6.44)$$

changing, slightly, the point of view, in comparison to the beginning of § 6.1. Here, we assume to know, already, the general solution of the homogeneous equation associated to (6.44). Aim of this section is, indeed, to describe the relation between solutions of $L\,y = 0$ and solutions of $L\,y = r(x)$, being $r(x)$ a given continuous function. The first and probably most important step in this direction is represented by the following Theorem 6.18.

Theorem 6.18. Let y_1 and y_2 be independent solutions of $L\,y = 0$, and let y_p be a solution of $L\,y = r(x)$. Then, any solution of the latter non–homogeneous equation has the form:

$$y(x) = c_1\, y_1(x) + c_2\, y_2(x) + y_p(x), \qquad (6.45)$$

where $c_1, c_2 \in \mathbb{R}$ are constant.

Proof. Using the linearity of the operator L, we see that:

$$L\,(y - y_p) = L\,y - L\,y_p = r - r = 0.$$

This means that $y - y_p$ can be express by a linear combination of y_1 and y_2. Hence, thesis (6.45) is proved. $\qquad \square$

Formula (6.45) is called *general* solution of the non–homogeneous equation (6.44).

6.2.1   Variation of parameters

Theorem 6.18 indicates that, to describe the general solution of the linear non–homogeneous equation (6.44), we need to know a particular solution to (6.44) and two independent solutions of the associated homogeneous equation. As a matter of fact, the knowledge of two independent solutions to $Ly = 0$ allows to individuate a particular solution of (6.44), using the method of the *variation of parameters*, introduced by Lagrange in 1774.

Theorem 6.19. Let y_1 and y_2 be two independent solutions of the homogeneous equation associated to (6.44). Then, a particular solution to (6.44) has the form:

$$y_p(x) = k_1(x)\, y_1(x) + k_2(x)\, y_2(x)\,, \tag{6.46}$$

where:

$$k_1(x) = -\int \frac{y_2(x)\, r(x)}{W(y_1, y_2)(x)}\, \mathrm{d}x\,, \qquad k_2(x) = \int \frac{y_1(x)\, r(x)}{W(y_1, y_2)(x)}\, \mathrm{d}x\,. \tag{6.47}$$

Proof. Assume that y_1 and y_2 are independent solutions of the homogeneous equation associated to (6.44), and look for a particular solution of (6.44) in the desired form:

$$y_p = k_1\, y_1 + k_2\, y_2\,,$$

where k_1, k_2 are two C^1 functions to be determined. Computing the first derivative of y_p yields:

$$y_p' = k_1'\, y_1 + k_2'\, y_2 + k_1\, y_1' + k_2\, y_2'\,. \tag{6.48}$$

Let us impose a first condition on y_1 and y_2, i.e., impose that they verify:

$$k_1'\, y_1 + k_2'\, y_2 = 0\,, \tag{6.49}$$

so that (6.48) reduces to:

$$y_p' = k_1\, y_1' + k_2\, y_2'\,. \tag{6.48a}$$

Now, compute y_p'', by applying differentiation to (6.48a):

$$y_p'' = k_1\, y_1'' + k_2\, y_2'' + k_1'\, y_1' + k_2'\, y_2'\,.$$

At this point, imposing that y_p solves equation (6.44) leads to forming the following expression (in which variable x is discarded, to ease the notation):

$$y_p'' + p\, y_p' + q\, y_p = (k_1\, y_1'' + k_2\, y_2'' + k_1'\, y_1' + k_2'\, y_2') + p\, (k_1\, y_1' + k_2\, y_2') + q\, (k_1\, y_1 + k_2\, y_2)$$

$$= k_1\, (y_1'' + p\, y_1' + q\, y_1) + k_2\, (y_2'' + p\, y_2' + q\, y_2) + (k_1'\, y_1' + k_2'\, y_2')\,.$$

In this way, a second condition on y_1 and y_2 is obtained:

$$k_1'\, y_1' + k_2'\, y_2' = r\,. \tag{6.50}$$

Equations (6.49) and (6.50) form a 2×2 linear system, in the variables k_1', k_2', that admits a unique solution:

$$k_1' = -\frac{y_2\, r}{W(y_1, y_2)}, \qquad k_2' = \frac{y_1\, r}{W(y_1, y_2)}, \qquad (6.51)$$

since its coefficient matrix is the Wronskian $W(y_1, y_2)$, which does not vanish, given the assumption that y_1 and y_2 are independent. Thesis (6.47) follows by integration of k_1', k_2' in (6.51).

$\square$

Example 6.20. In Example 6.4, we showed that the general solution of the homogeneous equation (6.18) has the form (6.19), i.e., $c_1 x + c_2 x^2 e^x$, $c_1, c_2 \in \mathbb{R}$. Here, we use Theorem 6.19 to find the general solution of the non–homogeneous equation:

$$x^2 (1+x)\, y'' - x\, (2 + 4x + x^2)\, y' + (2 + 4x + x^2)\, y = \left(x^2 (1+x) \right)^2. \quad (6.52)$$

As a first step, let us rewrite (6.52) in explicit form, namely:

$$y'' - \frac{x\,(2 + 4x + x^2)}{x^2\,(1+x)}\, y' + \frac{(2 + 4x + x^2)}{x^2\,(1+x)}\, y = x^2\,(1+x).$$

Then, using equation (6.47), we obtain:

$$k_1(x) = -\int x^2\, dx = -\frac{x^3}{3}, \qquad k_2(x) = \int x\, e^{-x}\, dx = -(1+x)\, e^{-x}.$$

Hence, the general solution of (6.52) is:

$$y(x) = c_1\, x + c_2\, x^2\, e^x - \frac{x^4}{3} - x^2\,(1+x).$$

Example 6.21. Consider the initial value problem:

$$\begin{cases} x^2\, y'' - x\, y' + 5\, y = x^2, \\ y(1) = y'(1) = 0. \end{cases}$$

In Example 6.12, the associated homogeneous equation was considered, of which two independent solutions were found, namely $y_1 = x\, \cos(2 \ln x)$ and $y_2 = x\, \sin(2 \ln x)$, whose Wronskian is $2x$. Now, writing the given non–homogenous differential equation in explicit form:

$$y'' - \frac{1}{x}\, y' + \frac{5}{x^2}\, y = 1,$$

and using (6.47), we find:

$$k_1(x) = -\frac{1}{2} \int \sin(2 \ln x)\, dx, \qquad k_2(x) = \frac{1}{2} \int \cos(2 \ln x)\, dx.$$

Evaluating the integrals:

$$k_1(x) = \frac{1}{2} \left(\frac{2}{5} x \cos(2 \ln x) - \frac{1}{5} x \sin(2 \ln x) \right),$$

$$k_1(x) = \frac{1}{2} \left(\frac{2}{5} x \sin(2 \ln x) - \frac{1}{5} x \cos(2 \ln x) \right).$$

Hence, a particular solution of the non–homogenous equation is

$$y_p = k_1 \, y_1 + k_2 \, y_2 = \frac{x^2}{5},$$

while the general solution is:

$$y = c_1 \, x \cos(2 \ln x) + c_2 \, x \sin(2 \ln x) + \frac{x^2}{5}.$$

To solve the initial value problem, c_1 and c_2 need to be determined, imposing the initial conditions:

$$y(1) = c_1 + \frac{1}{5} = 0, \qquad y'(1) = c_1 + 2 c_2 + \frac{2}{5} = 0,$$

yielding $c_1 = -\frac{1}{5}$, $c_2 = -\frac{1}{10}$. In conclusion, the solution of the given initial value problem is:

$$y(x) = \frac{1}{10} x \, (2 \, x - \sin(2 \ln x) - 2 \, \cos(2 \ln x)).$$

6.2.2 Non–homogeneous equations with constant coefficients

While studying a second–order differential equation, the easiest situation occurs, probably, when the equation coefficients are constant. In this case, the application of the variation of parameters method can be performed systematically. Here, though, we only provide the results, that is, we indicate how to search for some particular solution of a given constant–coefficient equation; the interested Reader can, then, apply the variation of parameters to validate our statements.

Assume that a constant–coefficient, non–homogeneous differential equation of second order is given:

$$M \, y = a \, y'' + b \, y' + c \, y = r(x), \tag{6.53}$$

where $r(x)$ is a continuous real function. Denote with $K(\lambda)$ the characteristic polynomial associated to the differential equation (6.53). A list is provided below, taken from [21], of particular functions $r(x)$, together with a recipe to find a relevant particular solution of (6.53).

(1) Let $r(x) = e^{\alpha x} P_n(x)$, being $P_n(x)$ a given polynomial of degree n.

(a) If $K(\alpha) \neq 0$, it means that α is not a root of the characteristic equation. Then, a particular solution of (6.53) has the form:

$$y_p = e^{\alpha x} Q_n(x), \qquad (6.54)$$

where $Q_n(x)$ is a polynomial of degree n to be determined.

(b) If $K(\alpha) = 0$, with multiplicity $s \geq 1$, it means that α is a root of the characteristic equation. Then, a particular solution of (6.53) has the form:

$$y_p = x^s \, e^{\alpha x} \, R_n(x), \qquad (6.55)$$

where $R_n(x)$ is a polynomial of degree n to be determined.

(2) Let $r(x) = e^{\alpha x} \left(P_n(x) \cos(\beta x) + Q_m(x) \sin(\beta x) \right)$, being P_n and Q_m given polynomials of degree n and m, respectively.

(a) If $K(\alpha + i\beta) \neq 0$, it means that $\alpha + i\beta$ is not a root of the characteristic equation. In this case, a particular solution of (6.53) has the form:

$$y_p = e^{\alpha x} \left(R_p(x) \cos(\beta x) + S_p(x) \cos(\beta x) \right), \qquad (6.56)$$

with $R_p(x)$ and $S_p(x)$ polynomials of degree $p = \max\{n, m\}$ to be determined.

(b) if $K(\alpha + i\beta) = 0$, it means that $\alpha + i\beta$ is a root of the characteristic equation, with multiplicity $s \geq 1$. Then, a particular solution of (6.53) has the form:

$$y_p = x^s \, e^{\alpha x} \left(R_p(x) \cos(\beta x) + S_p(x) \cos(\beta x) \right), \qquad (6.57)$$

where $R_p(x)$ and $S_p(x)$ are polynomials of degree $p = \max\{n, m\}$ to be determined.

(3) Let $r(x) = r_1(x) + \ldots + r_n(x)$ and let y_k be solution of $M \, y_k = r_k$. Then, $y = y_1 + \cdots + y_n$ solves $L \, y = r$. This fact is known as *super–position principle*.

Example 6.22. Consider the differential equation:

$$y''(x) + 2 \, y'(x) + y(x) = x^2 + x.$$

Here $\alpha = 0$ is not a root of the characteristic equation:

$$K(\lambda) = \lambda^2 + 2\lambda + 1 = 0,$$

thus, we are in situation (1a) and we look for a solution of the form (6.54), that is

$$y_p(x) = s_0\,x^2 + s_1\,x + s_2\,.$$

Differentiating:

$$\begin{cases} y_p'(x) = 2\,s_0\,x + s_1\,, \\ y_p''(x) = 2\,s_0\,, \end{cases}$$

and imposing that y_p solves the differential equation, we obtain:

$$s_0\,x^2 + (4\,s_0 + s_1)\,x + 2\,s_0 + 2\,s_1 + s_2 = x^2 + x\,.$$

Therefore, it must be:

$$\begin{cases} s_0 = 1\,, \\ 4\,s_0 + s_1 = 1\,, \\ 2\,s_0 + 2\,s_1 + s_2 = 0\,, \end{cases} \qquad \Longrightarrow \qquad \begin{cases} s_0 = 1\,, \\ s_1 = -3\,, \\ s_2 = 4\,. \end{cases}$$

Hence, a particular solution of the given equation is

$$y_p = x^2 - 3\,x + 4\,.$$

Finally, solving the associated homogeneous equation, we obtain the required general solution:

$$y(x) = x^2 - 3\,x + 4 + c_1\,e^{-x} + c_2\,x\,e^{-x}\,.$$

Example 6.23. Consider the differential equation:

$$y''(x) + y(x) = \sin x + \cos x\,.$$

Observe, first, that the general solution of the associated homogeneous equation is:

$$y_0(x) = c_1\,\cos x + c_2\,\sin x\,, \qquad c_1, c_2 \in \mathbb{R}\,.$$

Observe, further, that the characteristic equation has roots $\pm i$, Hence, we are in situation (2b) and we look for a solution of the form (6.57), that is:

$$y_p(x) = s_1\,x\,\cos x + s_2\,x\,\sin x\,, \qquad s_1, s_2 \in \mathbb{R}\,.$$

Imposing that $y_p(x)$ solves the given non–homogeneous equation, we find:

$$2\,s_2\,\cos x - 2\,s_1\,\sin x = \cos x + \sin x\,.$$

Solving the system:

$$s_1 = -\frac{1}{2}\,, \qquad s_2 = \frac{1}{2}\,,$$

leads to the general solution of the given equation:

$$y(x) = \frac{1}{2}\,x\,\sin x - \frac{1}{2}\,x\,\cos x + c_1\,\cos x + c_2\,\sin x\,, \qquad c_1, c_2 \in \mathbb{R}\,.$$

6.2.3 Exercises

1. Solve the following variable–coefficient, linear differential equations of second order:

   (a) $\quad y'' - \left(1 + \dfrac{2}{x}\right) y' + \dfrac{2}{x} y = 0$;

   (b) $\quad y'' - \dfrac{2x}{1+x^2} y' + \dfrac{2}{1+x^2} y = 0$;

   (c) $\quad y'' - \dfrac{1}{x} y' - 4 x^2 y = 0$.

   Hint: $y = e^{x^m}$.

2. Solve the following initial value problems, using the transformation of the given differential equation in normal form:

   (a) $\quad\begin{cases} y'' + (2 \sin x) y' + \left((\sin x)^2 + \cos x - \dfrac{6}{x^2}\right) y = 0, \\ y(1) = 0, \\ y'(1) = 1, \end{cases}$

   (b) $\quad\begin{cases} y'' + \dfrac{2}{x} y' + y = 0, \\ y(1) = 1, \\ y'(1) = 0. \end{cases}$

3. Find the general solution of the differential equation:

   $$y'' + (\cot x) y' - \dfrac{1}{(\sin x)^2} y = 0,$$

   using the fact that $y_1 = \dfrac{1}{\sin x}$ is a solution.

   Then, find, if it exists, the particular solution that vanishes for $x \to 0^+$.

   Say whether there exists a solution such that $\lim\limits_{x \to 0^+} \dfrac{y(x)}{x} = \dfrac{1}{2}$.

4. Solve the non–homogeneous equation:

   $$y'' - \dfrac{1}{x} y' - 4 x^2 y = -4 x^4,$$

   using the fact that $y_1(x) = e^{x^2}$ is a solution of the associated homogeneous equation.

Chapter 7

Prologue to Measure theory

Some basic notions are presented in this chapter, that are needed as introduction to *Measure theory*. Some familiarity with the concepts presented, here and in the following Chapters 8 to 10, is also assumed and recalled, briefly, for the sake of completeness.

7.1 Set theory

The concept of *set* is considered as given, we do not provide a definition for it. We are only concerned with set membership and operations with sets, where we shall always deal with collections of subsets of some universal set Ω.

7.1.1 Sets

Sets are usually denoted with capital letters. Set membership is denoted with the symbol "$\in$", therefore $x \in A$ means that x belongs to A. Set inclusion is denoted as $A \subset B$, which means that every member of A is a member of B. The case $A = B$ is a particular case of set inclusion. The inclusion is said to be *strict* if at least one element $b \in B$ exists such that $b \notin A$; in such a case, we may write $A \subsetneq B$; obviously, if $A \subsetneq B$, then $A \subset B$ is also true. The set of subsets of A is denoted by $\mathcal{P}(A)$, which is sometimes called the *power set* of A, since, if A has n elements, then $\mathcal{P}(A)$ has 2^n elements. Intersection and union are defined as:

$$A \cap B = \{x \mid x \in A \text{ and } x \in B\}, \qquad A \cup B = \{x \mid x \in A \text{ or } x \in B\}.$$

The complement A^c of a set A consists of the elements of Ω which are not members of A, denoted with:

$$A^c = \Omega \setminus A.$$

The difference is:

$$B \setminus A = \{x \in B \mid x \notin A\} = B \cap A^c,$$

while the symmetric difference is defined by:

$$A \Delta B = (A \setminus B) \cup (B \setminus A).$$

Symbol $\emptyset$ denotes the empty set, i.e. the set with no elements:

$$\emptyset = \{ x \in \Omega \mid x \neq x \}.$$

Note that $A \Delta B = \emptyset \iff A = B$.

7.1.2   Indexes and Cartesian product

Here, a rigorous extension is provided of the definitions of intersection and union to arbitrary collections of sets. Let us consider a finite family of sets $A_1, \ldots, A_n$; their list can be understood as the *image set* of a function $x : \{1, 2, \ldots, n\} \to \mathcal{X}$, where $\mathcal{X}$ actually indicates the set of n objects we want to index.

Assume, in general, that two non–empty sets $\mathcal{I}$ and $\mathcal{X}$ are given, together with a function $x : \mathcal{I} \to \mathcal{X}$, $x : i \mapsto x(i)$. We refer to $\mathcal{I}$ as the *index set*, while the triad $(\mathcal{X}, \mathcal{I}, x)$ is called the *indexing* of $\mathcal{X}$. The following conventions are adopted:

- the notation x_i indicates $x(i)$;

- the notation $(x_i)_{i \in \mathcal{I}}$ represents the mapping $x : \mathcal{I} \to \mathcal{X}$, and is said to be a collection of elements in $\mathcal{X}$ indexed by $\mathcal{I}$.

In other words, x establishes an *indexed family* of elements in $\mathcal{X}$ indexed by $\mathcal{I}$, and the elements of $\mathcal{X}$ are referred to as forming the family, i.e., the indexed family $(x_i)_{i \in \mathcal{I}}$ is interpreted as a collection, rather than a function. When $\mathcal{I} = \mathbb{N}$, we are obviously dealing with an usual sequence. When we, instead, consider a finite list of objects, then $\mathcal{I} = \{1, 2, \ldots, n\}$. It is also possible to consider as index set, for example, the power set of $\mathcal{X}$, that is, $\mathcal{I} = \mathcal{P}(\mathcal{X})$; in the latter case, the indexed family becomes the collection of the subsets of $\mathcal{X}$.

Union and intersection of arbitrary collections of sets are defined as:

$$\bigcap_{\alpha \in \mathcal{I}} A_\alpha = \{ x \mid x \in A_\alpha \text{ for all } \alpha \in \mathcal{I} \} = \{ x \mid \forall \alpha \in \mathcal{I}, \, x \in A_\alpha \},$$

$$\bigcup_{\alpha \in \mathcal{I}} A_\alpha = \{ x \mid x \in A_\alpha \text{ for some } \alpha \in \mathcal{I} \} = \{ x \mid \exists \alpha \in \mathcal{I}, \, x \in A_\alpha \},$$

and, recalling *De Morgan*[1] *Laws*, it holds that:

$$\left(\bigcup_{\alpha \in \mathcal{I}} A_\alpha \right)^c = \bigcap_{\alpha \in \mathcal{I}} A_\alpha^c, \qquad \left(\bigcap_{\alpha \in \mathcal{I}} A_\alpha \right)^c = \bigcup_{\alpha \in \mathcal{I}} A_\alpha^c. \qquad (7.1)$$

If $A \cap B = \emptyset$, then A and B are *disjoint*. A family of sets $(A_\alpha)_{\alpha \in \mathcal{I}}$ is *pairwise disjoint* if $A_\alpha \cup A_\beta = \emptyset$ whenever $\alpha \neq \beta$, for $\alpha, \beta \in \mathcal{I}$.

---

[1] Augustus De Morgan (1806–1871), British mathematician and logician.

7.1.3 Cartesian product

The Cartesian product $A \times B$ of a couple of sets A and B is defined as the set of ordered pairs:

$$A \times B = \{ (a, b) \mid a \in A, \, b \in B \} \,,$$

while, more in general, the Cartesian product:

$$\prod_{i \in \mathcal{I}} A_i$$

of an arbitrary, indexed family of sets $(A_i)_{i \in \mathcal{I}}$, is the collection of all functions:

$$a : \mathcal{I} \to \bigcup_{i \in \mathcal{I}} A_i \,, \qquad a : i \mapsto a_i \,, \qquad \text{such that } a_i \in A_i \text{ for any } i \in I \,.$$

When $A_i = A$ for any $i \in \mathcal{I}$, the Cartesian product is denoted as $A^{\mathcal{I}}$, which reduces to A^n in the case $\mathcal{I} = \{1, \dots, n\}$.

The Cartesian plane is $\mathbb{R}^2 = \mathbb{R} \times \mathbb{R}$, while $\mathbb{R}^n$ is the set of all n–tuples $(x_1, \dots, x_n)$ composed of real numbers. A *rectangle* is the Cartesian product of two intervals.

7.1.4 Functions

A function $f : A \to B$ can be intepreted as a subset of $A \times B$, in which each first coordinate determines the second one:

$$(a, b), (a, c) \in f \quad \implies \quad b = c \,.$$

Domain and Range of a function are, thus, respectively defined as:

$$\mathcal{D}_f = \{ a \in A \mid \exists b \in B, \, (a, b) \in f \} \,, \qquad \mathcal{R}_f = \{ b \in B \mid \exists a \in A, \, (a, b) \in f \} \,.$$

Informally, f associates elements of B with elements of A such that each $a \in A$ has at most one image $b \in B$, that is, $b = f(a)$. The Image of $X \subset A$ is:

$$f(X) = \{ b \in B \mid b = f(a) \text{ for some } a \in X \} \,,$$

and the Inverse image of $Y \subset B$ is:

$$f^{-1}(Y) = \{ a \in A \mid f(a) \in Y \} \,.$$

Given two functions g and f, such that $\mathcal{D}_f \subset \mathcal{D}_g$ and $g = f$ on $\mathcal{D}_f$, then we say that g *extends* f to $\mathcal{D}_g$ and, vice versa, f *restricts* g to $\mathcal{D}_f$.

The algebra of real functions is defined pointwise. The sum $f + g$ is defined as $(f + g)(x) = f(x) + g(x)$. The product fg is given by $(f g)(x) = f(x) g(x)$. The indicator function 1_A of the set A is the function:

$$1_A(x) = \begin{cases} 1 & \text{for} \quad x \in A \\ 0 & \text{for} \quad x \notin A \end{cases}$$

that verifies $1_{A \cap B} = 1_A \cdot 1_B \,, \quad 1_{A \cup B} = 1_A + 1_B - 1_A \cdot 1_B \,, \quad 1_{A^c} = 1 - 1_A \,.$

7.1.5   Equivalences

A *relation* between two sets, A and B, is a subset $\mathcal{R}$ of $A \times B$; we write $x \sim y$ to indicate that $(x, y) \in \mathcal{R}$. An *equivalence* relation on a set E is a relation $\mathcal{R}$ of $E \times E$, with the following properties, valid for any $x, y, z \in E$:

(i) reflexivity: $x \sim x$;

(ii) symmetry: $x \sim y \implies y \sim x$;

(iii) transitivity: $x \sim y$ and $y \sim z \implies x \sim z$.

For any $x \in E$, the subset $[x] := \{ e \in E \mid e \sim x \}$ is called *equivalence class* of the element $x \in E$. Clearly, when $x \sim y$, then $[x] = [y]$.

An equivalence relation on E subdivides E itself into disjoint equivalence classes. A *partition* of a set E is a family $\mathcal{F}$ of subsets of E such that:

(i) $\bigcup_{A \in \mathcal{F}} A = E$,

(ii) for any $A, B \in \mathcal{F}$, $A \neq B$, then $A \cap B = \emptyset$.

Consider $x, y \in E$ such that $x \neq y$; then, it is either $[x] = [y]$ or $[x] \neq [y]$. In the first case, we have already observed that it is $x \sim y$. In the second case, x and y are not equivalent, i.e., $x \nsim y$, therefore $[x] \cap [y] = \emptyset$. Hence, the equivalence classes of E partition the set E itself. The collection of the equivalence classes of E is called *quotient* set of E and is denoted by $E/\sim$. For instance, if $E = \mathbb{Z}$, the relation $x \sim y \iff x - y = 2k$, $k \in \mathbb{Z}$, is an equivalence in $\mathbb{Z}$, which partitions $\mathbb{Z}$ in the classes of even and odd numbers. In general, for any given $m \in \mathbb{Z}$, an equivalence relation $x \sim y$ can be defined, denoted by the symbol $\equiv_m$:

$$x \equiv_m y \iff x - y = m k, \quad \text{for a certain } k \in \mathbb{Z}.$$

The quotient set, obtained in this way, is called the set of *residue classes modulus m*, and is denoted by $\mathbb{Z}_m$.

7.1.6   Real intervals

Let us introduce some notation. $\mathbb{N}$ denotes the set of natural numbers, with the convention $0 \notin \mathbb{N}$, while $\mathbb{Z}$ is the set of integers, $\mathbb{Q}$ is the set of rational numbers, $\mathbb{R}$ denotes the set of real numbers and $\mathbb{C}$ is the set of complex numbers.

Intervals in $\mathbb{R}$ are denoted via endpoints, a square bracket indicating their inclusion, while an open bracket means exclusion, so that, for example:

$$(a, b] = \{ x \in \mathbb{R} \mid a < x \leq b \}.$$

Symbols ∞ and $-\infty$ are used to describe unbounded intervals, such as $(-\infty, b]$. Later on, we will define operations, in the extended real number system, involving these symbols.

7.1.7 Cardinality

Two non–empty sets, X and Y, share common *cardinality*, if there exists a bijection $f : X \to Y$. An empty set is finite; a non–empty set is finite if it shares cardinality with the set $I_n = \{1,\dots,n\}$, for some $n \in \mathbb{N}$. A set is infinite if it shares cardinality with one of its proper subsets. A set A is *countable*, or *numerable* (also *denumerable*, *enumerable*), if there exists a one–to–one correspondence between A and a subset of $\mathbb{N}$.

◇ $\mathbb{Q}$ is countable;   ◇ $\mathbb{R} \setminus \mathbb{Q}$ is not countable;

◇ $\mathbb{R}$ is not countable;   ◇ $[a,b]$ is not countable.

7.1.8 The Real Number System

Definition 7.1. Let X be a non–empty subset of $\mathbb{R}$.

(a) An element $u \in \mathbb{R}$ is called *upper bound* of X if $x \le u$ for any $x \in X$; in this case, X is said to be *bounded from above*.

(b) An element $\ell \in \mathbb{R}$ is called *lower bound* of X if $x \ge \ell$ for any $x \in X$; in this case X is said to be *bounded from below*.

(c) An element $u^\star$ is called *supremum*, or least lower bound, of X if $u^\star \le \mu$ for any upper bound u of X.

(d) An element $\ell_\star$ is called *infimum*, or greatest lower bound, of X if $\ell_\star \ge \nu$ for any lower bound ℓ of X.

Supremum and infimum of a set X are denoted by:

$$u^\star = \sup X, \qquad \ell_\star = \inf X.$$

7.1.9 The extended Real Number System

The set of the extended real numbers is denoted as:

$$\overline{\mathbb{R}} := \mathbb{R} \cup \{-\infty, +\infty\} = [-\infty, +\infty], \tag{7.2}$$

where we assume that, for any $x \in \mathbb{R}$, it holds $-\infty < x < +\infty$. Rules for computations with $-\infty$ and $+\infty$ are introduced in the following list.

1. $x + (+\infty) = +\infty = +\infty + x$,

2. $x + (-\infty) = -\infty = -\infty + x$,

3. $x > 0 \implies x \cdot (+\infty) = +\infty = +\infty \cdot x$,

4. $x > 0 \implies x \cdot (-\infty) = -\infty = -\infty \cdot x$,

5. $x < 0 \implies x \cdot (+\infty) = -\infty = -\infty \cdot x$,

6. $x < 0 \implies x \cdot (-\infty) = +\infty = -\infty \cdot x$,

7. $(+\infty) + (+\infty) = +\infty$,

8. $(-\infty) + (-\infty) = -\infty$,

9. $(+\infty) \cdot (+\infty) = +\infty = (-\infty) \cdot (-\infty)$,

10. $(+\infty) \cdot (-\infty) = -\infty = (-\infty) \cdot (+\infty)$,

11. $(+\infty) \cdot 0 = 0 = 0 \cdot (+\infty)$,

12. $(-\infty) \cdot 0 = 0 = 0 \cdot (-\infty)$.

Note that the following operations remain undefined:

$$(+\infty) + (-\infty) , \qquad (-\infty) + (+\infty) .$$

7.2   Topology

Basic notions of Topology are shortly resumed, now, which are needed to develop Measure theory, as well as to generalise the concepts studied in § 1.2. A complete reference to Topology is represented, for example, by [34]. The general definition of a Topology, in a non–empty set Ω, is given below.

Definition 7.2 (Topology). Let Ω be any non–empty set. A collection $\mathcal{T}$ of subsets of Ω is called *topology* if it verifies the four following properties:

(i) $\emptyset \in \mathcal{T}$;

(ii) $\Omega \in \mathcal{T}$;

(iii) closure under union: if $\mathcal{C} \subset \mathcal{T}$, then $\bigcup_{C \in \mathcal{C}} \in \mathcal{T}$;

(iv) closure under finite intersection: if $O_1, \ldots, O_n \in \mathcal{T}$, then $\bigcap_{k=1}^{n} O_k \in \mathcal{T}$.

The pair $(\Omega, \mathcal{T})$ is called a *topological space*, and the sets in $\mathcal{T}$ are called *open sets*.

Remark 7.3. Many of the topological spaces used in applications verify a further axiom, known as *Hausdorff* [2] property or *separation* or T_2 property:

---

[2]Felix Hausdorff (1868–1942), German mathematician.

(v) for any $x_1, x_2 \in \Omega$, $x_1 \neq x_2$, there exist $O_1, O_2 \in \mathcal{T}$, $O_1 \cap O_2 = \emptyset$, such that $x_1 \in O_1$ and $x_2 \in O_2$.

The idea of topological space is inspired by the open sets in $\mathbb{R}^n$, introduced in Definition 1.13 in § 1.2. We provide here further examples of topological spaces.

Example 7.4. Let Ω be any non–empty set. Then, $\mathcal{T} = \{\emptyset, \Omega\}$ is a topology, called *trivial* or *indiscrete* topology. In this situation, all points of the space cannot be distinguished by topological means.

Example 7.5. Let Ω be any non–empty set. Then $\mathcal{T} = \mathcal{P}(\Omega)$ is a topology, called *discrete* topology. Here, every set is open.

Example 7.6. If $\Omega = \mathbb{N}$, the family $\mathcal{T}$ of all the finite initial segments of $\mathbb{N}$, that is to say, the collection of sets $J_n = \{1, 2, \ldots, n\}$, is a topology.

Example 7.7. Let Ω be any non–empty set, and let $\mathcal{F}$ be a partition of Ω. The collection $\mathcal{T}$ of the subsets of Ω, obtained as union of elements of $\mathcal{F}$, is a topology, induced by the partition $\mathcal{F}$.

7.2.1 Closed sets

Let $(\Omega, \mathcal{T})$ be a topological space. The set $C \subset \Omega$ is said to be *closed* if $C^c \in \mathcal{T}$, that is, if the complement of C is an open set in $\mathcal{T}$.

From the topology axioms in Definition 7.2 and De Morgan Laws (7.1), the collection $\mathcal{K}$, formed by the closed sets in a topological space, verifies the following four properties:

(i) $\emptyset \in \mathcal{K}$;

(ii) $\Omega \in \mathcal{K}$;

(iii)   closure under intersection: if $\mathcal{C} \subset \mathcal{K}$, then $\bigcap_{C \in \mathcal{C}} \in \mathcal{K}$;

(iv)   closure under finite union: if $O_1, \ldots, O_n \in \mathcal{K}$, then $\bigcup_{k=1}^{n} O_k \in \mathcal{K}$.

7.2.2 Limit

Let $(\Omega, \mathcal{T})$ be a topological space, and consider $S \subset \Omega$. Then, $x \in \Omega$ is:

(a) *separated* from S if and only if there exists $A \in \mathcal{T}$ such that $x \in A$ and $A \cap S = \emptyset$;

(b) an *adherent* point for S if and only if $A \cap S \neq \emptyset$ for any $A \in \mathcal{T}$ such that $x \in A$;

(c) an *accumulation* point for S if and only if x is adherent to $S \setminus \{x\}$;

(d) an *isolated* point in S if and only if x is not an accumulation point for S.

Remark 7.8. A few facts should be noticed.

■ If x is an accumulation point for S, then x is also an adherent point for S.

■ If $x \notin S$ is adherent for S, then x is an accumulation point for S.

■ There exist adherent points for S that are not accumulation points for S.

An *open neighbourhood* of $x \in \Omega$ is an open set $U \in \mathcal{T}$ such that $x \in U$; for simplicity, it will be referred to as *neighbourhood*. It is possible to express the notions of separated, adherent and accumulation points using the idea of neighbourhood. The concept of limit can also be generalised.

Definition 7.9. Let $(\Omega_1, \mathcal{T}_1)$ and $(\Omega_2, \mathcal{T}_2)$ be topological spaces, and consider $A \subset \Omega_1$ and $f : A \to \Omega_2$. Furthemore, let $x_0 \in A$ be an accumulation point for A, and $\ell \in \Omega_2$. Then:

$$\lim_{x \to x_0} f(x) = \ell$$

if, for any $\mathcal{T}_2$–neighbourhood V of ℓ, there exists a $\mathcal{T}_1$–neighbourhood U of x_0 such that

$$x \in A \cap (U \setminus \{x_0\}) \quad \Longrightarrow \quad f(x) \in V.$$

7.2.3   Closure

Consider a topological space $(\Omega, \mathcal{T})$ and a set $S \subset \Omega$. The *closure* $\overline{S}$ is the collection of all adherent points of S. By construction, it holds $S \subset \overline{S}$. The closure of a closed set is the set itself, and we can indeed state the following Theorem 7.10.

Theorem 7.10. In the topological space $(\Omega, \mathcal{T})$, a set $S \subset \Omega$ is closed if and only if $S = \overline{S}$.

Theorem 7.10 has a few implications. First of all, $\overline{\emptyset} = \emptyset$ and $\overline{\Omega} = \Omega$. Moreover, given $S \subset \Omega$, since $\overline{S}$ is a closed set, then $\overline{\overline{S}} = \overline{S}$. Furthermore, if $S_1 \subset S_2$, then $\overline{S_1} \subset \overline{S_2}$. Finally, $\overline{S_1 \cup S_2} = \overline{S_1} \cup \overline{S_2}$.

The main property of the closure of a set S, though, is that $\overline{S}$ is the smallest closed set containing S. In fact, the following Theorem 7.11 holds.

Theorem 7.11. Let $(\Omega, \mathcal{T})$ be a topological space, and consider $S \subset \Omega$. Denote with $\mathcal{K}$ the collection of closed sets in $(\Omega, \mathcal{T})$. Furthermore, denote with $\mathcal{I}$ the family of subsets of Ω such that:

$$\mathcal{I} = \{ K \in \mathcal{K} \mid S \subset K \}.$$

Then

$$\overline{S} = \bigcap_{K \in \mathcal{I}} K.$$

7.2.4 Compactness

Definition 7.12. Let Ω be a given set and let $S \subset \Omega$. A *covering* of S in Ω is an indexed family $(A_i)_{i \in \mathcal{I}}$ in Ω such that:

$$S \subset \bigcup_{i \in \mathcal{I}} A_i.$$

In a topological space, a covering is called *open* if A_i is an open set for any $i \in \mathcal{I}$.

The notion of open covering leads to the fundamental definition of compactness in a topological space.

Definition 7.13. Let $(\Omega, \mathcal{T})$ be a topological space. A set $K \subset \Omega$ is called *compact* if any open covering of K admits a finite *sub–covering*, meaning that there exists a finite subset $\mathcal{I}_0 \subset \mathcal{I}$ such that

$$K \subset \bigcup_{i \in \mathcal{I}_0} A_i.$$

In the familiar context of the Euclidean, open set, topology in $\mathbb{R}^n$, it is possible to show that a subset is compact if and only if it is closed and bounded. This property of compactness does not hold for general topological spaces. In the case of the real line $\mathbb{R}$, intervals of the form $[a, b]$ constitute examples of compact sets. It is further possible to show that, in a T_2–space (where the T_2 property is defined in Remark 7.3), any compact set is also a closed set. Finally, finite subsets in a topological space are compact.

7.2.5 Continuity

When, in a set, a topology is available, it is possible to introduce the concept of continuity. Let $(\Omega_1, \mathcal{T}_1)$ and $(\Omega_2, \mathcal{T}_2)$ be topological spaces and consider $f : \Omega_1 \to \Omega_2$. Function f is *continuous* if, for any open set $A_2 \in \Omega_2$, the inverse image $f^{-1}(A_2)$ is an open set in the space $(\Omega_1, \mathcal{T}_1)$, that is, in formal words:

$$A_2 \in \mathcal{T}_2 \implies f^{-1}(A_2) \in \mathcal{T}_1.$$

It is possible to show that f is continuous if and only if, for any closed set K_2 in Ω_2, the inverse image $f^{-1}(K_2)$ is a closed set in Ω_1. It is also possible to formulate a *local* notion of continuity, at a given point x_0, using appropriate neighbourhoods; as we do not need to analyse this problem here, we leave it to the interested Reader as an exercise.

It is interesting to remark that Weierstrass Theorem 1.30 can be generalised to the case of topological spaces.

Theorem 7.14 (Weierstrass theorem on compactness in topological spaces). If $f : \Omega_1 \to \Omega_2$ is a continuous map, and if K_1 is a compact subset of Ω_1, then $f(K_1)$ is compact in Ω_2.

Chapter 8

Lebesgue integral

Here and in Chapters 9 and 10, we deal with function μ, called *measure*, which returns area, or volume, or probability, of a given set. We assume that μ is already defined, adopting an axiomatic approach which turns out advantageous, as the same theoretical results apply to other situations, besides area in $\mathbb{R}^2$ or volume in $\mathbb{R}^3$, and which is particularly fruitful in Probability theory. A general domain that can be assumed for μ is a σ-algebra, defined in § 8.1. For completeness, a few basic concepts are recalled, for which some familiarity is assumed.

8.1 Measure theory

Measure theory is introduced in an axiomatic way, to keep its exposition to a minimum. Some ideas are provided, later, to explain the construction of the *Lebesgue*[1] *measure*, which represents the most important measure on the real line.

8.1.1 σ-algebras

Let us introduce, first, the notion of σ-algebra of sets.

Definition 8.1. Let Ω be any non–empty set. A collection $\mathcal{A}$ of subsets of Ω is called *σ-algebra on Ω* if:

(i) $\Omega \in \mathcal{A}$;

(ii)   closure under complement: $A \in \mathcal{A} \implies A^c \in \mathcal{A}$;

(iii) closure under countable union: $(A_n)_{n \in \mathbb{N}} \subset \mathcal{A} \implies \displaystyle\bigcup_{n=1}^{\infty} A_n \in \mathcal{A}$.

The pair $(\Omega, \mathcal{A})$ is called *measurable space*, and the sets in $\mathcal{A}$ are called *measurable sets*.

---

[1] Henri Léon Lebesgue (1875–1941), French mathematician.

From Definition 8.1 and De Morgan Laws (7.1), it follows a fourth property of *closure under countable intersections* of the collection of measurable sets $\mathcal{A}$. When the *countable union* property (iii) of Definition 8.1 is replaced with the weaker assumption of *finite union*, the family of sets $\mathcal{A}$ becomes an *algebra of sets*.

Definition 8.2. Let Ω be any non–empty set. A collection $\mathcal{A}$ of subsets of Ω is called *algebra of sets* if:

(i) $\Omega \in \mathcal{A}$;

(ii)   closure under complement: $A \in \mathcal{A} \implies A^c \in \mathcal{A}$;

(iii)   closure under finite union: $(A_n)_{n \in \mathbb{N}} \subset \mathcal{A} \implies \bigcup_{n=1}^{m} A_n \in \mathcal{A}$.

We do not develop Measure theory in the context of algebra of sets, as it is beyond the purpose of our introductory treatment.

Example 8.3. Let Ω be any non–empty set.

- $\mathcal{A} = \{\Omega, \emptyset\}$ is a σ–algebra. Since it is the simplest possible σ–algebra, it is called *trivial*.

- $\mathcal{A} = \mathcal{P}(\Omega)$ (i.e., the power set of Ω) is a σ–algebra.

- If A is a non–empty subset of Ω, then $\mathcal{A} = \{\emptyset, A, A^c, \Omega\}$ is a σ–algebra.

- If Ω is an infinite set, then the collection $\mathcal{A}$ of subsets $A \subset \Omega$, with A countably infinite or A^c countably infinite, is a σ–algebra.

- If Ω is an infinite set, then the collection $\mathcal{A}$ of subsets $A \subset \Omega$, with A finite or A^c finite, is an algebra, but it is not a σ–algebra.

It may happen that a given collection of sets, $\mathcal{X}$, is not a σ–algebra, but there exists the minimal σ–algebra, which contains $\mathcal{X}$. Hence, to obtain non–trivial σ–algebras, we need to consider the following abstract construction.

Lemma 8.4. Consider a family of σ–algebras on Ω. Then, the intersection of all the σ–algebras from this family is also a σ–algebra on Ω.

Proof. Denote by $\mathcal{H}$ a non–empty collection of σ–algebras on Ω, and define:

$$\mathcal{A}_0 = \bigcap_{A \in \mathcal{H}} \mathcal{A}.$$

To prove Lemma 8.4, we need to prove that $\mathcal{A}_0$ is a σ–algebra. Observe, first, that $\Omega \in \mathcal{A}_0$, since $\Omega \in \mathcal{A}$ for any $\mathcal{A} \in \mathcal{H}$, by definition. Then, choose a set $A \in \mathcal{A}_0$, so that A belongs to any σ–algebra in $\mathcal{A} \in \mathcal{H}$ and, therefore,

$A^c \in \mathcal{A}_0$. Finally, if (A_n) is a sequence of sets in $\mathcal{A}_0$, it is clear that such a sequence belongs to any σ–algebra in $\mathcal{H}$, implying:

$$\bigcup_{n \in \mathbb{N}} A_n \in \mathcal{A}$$

for each $\mathcal{A} \in \mathcal{H}$, so that:

$$\bigcup_{n \in \mathbb{N}} A_n \in \mathcal{A}_0 .$$

$\square$

Using Lemma 8.4, we can define the σ–algebra generated by a family of arbitrary sets.

Definition 8.5. Let $\mathcal{X}$ be a collection of subsets of Ω. Denote with $\mathcal{H}$ the collection of σ–algebras in Ω, including $\mathcal{X}$. Then:

$$\sigma(\mathcal{X}) = \bigcap_{\mathcal{A} \in \mathcal{H}} \mathcal{A}$$

is a σ–algebra, called *σ–algebra generated by $\mathcal{X}$*.

8.1.2 Borel sets

This construction is of great interest when, working on the real line, we consider the collection $\mathcal{X}$ of open sets in $\mathbb{R}$. In this situation, the generated σ–algebra, $\sigma(\mathcal{X})$, is called *Borel σ–algebra*, and it is denoted by $\mathcal{A}(\mathbb{R})$, while every member of $\mathcal{A}(\mathbb{R})$ is called *Borel set*.

Let $\mathcal{I}$ be a collection of intervals $[a, b)$ with $a < b$, i.e.,

$$\mathcal{I} = \{ \, [a, b) \mid a, b \in \mathbb{R}, \ a < b \} \, ,$$

then:

$$\sigma(\mathcal{I}) = \mathcal{A}(\mathbb{R}) \, . \tag{8.1}$$

Equality (8.1) also holds when:

$$\mathcal{I} = \{ \, (a, b) \mid a, b \in \mathbb{R}, \ a < b \} \, ,$$
$$\mathcal{I} = \{ \, (a, +\infty) \mid a \in \mathbb{R} \} \, ,$$
$$\mathcal{I} = \{ \, (-\infty, a) \mid a \in \mathbb{R} \} \, .$$

The n–dimensional Borel[2] σ–algebra in $\mathbb{R}^n$ is the σ–algebra generated by all open subsets of $\mathbb{R}^n$, and it is denoted by $\mathcal{A}(\mathbb{R}^n)$. By construction, $\mathcal{A}(\mathbb{R}^n)$ contains all open/closed sets and all countable unions/intersections of open/closed sets. The Borel σ–algebra does not represent the whole of $\mathcal{P}(\mathbb{R}^n)$; this result is known as Vitali[3] covering theorem ([10], Theorem 2.1.4).

---

[2] Félix Édouard Justin Émile Borel (1871–1956), French mathematician and politician.
[3] Giuseppe Vitali (1875–1932), Italian mathematician.

8.1.3   Measures

Definition 8.6. Let $\mathcal{A}$ be a σ–algebra on Ω. Function $\mu : \mathcal{A} \to [0, +\infty]$ is said to be a *measure* on $\mathcal{A}$ if:

(i) $\mu(\emptyset) = 0$

(ii) the property of *countable additivity* holds, that is, for any disjoint sequence $(A_n) \subset \mathcal{A}$:

$$\mu\left(\bigcup_{n=1}^{\infty} A_n\right) = \sum_{n=1}^{\infty} \mu(A_n).$$

The triple $(\Omega, \mathcal{A}, \mu)$ is called *measure space*, and measure μ is called:

- *finite*   if $\mu(\Omega) < +\infty$; in particular, μ is called *probability* measure if $\mu(\Omega) = 1$;

- *σ–finite*   if there exists a sequence $(A_n) \subset \mathcal{A}$, such that $\bigcup_{n=1}^{\infty} A_n = \Omega$, where $\mu(A_n) < +\infty$ for any $n \in \mathbb{N}$;

- *complete*   if $A \in \mathcal{A}$, with $\mu(A) = 0$, and $B \subset A \implies B \in \mathcal{A}$, and $\mu(B) = 0$;

- *concentrated* on $A \in \mathcal{A}$ if $\mu(A^c) = 0$; in this case, A is said to be a *support* of μ.

Example 8.7. Here, the so–called *counting measure* is introduced. Let Ω be an arbitrary set, and $\mathcal{A}$ be a σ–algebra on Ω. Define $\mu^{\#} : \mathcal{A} \to [0, \infty]$ as:

$$\mu^{\#}(A) = \begin{cases} \#A & \text{if } A \text{ is finite,} \\ +\infty & \text{if } A \text{ is infinite.} \end{cases}$$

where $\#A$ indicates the cardinality of A. Function $\mu^{\#}$ is a measure on $\mathcal{A}$.

Example 8.8. Let $x \in \Omega$. For any $A \in \mathcal{P}(\Omega)$, define:

$$\delta_x(A) = \begin{cases} 1 & \text{if } x \in A, \\ 0 & \text{if } x \notin A. \end{cases}$$

δ_x is a measure on $\mathcal{P}(\Omega)$ called *Dirac x–measure*. It is a measure concentrated on the singleton set $\{x\}$.

The following intuitive results, stated in Proposition 8.9, are known as *monotonicity* and *subtractivity* of the measure.

Proposition 8.9. If $A, B \in \mathcal{A}$ and $A \subset B$, then:

$$\mu(A) \leq \mu(B).$$

Moreover, if $\mu(A) < \infty$, then:

$$\mu(B \setminus A) = \mu(B) - \mu(A).$$

Proof. Equality $B = (B \setminus A) \cup A$ implies:

$$\mu(B) = \mu(B \setminus A) + \mu(A). \tag{8.2}$$

Since $\mu(B \setminus A) \geq 0$, it follows that $\mu(B) \geq \mu(A)$, thus *monotonicity* is proved. Now, the assumption $\mu(A) < \infty$ means that $\mu(A)$ is finite and, thus, it can be substracted from both sides of equality (8.2), proving *subtractivity*. □

Proposition 8.10. If $A, B \in \mathcal{A}$ and if $\mu(A \cap B) < \infty$, then:

$$\mu(A \cup B) = \mu(A) + \mu(B) - \mu(A \cap B).$$

Proof. Notice that $(A \cap B) \subset A$, $(A \cap B) \subset B$, and:

$$A \cup B = (A \setminus (A \cap B)) \cup B.$$

Since $\mu(A \cap B) < \infty$, it holds:

$$\mu(A \cup B) = \mu(A) - \mu(A \cap B) + \mu(B).$$

□

The following Lemma 8.11 shows that, when dealing with a sequence of non–disjoint sets, it is always possible to rearrange it and treat, instead, a pairwise disjoint sequence of sets, equivalent to the original sequence.

Lemma 8.11. Consider a measure space $(\Omega, \mathcal{A}, \mu)$. Let $(A_k)_{k \in \mathbb{N}}$ be a countable sequence of measurable sets in $\mathcal{A}$. Then, there exists a sequence of measurable sets $(B_k)_{k \in \mathbb{N}}$ such that:

1. $B_k \in \mathcal{A}$ for any $k \in \mathbb{N}$;

2. $B_k \subset A_k$ for any $k \in \mathbb{N}$;

3. $B_k \cap B_j = \emptyset$ for any $k \neq j$, i.e., sequence $(B_k)_{k \in \mathbb{N}}$ is pairwise disjoint;

4. $\displaystyle\bigcup_{k=1}^{\infty} A_k = \bigcup_{k=1}^{\infty} B_k$.

Proof. The sequence $(B_k)_{k \in \mathbb{N}}$ can be defined inductively, setting:

$$B_1 = A_1, \qquad B_k = A_k \setminus \bigcup_{i=1}^{k-1} A_i \quad \text{for } k > 1.$$

Properties (1)–(2) are straightforward, both being a consequence of the construction of each set B_k. To demonstrate property (3), let us fix $k, j \in \mathbb{N}$, assuming $k < j$ without loss of generality, and consider the intersection:

$$B_k \cap B_j = \left(A_k \setminus \bigcup_{i=1}^{k-1} A_i \right) \cap \left(A_j \setminus \bigcup_{i=1}^{j-1} A_i \right).$$

Now, using De Morgan Laws (7.1):

$$B_k \cap B_j = \left(A_k \cap A_1^c \cap A_2^c \ldots \cap A_{k-1}^c \right) \cap \left(A_j \cap A_1^c \cap A_2^c \ldots \cap A_{j-1}^c \right).$$

The assumption $k < j$ implies that, in the second group of intersections, it appears the set A_k^c. Thus, we can infer that:

$$B_k \cap B_j = \emptyset.$$

We now demonstrate property (4). Notice first that, from property (2), it is immediate the inclusion:

$$\bigcup_{k=1}^{\infty} B_k \subset \bigcup_{k=1}^{\infty} A_k.$$

To complete the proof, the reverse inclusion must be shown. To this purpose, let us choose:

$$x \in \bigcup_{k=1}^{\infty} A_k.$$

Hence, x must belong to one of the sets A_k, at least, implying that the following definition is well–posed:

$$m = \min\{ k \in \mathbb{N} \mid x \in A_k \}.$$

In other words, $x \in A_m$ and $x \notin A_k$ for $k = 1, \ldots, m - 1$. Therefore, it must be $x \in B_m$, so that:

$$x \in \bigcup_{k=1}^{\infty} B_k.$$

which completes the proof. $\square$

While working with countable families of sets, not necessarily pairwise disjoint, we meet the so–called property of *countable subadditivity*.

Proposition 8.12. If $(A_k)_{k \geq 1} \subset \mathcal{A}$, then:

$$\mu\left(\bigcup_{k=1}^{\infty} A_k \right) \leq \sum_{i=k}^{\infty} \mu(A_k).$$

Proof. Denote with $(B_k)_{k \in \mathbb{N}}$ the sequence of measurable sets, obtained from the sequence $(A_k)_{k \in \mathbb{N}}$ using the procedure of Lemma 8.11. In this way, for any $k \in \mathbb{N}$, it holds that $B_k \in \mathcal{A}$ and $B_k \subset A_k$. It also holds that the sequence $(B_k)_{k \in \mathbb{N}}$ is pairwise disjoint, that is, $B_k \cap B_j = \emptyset$ for any $k \neq j$. Moreover:

$$\bigcup_{k=1}^{\infty} A_k = \bigcup_{k=1}^{\infty} B_k .$$

We can, then, infer:

$$\mu \left(\bigcup_{k=1}^{\infty} A_k \right) = \mu \left(\bigcup_{k=1}^{\infty} B_k \right) = \sum_{k=1}^{\infty} \mu(B_k) \leq \sum_{k=1}^{\infty} \mu(A_k) .$$

$\square$

The next Theorem 8.13 states two very interesting results, related to the situation of increasing or decreasing families of nested sets.

Theorem 8.13. Let $(A_k)_{k \geq 1} \subset \mathcal{A}$ be an increasing sequence of sets, i.e., $A_k \subset A_{k+1}$. Then:

$$\mu \left(\bigcup_{k=1}^{\infty} A_k \right) = \lim_{n \to \infty} \mu(A_n). \tag{8.3}$$

Let $(A_k)_{k \geq 1} \subset \mathcal{A}$ be a decreasing sequence of sets, i.e., $A_{k+1} \subset A_k$, with $\mu(A_1) < \infty$. Then:

$$\mu \left(\bigcap_{k=1}^{\infty} A_k \right) = \lim_{n \to \infty} \mu(A_n). \tag{8.4}$$

Proof. Let us first prove (8.3). By assumption, $A_k \subset A_{k+1}$, hence, the non-negative sequence $(\mu(A_k))_{k \in \mathbb{N}}$ is monotonically increasing, which implies the existence of the limit:

$$\ell = \lim_{m \to \infty} \mu(A_m).$$

On the other hand, the inclusion $A_m \subset \bigcup_{k=1}^{\infty} A_k$ holds true, for any $m \in \mathbb{N}$.
Passing to the limit:

$$\ell = \lim_{m \to \infty} \mu(A_m) \leq \lim_{k \to \infty} \mu \left(\bigcup_{k=1}^{\infty} A_k \right) . \tag{8.5}$$

If $\ell = +\infty$, there is nothing more to proof. If, instead, $\ell < +\infty$, then, due to the monotonicity of the sequence $(\mu(A_k))_{k \in \mathbb{N}}$, we infer that, for any $k \in \mathbb{N}$:

$$\mu(A_k) \leq \ell < +\infty. \tag{8.6}$$

Now, observe that the union of sets can be rewritten as:

$$\bigcup_{k=1}^{\infty} A_k = A_1 \cup \bigcup_{k=1}^{\infty} (A_{k+1} \setminus A_k). \tag{8.7}$$

The union in the right–hand side of (8.7) is, by construction, disjoint. Since condition (8.6) also holds, we can use Proposition 8.9 to obtain:

$$\mu\left(\bigcup_{k=1}^{\infty} A_k\right) = \mu(A_1) + \sum_{k=1}^{\infty} \mu\big((A_{k+1} \setminus A_k)\big)$$

$$= \mu(A_1) + \sum_{k=1}^{\infty} \big(\mu(A_{k+1}) - \mu(A_k)\big).$$

In the previous chain of equalities, the last series is telescopic, therefore:

$$\mu\left(\bigcup_{k=1}^{\infty} A_k\right) = \mu(A_1) + \lim_{n\to\infty} \sum_{k=1}^{n} \big(\mu(A_{k+1}) - \mu(A_k)\big) = \lim_{n\to\infty} \mu(A_n) = \ell,$$

which proves (8.3).

We now prove the second relation (8.4). Define the measurable set $B = \bigcap_{k=1}^{\infty} A_k$, and form:

$$A_1 \setminus B = A_1 \cap B^c = A_1 \cap \bigcup_{k=1}^{\infty} A_k^c = \bigcup_{k=1}^{\infty} (A_1 \cap A_k^c) = \bigcup_{k=1}^{\infty} (A_1 \setminus A_k).$$

The sequence of sets $(A_1 \setminus A_k)_{k\in\mathbb{N}}$ is increasing, thus, we can apply relation (8.3) and infer that:

$$\mu(A_1 \setminus B) = \lim_{k\to\infty} \mu(A_1 \setminus A_k). \tag{8.8}$$

Now, $\mu(A_1) < +\infty$ by hypothesis. Hence, from Proposition 8.9, it follows that $\mu(A_1 \setminus B) = \mu(A_1) - \mu(B)$, which can be inserted into (8.8) to yield, recalling the definition of B:

$$\mu(A_1) - \mu\left(\bigcap_{k=1}^{\infty} A_k\right) = \lim_{k\to\infty} \mu(A_1 \setminus A_k) = \mu(A_1) - \lim_{k\to\infty} \mu(A_k). \tag{8.9}$$

Thesis (8.4) follows by eliminating $\mu(A_1)$ form both sides of (8.9). $\qquad\square$

8.1.4 Exercises

1. Consider $\Omega = \{1,2,3\}$. Find necessary and sufficient conditions on the real numbers x, y, z, such that there exists a probability measure μ on the σ–algebra $\mathcal{A} = \mathcal{P}(\Omega)$, where:

$$x = \mu(\{1,2\}), \qquad y = \mu(\{2,3\}), \qquad z = \mu(\{1,3\}).$$

2. Consider a measure space $(\Omega, \mathcal{A}, \mu)$.

   (a) Let $E_1, E_2 \in \mathcal{A}$. The *symmetric difference* $E_1 \Delta E_2$ is defined as:

   $$E_1 \Delta E_2 := (E_1 \setminus E_2) \cup (E_2 \setminus E_1).$$

   Suppose that $\mu(E_1 \Delta E_2) = 0$. Show that $\mu(E_1) = \mu(E_2)$.

   (b) Show that, under the assumption that μ is a complete measure, if $E_1 \in \mathcal{A}$ and $\mu(E_1 \Delta E_2) = 0$, then $E_2 \in \mathcal{A}$.

Solution to Exercise 2.

(a) To prove the first statement, we proceed as follows.

$$\mu(E_1 \Delta E_2) = 0 \implies \mu\Big((E_1 \setminus E_2) \cup (E_2 \setminus E_1)\Big) = 0$$
$$\implies \mu(E_1 \setminus E_2) + \mu(E_2 \setminus E_1) = 0$$
$$\implies \mu\Big(E_1 \setminus (E_1 \cap E_2)\Big) + \mu\Big(E_2 \setminus (E_1 \cap E_2)\Big) = 0.$$

Since $\mu : \Omega \to [0, +\infty]$, it follows that $\mu(E_1 \setminus (E_1 \cap E_2)) = 0$ and $\mu(E_2 \setminus (E_1 \cap E_2)) = 0$. Moreover, observing that $(E_1 \cap E_2) \subset E_1$ and $(E_2 \cap E_1) \subset E_2$, we can write:

$$\begin{cases} \mu(E_1) - \mu(E_1 \cap E_2) = 0, \\ \mu(E_2) - \mu(E_1 \cap E_2) = 0. \end{cases}$$

Thus:

$$\mu(E_1) - \mu(E_1 \cap E_2) = \mu(E_2) - \mu(E_1 \cap E_2) \implies \mu(E_1) = \mu(E_2).$$

(b) For this second point, we have:

$$\mu(E_1 \Delta E_2) = 0 \implies \mu\Big((E_1 \setminus E_2) \cup (E_2 \setminus E_1)\Big) = 0.$$

Since μ is complete, then $(E_1 \setminus E_2)$ and $(E_2 \setminus E_1)$ are measurable, i.e., they belong to $\mathcal{A}$. In this way, since the σ-algebra is closed with respect to union, intersection and set complementation, we have:

$$E_1 \setminus (E_1 \Delta E_2) = E_1 \cap (E_1 \Delta E_2)^c \in \mathcal{A},$$

hence:

$$E_2 = (E_2 \setminus E_1) \cup \Big(E_1 \setminus (E_1 \Delta E_2)\Big) \in \mathcal{A}.$$

8.2   Translation invariance

Consider the situation of a universal set, where the measure space is established, that coincides with the real line $\mathbb{R}$, or with the Euclidean m–dimensional space $\mathbb{R}^m$: it is natural, then, to relate the property of measure with the algebraic structure of the universal set. The main property that a measure may verify is its invariance with respect to translation.

Definition 8.14. Consider $x \in \mathbb{R}^m$ and $A \in \mathcal{P}(\mathbb{R}^m)$. The *translate* of A with respect to x is the set:

$$A + x := \{u \in \mathbb{R}^m \mid u = x + a \text{ for some } a \in A\}.$$

In Measure theory, it makes sense to compare the measure of the two sets A and $A + x$: for geometric reasons, we may expect them to be the same; this holds for an important class of measures. We have, indeed, the following result, that we state without proof.

Theorem 8.15. There exists a unique complete measure, defined on $\mathbb{R}$ and denoted by ℓ, and there exists a unique σ–algebra $\mathcal{M}(\mathbb{R})$, which are translation invariant, i.e.:

$$\ell(A) = \ell(A + x), \quad \text{for any } x \in \mathbb{R}, \ A \in \mathcal{M}(\mathbb{R}).$$

Moreover, for any interval with endpoints $a, b \in \mathbb{R}$, with $a < b$, it holds:

$$\ell([a, b]) = \ell([a, b)) = \ell((a, b]) = \ell((a, b)) = b - a.$$

ℓ is called *Lebesgue measure* in $\mathbb{R}$.

The relation between the σ–algebra $\mathcal{M}(\mathbb{R})$ of the Lebesgue measurable sets and the Borel σ–algebra is stated in the following Remark 8.16.

Remark 8.16. The chain of strict inclusions is true:

$$\mathcal{A}(\mathbb{R}) \subsetneq \mathcal{M}(\mathbb{R}) \subsetneq \mathcal{P}(\mathbb{R}).$$

An extensive treatment and a construction of the Lebesgue measure and its properties can be found in [52, 51, 6]. For our purposes, our axiomatic approach suffices: there exists a unique translation invariant measure, which associates the measure μ of every finite interval $[a, b]$ with its length $b - a$.

8.2.1   Exercises

1. Find the Lebesgue measure of the set:

$$\bigcup_{n=1}^{\infty} \left\{ x \mid \frac{1}{n+1} \leq x < \frac{1}{n} \right\}.$$

8.3 Simple functions

Definition 8.17. Let $(\Omega, \mathcal{A}, \mu)$ be a measure space, and consider $A \in \mathcal{A}$. The function $\varphi : A \to [-\infty, +\infty]$ is called *simple* if:

(i) $\varphi(A)$ is a finite set, i.e., $\varphi(A) = \{\varphi_1, \varphi_2, \dots, \varphi_m\}$;

(ii) there exist $A_1, \dots, A_m \in \mathcal{A}$ such that $A_i \cap A_j = \emptyset$ for $i \neq j$, and
$$A = \bigcup_{i=1}^{m} A_i \; ;$$

(iii) for $x \in A_i$, with $i = 1, \dots, m$, it holds $\varphi(x) = \varphi_i$.

The plainest simple function is the *characteristic* function of a given set.

Example 8.18. If A is a subset of the set Ω, the characteristic function of A is defined by:
$$\mathbf{1}_A(x) = \begin{cases} 1 & \text{if } x \in A, \\ 0 & \text{if } x \notin A. \end{cases}$$

Remark 8.19. Any simple function can be represented as linear combination of characteristic functions, since it holds:
$$\varphi(x) = \sum_{i=1}^{m} \varphi_i \, \mathbf{1}_{A_i}(x).$$

8.3.1 Integral of simple functions

Let $(\Omega, \mathcal{A}, \mu)$ be a measure space, let $A \in \mathcal{A}$ be a measurable set, and let φ be a simple function defined on A; assume further that φ is non–negative.

Definition 8.20. The integral of φ, on the set A, with respect to measure μ, is defined by:
$$\int_A \varphi \, d\mu := \sum_{i=1}^{m} \varphi_i \, \mu(A_i).$$

When a simple function is defined on the real line, the idea of its integration is intuitive and it is inspired by the geometric concept of area of a family of rectangles, as illustrated in Figure 8.1.

Observe that, to define a simple function correctly, we have to adopt a Measure theory convention, that concerns ∞ in the extended real number system, namely:
$$\pm\infty \cdot 0 = 0.$$

We now state, without proof, the main properties of the integral of a simple function, starting with the properties of *positivity* and *linearity*.

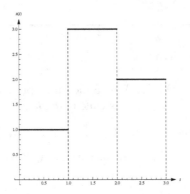

Figure 8.1: The integral of a simple real function.

Proposition 8.21. Let $(\Omega, \mathcal{A}, \mu)$ be a measure space, let $A \in \mathcal{A}$ be a measurable set, and let φ_1, φ_2 be simple functions, defined on A. Then, the following properties hold.

Linearity: if α_1, $\alpha_2 \in \mathbb{R}$, then

$$\int_A (\alpha_1 \varphi_1 + \alpha_2 \varphi_2) \, d\mu = \alpha_1 \int_A \varphi_1 \, d\mu + \alpha_2 \int_A \varphi_2 \, d\mu.$$

Positivity: if $\varphi_1 \geq 0$, then

$$\int_A \varphi_1 \, d\mu \geq 0.$$

Even though it is evident and easy to show, the following Proposition 8.22 has a fundamental importance in Measure theory, and it follows from the concept of integral of a simple function.

Proposition 8.22. Let $(\Omega, \mathcal{A}, \mu)$ be a measure space, and let $A \in \mathcal{A}$ be a measurable set. Then

$$\mu(A) = \int_\Omega \mathbf{1}_A \, d\mu.$$

Proof. Observe that the characteristic function of a set A is a simple function. Hence, since $\mathbf{1}_A(x) = 0$ when $x \in A^c$, and since $\mathbf{1}_A(x) = 1$ when $x \in A$, it follows:

$$\int_\Omega \mathbf{1}_A \, d\mu = 0 \cdot \mu(A^c) + 1 \cdot \mu(A) = \mu(A).$$

□

8.4 Measurable functions

The notion of measurable functions is of great interest in Measure theory, but also in Probability theory, since measurable functions can be interpreted as random variables.

Definition 8.23. Let $(\Omega, \mathcal{A}, \mu)$ be a measure space, and let $A \in \mathcal{A}$ be a measurable set. Function $f : A \to [-\infty, +\infty]$ is called measurable if, for any $\alpha \in \mathbb{R}$:

$$\{x \in A \mid f(x) \leq \alpha\} \in \mathcal{A}.$$

The choice of "$\leq$" is not restrictive, as shown in Proposition 8.24.

Proposition 8.24. Let $(\Omega, \mathcal{A}, \mu)$ be a measure space, and let $A \in \mathcal{A}$ be a measurable set. The following statements for function $f : A \to [-\infty, +\infty]$ are equivalent:

(i) $\{x \in A \mid f(x) > \alpha\} \in \mathcal{A}$;   (iii) $\{x \in A \mid f(x) < \alpha\} \in \mathcal{A}$;

(ii) $\{x \in A \mid f(x) \geq \alpha\} \in \mathcal{A}$;   (iv) $\{x \in A \mid f(x) \leq \alpha\} \in \mathcal{A}$.

Proof. The proof follows from the equalities recalled below, for which we refer to Theorem 11.15 of [52]:

$$\{x \in A \mid f(x) \leq \alpha\} = \bigcap_{n=1}^{\infty} \{x \in A \mid f(x) < \alpha + \frac{1}{n}\},$$

$$\{x \in A \mid f(x) \geq \alpha\} = \bigcap_{n=1}^{\infty} \{x \in A \mid f(x) > \alpha - \frac{1}{n}\},$$

and

$$\{x \in A \mid f(x) < \alpha\} = A \setminus \{x \in A \mid f(x) \geq \alpha\},$$
$$\{x \in A \mid f(x) > \alpha\} = A \setminus \{x \in A \mid f(x) \leq \alpha\}.$$

They show, in a chain, that statement (i) $\implies$ (ii), statement (ii) $\implies$ (iii), statement (iii) $\implies$ (iv), and, finally, statement (iv) $\implies$ (i). $\qquad\square$

The property of measurability is well related with the basic algebraic operations between functions, hence, we can state, without proof, the following Proposition 8.25.

Proposition 8.25. Let $(\Omega, \mathcal{A}, \mu)$ be a measure space, and let $A \in \mathcal{A}$ be a measurable set. Let, further, $f, g : A \to \mathbb{R}$ be measurable functions. Then, the following functions are measurable:

(a) αf,   with $\alpha \in \mathbb{R}$;

(e) $\max \{f,g\}$;

(b) $f + g$;

(f) $f^+ := \max \{f,0\}$;

(c) $f g$;

(g) $f^- := \max \{-f,0\}$;

(d) $\dfrac{f}{g}$;

(h) $|f| = f^+ + f^-$.

Continuous and monotonic functions are indeed measurable, as illustrated in Proposition 8.26.

Proposition 8.26. Let $(\Omega,\mathcal{A},\mu)$ be a measure space, and let $A \in \mathcal{A}$ be a measurable set. Then, any continuous function and any monotonic function, defined on A, is measurable.

In Proposition 8.27, we analyse how measurability interacts with sequences of functions.

Proposition 8.27. Let $(\Omega,\mathcal{A},\mu)$ be a measure space, and let $A \in \mathcal{A}$ be a measurable set. Let, further, (f_n) be a sequence of measurable functions, defined on A. Then, the following are also measurable functions:

$$\max_{n \leq k} f_n, \qquad \min_{n \leq k} f_n, \qquad \sup_{n \in \mathbb{N}} f_n, \qquad \inf_{n \in \mathbb{N}} f_n.$$

Recall the definition of upper and lower limit of a sequence (x_n) :

$$\limsup_{n \to \infty} x_n := \inf_{k \in \mathbb{N}} \sup_{n \geq k} x_n, \qquad \liminf_{n \to \infty} x_n := \sup_{k \in \mathbb{N}} \inf_{n \geq k} x_n, \qquad (8.10)$$

and notice that upper and lower limits are always well defined, since the following sequences are decreasing and increasing, respectively:

$$u_n := \inf_{k \in \mathbb{N}} \sup_{n \geq k} x_n, \qquad l_n := \sup_{k \in \mathbb{N}} \inf_{n \geq k} x_n.$$

We are now in the position to state the following Corollary 8.28.

Corollary 8.28. Let $(\Omega,\mathcal{A},\mu)$ be a measure space, and let $A \in \mathcal{A}$ be a measurable set. Let, further, (f_n) be a sequence of measurable functions, defined on A. Then, the following are also measurable functions:

$$\limsup_{n \to \infty} f_n, \qquad \liminf_{n \to \infty} f_n, \qquad \lim_{n \to \infty} f_n.$$

Reasoning with sequences of functions can yield interesting results, like the one stated in the following Theorem 8.29.

Theorem 8.29. Let $f : \mathbb{R} \to \mathbb{R}$ be a differentiable function. Then, the derivative f' is a measurable function.

Proof. For any $n \in \mathbb{N}$, define:

$$g_n(x) = n\left(f\left(x + \frac{1}{n}\right) - f(x)\right) = \frac{f\left(x + \frac{1}{n}\right) - f(x)}{\frac{1}{n}}.$$

The thesis follows, observing that (g_n) is a sequence of measurable functions and that $g_n \longrightarrow f'(x)$. $\square$

From their Definition 8.17, simple functions are measurable. Their importance lies in the fact that simple functions can approximate measurable functions, as shown by the following Theorem 8.30.

Theorem 8.30. Let $(\Omega, \mathcal{A}, \mu)$ be a measure space, and let $A \in \mathcal{A}$ be a measurable set. Let, further, $f : A \to \overline{R}$ be a measurable function. Then, there exists a sequence of simple functions (f_n), defined on A, such that, for any $x \in A$, it holds $|f_n(x)| \leq |f(x)|$ and:

$$\lim_{n \to \infty} f_n(x) = f(x).$$

Remark 8.31. The quite technical proof of Theorem 8.30 can be found, for instance, in Theorem 11.20 of [52]. Here, we only provide the interesting result that the sequence of simple functions, approximating the given measurable function f, can be defined as follows, for any $n \in \mathbb{N}$:

$$f_n(x) = \begin{cases} n & \text{if} \quad f(x) \geq n, \\ \dfrac{k-1}{2^n} & \text{if} \quad \dfrac{k-1}{2^n} \leq f(x) < \dfrac{k}{2^n}, \quad 1 \leq k \leq n\,2^n. \end{cases}$$

8.5 Lebesgue integral

We introduce the notion of integral, using the approximation of a measurable function with sequences of simple functions, in the spirit of Theorem 8.30. We start with working within the context of non–negative functions.

Definition 8.32. Let $(\Omega, \mathcal{A}, \mu)$ be a measure space, and let $A \in \mathcal{A}$ be a measurable set. Let, further, $f : A \to [0, +\infty]$ be a measurable function. The integral of f on A, with respect to measure μ, is defined as:

$$\int_A f \, d\mu := \sup \left\{ \int_A \varphi \, d\mu \ \Big|\ 0 \leq \varphi \leq f, \quad \varphi \text{ simple} \right\}.$$

Here is a list of the main properties of the integral of a non–negative function. The three properties can be demonstrated by observing, first, that they hold for simple functions, and then forming the limit, via Theorem 8.30.

(P1) Positivity property: $\quad 0 \le \int_A f \, d\mu$.

(P2) Monotonicity property: $\quad 0 \le f \le g \quad \Longrightarrow \quad \int_A f \, d\mu \le \int_A g \, d\mu$.

(P3) $\quad \int_A f \, d\mu < +\infty \quad \Longrightarrow \quad \mu\left(\{x \in A \mid f(x) = +\infty\}\right) = 0$.

We are now in the position to define the integral for measurable functions which may change sign. In particular, the integral is defined for the class of *absolutely integrable* functions.

Definition 8.33. Let $(\Omega, \mathcal{A}, \mu)$ be a measure space, and let $A \in \mathcal{A}$ be a measurable set. Let, further, $f : A \to [0, +\infty]$ be a measurable function. We say that f is *summable* on Ω if both integrals:

$$\int_\Omega f^+ \, d\mu \qquad \text{and} \qquad \int_\Omega f^- \, d\mu \qquad\qquad (8.11)$$

are finite. In this case, the integral of f is defined as:

$$\int_\Omega f \, d\mu := \int_\Omega f^+ \, d\mu - \int_\Omega f^- \, d\mu . \qquad\qquad (8.12)$$

Moreover, the integral of f on $\overline{\mathbb{R}}$ exists if at least one of the two integrals (8.11) is finite. In the latter case, the integral is still defined by (8.12), but it may be infinite. The undefined situation is, obviously, $+\infty - \infty$.

Take notice that some Authors employ alternative notations:

$$\int_\Omega f(x) \, d\mu(x), \qquad\qquad \int_\Omega f(x) \, \mu(dx) .$$

Remark 8.34. From Definition 8.33, it can be easily inferred that:

$$\left| \int_\Omega f \, d\mu \right| \le \int_\Omega |f| \, d\mu .$$

The following Definition 8.35 adapts Definition 8.33 to functions defined on subsets of Ω, in the measure space $(\Omega, \mathcal{A}, \mu)$.

Definition 8.35. Let $(\Omega, \mathcal{A}, \mu)$ be a measure space. Consider $A \in \mathcal{A}$ and let $f : A \to [-\infty, +\infty]$ be a measurable function. Then, f is integrable on A if, following Definition 8.33, function $\hat{f} : \Omega \to [-\infty, +\infty]$ defined by:

$$\hat{f}(x) := \begin{cases} f(x) & \text{if } x \in A \\ 0 & \text{if } x \notin A \end{cases}$$

is integrable on Ω. The integral of f on A is, then, given by:

$$\int_A f \, d\mu := \int_\Omega \hat{f} \, d\mu .$$

The notation $\mathcal{L}(\Omega)$ represents the collection of all measurable functions, on Ω, which are summable. The properties of the integral imply that $\mathcal{L}(\Omega)$ is a vector space on $\mathbb{R}$.

8.6 Almost everywhere

There are cases in which the integral defined using the Lebesgue measure theory approach is more effective than the Riemann integral. This happens, in particular, when the Lebesgue integral allows to deal with a property that does not hold for all x, and the set, on which the property does not hold, has zero measure.

Definition 8.36. Let $(\Omega, \mathcal{A}, \mu)$ be a measure space. Consider a property $\mathcal{P}$, and consider the set formed by those elements of Ω for which $\mathcal{P}$ does not hold. If such a set has zero measure, then $\mathcal{P}$ *holds almost everywhere* in Ω.

In other words, a property is said to hold almost everywhere, if there exists a set $N \in \mathcal{A}$ such that $\mu(N) = 0$ and such that all the elements of Ω, where $\mathcal{P}$ does not hold, belong to N.

Example 8.37. Consider two measurable functions $f, g : \Omega \to [-\infty, +\infty]$. We say that $f = g$ almost everywhere if:

$$\mu\left(\{x \in \Omega \mid f(x) \neq g(x)\}\right) = 0.$$

Example 8.38. Let $f, g : \Omega \to [-\infty, +\infty]$ be measurable functions. We say that $f \leq g$ almost everywhere if:

$$\mu\left(\{x \in \Omega \mid f(x) > g(x)\}\right) = 0,$$

or that $f \geq g$ almost everywhere if:

$$\mu\left(\{x \in \Omega \mid f(x) < g(x)\}\right) = 0.$$

Example 8.39. The sequence of functions $f_n(x) = x^n$, defined on the interval $[0, 1]$, converges to 0 almost everywhere, if we take the Lebesgue measure. In fact, if $x \in [0, 1]$, then

$$\lim_{n \to \infty} x^n = \begin{cases} 0 & \text{if } x \in [0, 1[, \\ 1 & \text{if } x = 1. \end{cases}$$

This shows that the limit function is not the null function, but the set on which the limit function differs from the null function has zero measure.

The following Proposition 8.40 provides reasons of the relevance of the almost everywhere properties, when they hold for a complete measure space.

Proposition 8.40. Let $(\Omega, \mathcal{A}, \mu)$ be a measure space, where μ is a complete measure. Let $f, g : \Omega \to [-\infty, +\infty]$ be functions that are almost everywhere equal. If f is measurable, then g is measurable. Moreover, if $f \in \mathcal{L}(\Omega)$, then $g \in \mathcal{L}(\Omega)$ and:

$$\int_\Omega f \, d\mu = \int_\Omega g \, d\mu.$$

8.7   Connection with Riemann integral

The Lebesgue integral was introduced in order to generalise the Riemann[4] integral. In this § 8.7 the Lebesgue–Vitali theorem is stated, without proof, which explains the interplays between the two notions of integral. For completeness, we first recall the definition of Riemann integral.

8.7.1   The Riemann integral

A *partition* of a real interval $[a, b]$ is any finite subset σ of $[a, b]$, such that $a, b \in \sigma$. Hence, for a suitable $n \in \mathbb{N}$, we must have:

$$\sigma = \{ a = x_0, x_1, \dots, x_{n-1}, x_n = b \},$$

where the elements of σ are listed according to the convention that $x_{i-1} < x_i$, for any $i = 1, \dots, n$. The norm of σ is the positive number:

$$\|\sigma\| = \max_{1 \leq i \leq n} (x_i - x_{i-1}).$$

The collection of all partitions of $[a, b]$ is denoted by $\Omega([a, b])$. Given $\sigma_1, \sigma_2 \in \Omega([a, b])$, if $\sigma_1 \subset \sigma_2$ then $\|\sigma_1\| \geq \|\sigma_2\|$, and we say that σ_2 is a refinement of σ_1.

The notation $f \in B([a, b])$ indicates that $f : [a, b] \longrightarrow \mathbb{R}$ is a bounded function, i.e., for any $x \in [a, b]$:

$$-\infty < m := \inf_{t \in [a, b]} f(t) \leq f(x) \leq \sup_{t \in [a, b]} f(t) := M < +\infty.$$

The lower sum of f, induced by σ, is the real number:

$$s(f, \sigma) := \sum_{i=1}^{n} m_i (x_i - x_{i-1}),$$

---

[4]Georg Friedrich Bernhard Riemann (1826–1866), German mathematician.

where, for any $1 \leq i \leq n$:

$$m_i = \inf_{x_{i-1} \leq x \leq x_i} f(x) .$$

Similarly, the upper sum of f, induced by σ, is the real number:

$$S(f,\sigma) := \sum_{i=1}^{n} M_i \, (x_i - x_{i-1}) ,$$

where:

$$M_i = \sup_{x_{i-1} \leq x \leq x_i} f(x) .$$

These definitions have a plain geometrical meaning; in particular, when f is non–negative, the idea is to calculate the area of the plane region situated underneath the graph of f, over the interval $[a,b]$, as illustrated in Figure 8.2.

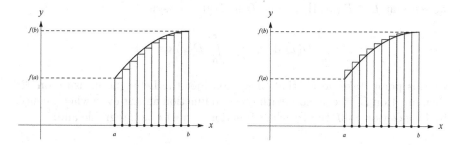

Figure 8.2: Lower sum (left) and upper sum (right).

The real numbers:

$$\sup_{\sigma \in \Omega([a,b])} s(f,\sigma) \qquad \text{and} \qquad \inf_{\sigma \in \Omega([a,b])} S(f,\sigma)$$

are called, respectively, lower integral and upper integral of $f \in B([a,b])$. They are represented with the notations:

$$\sup_{\sigma \in \Omega([a,b])} s(f,\sigma) := \underline{\int_a^b} f(x) \, \mathrm{d}x , \qquad \inf_{\sigma \in \Omega([a,b])} S(f,\sigma) := \overline{\int_a^b} f(x) \, \mathrm{d}x .$$

From the definitions, it follows immediately:

$$\underline{\int_a^b} f(x) \, \mathrm{d}x \leq \overline{\int_a^b} f(x) \, \mathrm{d}x .$$

A function $f \in B([a,b])$ is Riemann integrable if:

$$\underline{\int_a^b} f(x)\,\mathrm{d}x = \overline{\int_a^b} f(x)\,\mathrm{d}x\,.$$

In this case, we denote $f \in R([a,b])$, and the common value of upper and lower integrals is denoted by:

$$\int_a^b f(x)\,\mathrm{d}x\,.$$

The inclusion $R([a,b]) \subset B([a,b])$ is proper. In fact, in the particular case $[a,b] = [0,1]$, if we consider the following function, due to Dirichlet[5]:

$$D(x) = \begin{cases} 0 & \text{if } x \in [0,1] \cap (\mathbb{R}\backslash\mathbb{Q})\,, \\ 1 & \text{if } x \in [0,1] \cap \mathbb{Q}, \end{cases}$$

we see that $D \in B([0,1])$, while $D \notin R([0,1])$, since:

$$\underline{\int_0^1} D(x)\,\mathrm{d}x = 0 < \overline{\int_0^1} D(x)\,\mathrm{d}x = 1\,.$$

At this point, take notice that, if we consider the Lebesgue measure on $\mathbb{R}$, then function $D(x)$ coincides with the zero function almost everywhere; hence, by Proposition 8.40, the Dirichlet function turns to be integrable and:

$$\int_{[0,1]} D\,\mathrm{d}\ell = 0\,.$$

In other words, the Dirichlet function constitutes an example of a Lebesgue integrable function, that is not Riemann integrable.

8.7.2   Lebesgue–Vitali theorem

The full description of the relation between Riemann and Lebesgue integrals is due to Lebesgue and Vitali, and it is explained in the following Theorem 8.41, whose proof can be found, for example, in Theorem 2.5.4 of [12].

Theorem 8.41. Let $f : [a,b] \to \mathbb{R}$ be a bounded function. Then:

(1) $f \in R([a,b]) \quad \Longleftrightarrow \quad f \in C([a,b]) \quad$ for almost any $x \in [a,b]$;

(2) $f \in R([a,b]) \quad \Longrightarrow \quad f \in \mathcal{L}([a,b],\ell) \quad$ and

$$\int_a^b f(x)\,\mathrm{d}x = \int_{[a,b]} f\,\mathrm{d}\ell\,.$$

---

[5] Johann Peter Gustav Lejeune Dirichlet (1805–1859), German mathematician.

Remark 8.42. In view of Theorem 8.41, we will use the traditional Leibniz–Riemann notation:

$$\int_a^b f(x)\,\mathrm{d}x$$

to denote, also, the Lebesgue integral of the measurable function f, defined on the interval $[a,b]$.

8.7.3 An interesting example

One result, of classical theory of Riemann integration, states that any monotonic function $f : [a,b] \to \mathbb{R}$ is Riemann integrable. On the other hand, Theorem 8.41 ensures that the set of points, in which a Riemann integrable function is not continuous, has zero Lebesgue measure. The following Example 8.43 introduces a monotonic function f, defined on a bounded interval $[a,b]$, that has a countable set of points $(x_n)_{n \in \mathbb{N}} \subset [a,b]$ in which it is not continuous, being:

$$f^+(x_n) = \lim_{x \to x_n^+} f(x) \quad \neq \quad \lim_{x \to x_n^-} f(x) = f^-(x_n)$$

with $f^+(x_n), f^-(x_n) \in \mathbb{R}$.

Example 8.43. Consider a strictly increasing sequence $(x_n) \subset [0,1]$, and define:

$$x_\infty := \lim_{n \to \infty} x_n,$$

being the existence of x_∞ ensured by the hypothesis on (x_n). Now, introduce the function $f : [0,1] \to \mathbb{R}$, defined as:

$$f(x) = \begin{cases} x_n & \text{if} \quad x \in [\frac{n-1}{n}, \frac{n}{n+1}), \\ x_\infty & \text{if} \quad x = 1. \end{cases}$$

By construction, f is strictly increasing and, thus, integrable on $[0,1]$. Again by construction, f is discontinuous at any point $\xi_n = \dfrac{n}{n+1}$, where f jumps, as well as at $x = 1$. Moreover:

$$\int_0^1 f(x)\,\mathrm{d}x = \sum_{n=1}^{\infty} \int_{\frac{n-1}{n}}^{\frac{n}{n+1}} x_n\,\mathrm{d}x = \sum_{n=1}^{\infty} x_n \left(\frac{n}{n+1} - \frac{n-1}{n}\right) = \sum_{n=1}^{\infty} \frac{x_n}{n(n+1)},$$

which implies that the integral of f is strictly positive. For instance, as illustrated in Figure 8.3, if $x_n = 1 - \dfrac{1}{n}$, then:

$$\int_0^1 f(x)\,\mathrm{d}x = \sum_{n=1}^{\infty} \left(\frac{2}{n} - \frac{2}{n+1} - \frac{1}{n^2}\right) = 2\sum_{n=1}^{\infty} \left(\frac{1}{n} - \frac{1}{n+1}\right) - \sum_{n=1}^{\infty} \frac{1}{n^2} = 2 - \frac{\pi^2}{6}.$$

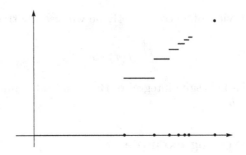

Figure 8.3: Example 8.43, with $x_n = 1 - \dfrac{1}{n}$.

8.8　Non Lebesgue integrals

8.8.1　Dirac measure

Recall the definition of the *Dirac*[6] *measure*, already met in Example 8.8. Given a non–empty set Ω, consider $x_0 \in \Omega$. For any $A \in \mathcal{P}(\Omega)$, the Dirac measure, concentrated in x_0, is defined as:

$$\delta_{x_0}(A) = \begin{cases} 1 & \text{if } x_0 \in A, \\ 0 & \text{if } x_0 \notin A, \end{cases}$$

thus, it can be described in terms of the characteristic function:

$$\delta_{x_0}(A) = \mathbf{1}_A(x_0).$$

Moreover, any real function on Ω is integrable with respect to the Dirac δ_{x_0}; from the general definition of integral, in fact, we have:

$$\int_\Omega f \, d\delta_{x_0} = f(x_0).$$

8.8.2　Discrete measure

Let us generate a measure in Ω, via the following construction. Given a sequence (x_n) in Ω, for any $A \subset \Omega$ define:

$$\mu(A) = \sum_{n \geq 1} \mathbf{1}_A(x_n)$$

In other words, μ counts how many elements of (x_n) belong to A.

---

[6]Paul Adrien Maurice Dirac (1902–1984), British theoretical physicist.

It is possible to show that a measurable function f is integrable if and only if the numeric series $\sum\limits_{n \geq 1} |f(x_n)|$ converges. In such a case, we also have:

$$\int_\Omega f \, d\mu = \sum_{n \geq 1} f(x_n).$$

The above result also means that infinite series can be seen as Lebesgue integrals; indeed, if $\Omega = \mathbb{R}$ and (a_n) is a sequence such that $\sum\limits_{n \geq 1} a_n$ converges absolutely, then we can write:

$$\sum_{n \geq 1} a_n = \int_\mathbb{R} a(x) \, d\mu(x),$$

where $a(n) = a_n$, and $a(x) = 0$ if $x \notin \mathbb{N}$, and where μ is the measure which counts non–negative numbers.

8.9 Generation of measures

Up to now, we exposed the theory of Lebesgue integral under the assumption that a measure is given. When Measure theory is employed in Probability theory, there are many measures to deal with. The way these measures are generated is expressed by the following Theorem 8.44, known as *measure generation* theorem: we present a proof of it inspired by [52].

Theorem 8.44. Let $(\Omega, \mathcal{A}, \mu)$ be a measure space, and let $f : \Omega \to [0, +\infty]$ be non–negative and measurable. Then, for any $A \in \mathcal{A}$:

$$\phi(A) := \int_A f \, d\mu$$

is a measure on $\mathcal{A}$. Moreover, ϕ is a finite measure, that is, $\phi(\Omega) < +\infty$, if and only if $f \in \mathcal{L}(\Omega, \mu)$.

Proof. It is obvious that $\phi(\emptyset) = 0$. To complete the proof, we have to show that, if $A_n \in \mathcal{A}$, $n \in \mathbb{N}$, is such that $A_i \cap A_j = \emptyset$ for $i \neq j$, then, setting $A = \bigcup\limits_{n=1}^{\infty} A_n$ yields:

$$\phi(A) = \sum_{n=1}^{\infty} \phi(A_n).$$

If f is a characteristic function of some measurable set E, then the countable additivity of ϕ follows from the countable additivity of μ; in such a case, in fact, we have:

$$\phi(A) = \int_A 1_E \, d\mu = \mu(A \cap E).$$

Therefore, if $A, B \in \mathcal{A}$, with $A \cap B = \emptyset$, then:

$$\phi(A \cup B) = \mu((A \cup B) \cap E) = \mu((A \cap E) \cup (B \cap E)).$$

Since $(A \cap E) \cap (B \cap E) = \emptyset$, it follows that:

$$\phi(A \cup B) = \mu((A \cap E) \cup (B \cap E)) = \mu(A \cap E) + \mu(B \cap E)$$
$$= \phi(A) + \phi(B).$$

The extension to a countable union of disjoint sets is straightforward.

If f is a simple function, the conclusion still holds, since:

$$f = \sum_i f_i \, 1_{E_i}.$$

In the general case of non–negative measurable f, for any simple function s such that $0 \le s \le f$, we have:

$$\int_A s \, d\mu = \sum_{n=1}^{\infty} \int_{A_n} s \, d\mu \le \sum_{n=1}^{\infty} \int_{A_n} f \, d\mu = \sum_{n=1}^{\infty} \phi(A_n).$$

Now, recalling the definition of integral of a measurable function, and considering the supremum, we find:

$$\sup_{0 \le s \le f} \int_A s \, d\mu = \int_A f \, d\mu = \phi(A) \le \sum_{n=1}^{\infty} \phi(A_n). \tag{8.13}$$

To obtain the thesis, we have to reverse inequality (8.13) and prove it. Now, if there exists an A_n such that $\phi(A_n) = +\infty$, then, since $A \supset A_n$, it follows that $\phi(A) \ge \phi(A_n)$ and the thesis is immediate. Hence, the proof can be limited to the case in which $\phi(A_n) < +\infty$ for any $n \in \mathbb{N}$. Then, for any $\varepsilon > 0$, a simple function s can be chosen such that $0 \le s \le f$ and:

$$\int_{A_1} s \, d\mu \ge \int_{A_1} f \, d\mu - \frac{\varepsilon}{2}, \qquad \int_{A_2} s \, d\mu \ge \int_{A_2} f \, d\mu - \frac{\varepsilon}{2}.$$

Thus:

$$\phi(A_1 \cup A_2) \ge \int_{A_1 \cup A_2} s \, d\mu = \int_{A_1} s \, d\mu + \int_{A_2} s \, d\mu \ge \phi(A_1) + \phi(A_2) - \varepsilon,$$

from which it can be inferred that:

$$\phi(A_1 \cup A_2) \ge \phi(A_1) + \phi(A_2).$$

Generalising to any $n \in \mathbb{N}$:

$$\phi(A_1 \cup \cdots \cup A_n) \geq \phi(A_1) + \cdots + \phi(A_n) \,. \tag{8.14}$$

Now, since $A \supset A_1 \cup \cdots \cup A_n$, inequality (8.14) implies:

$$\phi(A) \geq \sum_{n=1}^{\infty} \phi(A_n) \,. \tag{8.15}$$

Finally, the thesis follows from (8.13) and (8.15). □

8.10 Passage to the limit

In applications involving a sequence (f_n) of measurable functions, it is of great importance the *passage to the limit*, related to the possibility of switching between the operations of limit and integration. We addressed this problem in § 2.3, in particular in Theorem 2.15, for uniformly convergent sequences of functions. Here, we provide a more general, measure theoretical treatment.

8.10.1 Monotone convergence

We can surely state that passing to the limit represents one of the main aims of Lebesgue integration theory, in generalising the classic context of Riemann integration. In this § 8.10.1 we build the main tool for the passage–to–the–limit theory for the Lebesgue integral. The first and probably most important result, in this theory, is Theorem 8.45, on monotone convergence, due to Beppo Levi[7].

Theorem 8.45 (Beppo Levi – monotone convergence). Let $E \in \mathcal{A}$ and let (f_n) be a sequence of measurable functions, such that, for any $x \in E$:

$$0 \leq f_1(x) \leq f_2(x) \leq \cdots \leq f_n(x) \leq \cdots \cdots \,. \tag{8.16}$$

If:

$$f(x) = \lim_{n \to \infty} f_n(x) \,, \tag{8.17}$$

then:

$$\lim_{n \to \infty} \int_E f_n \, d\mu = \int_E f \, d\mu \,.$$

Remark 8.46. When f_n is decreasing, Levi Theorem 8.45 does not hold.

---

[7]Beppo Levi (1875–1961), Italian mathematician.

To see it, consider the counter–example $f_n(x) = \dfrac{1}{n}$, for any $x \in \mathbb{R}$. Here:

$$f_n(x) \searrow 0,$$

a notation indicating that the sequence decreases to zero, as $n \to \infty$. Thus, $f(x) = \lim_{n \to \infty} f_n(x) = 0$. At the same time, though, for any $n \in \mathbb{N}$, it holds:

$$\int_{-\infty}^{\infty} f_n(x)\, dx = \infty.$$

Proof. (*Theorem 8.45*) Relations (8.16) imply that there exists $\alpha \in [0, +\infty]$ such that:

$$\lim_{n \to \infty} \int_E f_n\, d\mu = \alpha.$$

Since $\int_E f_n\, d\mu \le \int_E f\, d\mu$, it is also:

$$\alpha \le \int_E f\, d\mu. \tag{8.18}$$

To obtain the thesis, we have to reverse inequality (8.18) and prove it. Let us fix $0 < c < 1$ and a simple function s such that $0 \le s \le f$. Define, further, for any $n \in \mathbb{N}$:

$$E_n = \{\, x \in E \mid f_n(x) \ge c\, s(x)\,\}.$$

From (8.16) we infer that $E_1 \subset E_2 \subset \cdots \subset E_n \subset \cdots\cdots$ and, from (8.17), we also have:

$$E = \bigcup_{n=1}^{\infty} E_n.$$

Now, given an $m \in \mathbb{N}$, the following inequalities are verified, since $E_m \subset E$:

$$\int_E f_n\, d\mu \ge \int_{E_m} f_n\, d\mu \ge c \int_{E_m} s\, d\mu. \tag{8.19}$$

At this point, evaluating the limit for $n \to \infty$ in (8.19) leads to:

$$\alpha \ge c \int_{E_m} s\, d\mu. \tag{8.20}$$

Due to the measure generation Theorem 8.44, the integral in (8.20) is a countable additive set function. Then, we can use Theorem 8.13, for increasing sequences of nested sets, to infer:

$$\alpha \ge c \int_E s\, d\mu.$$

Finally, we evaluate the limit $c \nearrow 1$, obtaining:

$$\alpha \geq \int_E s \, d\mu,$$

and we consider the supremum for s, that is:

$$\alpha \geq \int_E f \, d\mu. \tag{8.21}$$

Thesis follows from (8.18) and (8.21). □

8.10.1.1 Analytic functions

The monotone convergence Theorem 8.45 applies in a natural way to *analytic functions*, that are functions admitting a convergent expansion in power series. The connection to analytic functions is established by the following Corollary 8.47 to Theorem 8.45.

Corollary 8.47. Let $\sum_{n=1}^{\infty} f_n$ be a series of positive functions on Ω. Then:

$$\int_\Omega \sum_{n=1}^{\infty} f_n \, d\mu = \sum_{n=1}^{\infty} \int_\Omega f_n \, d\mu.$$

The next Example 8.48 aims to clarify how to deal with analytic functions using monotone convergence. Take notice of the importance of this example, which provides a further solution to the Basel problem presented in § 2.7.

Example 8.48. Compute the value of the so–called *Leibniz integral*, that is:

$$\int_0^1 \frac{\ln x}{x^2 - 1} \, dx = \frac{\pi^2}{8}. \tag{8.22}$$

As mentioned, integral (8.22) is connected to the Basel problem in § 2.7. Recalling the geometric series expansion (2.38), the following result can be inferred:

$$\frac{1}{1 - x^2} = \sum_{n=0}^{\infty} x^{2n}.$$

Then, we employ Levi Theorem 8.45:

$$\int_0^1 \frac{-\ln x}{1 - x^2} \, dx = \sum_{n=0}^{\infty} \int_0^1 -x^{2n} \ln x \, dx.$$

Integrating by parts:

$$\int_0^1 -x^{2n} \ln x \, dx = \left[-\frac{x^{2n+1}}{2n+1} \ln x \right]_0^1 + \int_0^1 \frac{x^{2n}}{2n+1} \, dx = \frac{1}{(2n+1)^2}. \tag{8.23}$$

In other words, the integral in the left–hand side of (8.22) can be expressed in terms of a numerical series:

$$\int_0^1 \frac{\ln x}{x^2 - 1}\, dx = \sum_{n=0}^{\infty} \frac{1}{(2\,n+1)^2} . \tag{8.24}$$

Formula (8.22) follows from identity (2.69).

The monotone convergence Theorem 8.45 allows, also, the computation of many other infinite series, a few of which will be presented in the following examples. Let us start with the most elementary cases.

Example 8.49. This interesting example, of infinite series summation, is taken from [17]. We show that:

$$\sum_{n=1}^{\infty} \frac{1}{n^2 + \dfrac{n}{2}} = 4\,(1 - \ln 2) . \tag{8.25}$$

If $a, b \geq 0$, $a \neq b$, consider the infinite series:

$$S(a, b) = \sum_{n=1}^{\infty} \frac{1}{(n+a)\,(n+b)} .$$

Making use of partial fraction decomposition, we get:

$$S(a, b) = \frac{1}{a - b} \sum_{n=1}^{\infty} \left(\frac{1}{n+b} - \frac{1}{n+a} \right) .$$

Recalling that, for any $\alpha > 0$, it holds:

$$\frac{1}{\alpha} = \int_0^{\infty} e^{-\alpha x}\, dx ,$$

we get further:

$$S(a, b) = \frac{1}{a - b} \sum_{n=1}^{\infty} \int_0^{\infty} \left[e^{-(n+b)\,x} - e^{-(n+a)\,x} \right] dx ,$$

$$= \frac{1}{a - b} \sum_{n=1}^{\infty} \int_0^{\infty} e^{-n\,x} \left[e^{-b\,x} - e^{-a\,x} \right] dx .$$

As in Example 8.48, the trick is to recognise the geometric series expansion (2.31) in the last integrand:

$$\sum_{n=0}^{\infty} \left(e^{-x} \right)^n = \frac{1}{1 - e^{-x}} , \quad \text{i.e.,} \quad \sum_{n=1}^{\infty} e^{-n\,x} = \frac{1}{1 - e^{-x}} - 1 = \frac{1}{e^x - 1} = \frac{e^{-x}}{1 - e^{-x}} ,$$

and, then, use the monotone convergence Theorem 8.45, which yields:

$$S(a,b) = \frac{1}{a-b} \int_0^\infty \left[e^{-bx} - e^{-ax} \right] \frac{e^{-x}}{1 - e^{-x}} dx.$$

At this point, considering the change of variable $t = e^{-x}$:

$$S(a,b) = \frac{1}{a-b} \int_0^1 \frac{t^b - t^a}{1 - t} dt. \tag{8.26}$$

The left–hand side of (8.25) corresponds to forming $S\left(\frac{1}{2}, 0\right)$:

$$\sum_{n=1}^\infty \frac{1}{n^2 + \dfrac{n}{2}} = S\left(\frac{1}{2}, 0\right) = 2 \int_0^1 \frac{1 - \sqrt{t}}{1 - t} dt = 2 \int_0^1 \frac{1}{1 + \sqrt{t}} dt.$$

Finally, recovering the variable $t = u^2$, with $dt = 2\,du$:

$$\sum_{n=1}^\infty \frac{1}{n^2 + \dfrac{n}{2}} = 4 \int_0^1 \frac{u}{1 + u} du.$$

Equality (8.25) follows by computing the last integral.

Remark 8.50. Consider again equation (8.26), but, this time, let $a = b > 0$. In this case, the infinite series becomes [17]:

$$S(a) := S(a,a) = \sum_{n=1}^\infty \frac{1}{(n + a)^2}.$$

To evaluate it, we can repeat the argument of Example 8.49, starting from the definite integral:

$$\frac{1}{\alpha^2} = \int_0^\infty x\, e^{-\alpha}\, dx, \qquad \alpha > 0.$$

Making use of the geometric expansion (2.31) and the monotone convergence Theorem 8.45:

$$S(a) = \sum_{n=1}^\infty \frac{1}{(n + a)^2} = \sum_{n=1}^\infty \int_0^\infty x\, e^{-(n+a)x}\, dx = \sum_{n=1}^\infty \int_0^\infty x\, e^{-ax} e^{-nx}\, dx$$

$$= \int_0^\infty x \frac{e^{-ax}}{e^x - 1} dx.$$

With the change of variable $e^{-x} = t$, we arrive at the identity:

$$S(a) = \sum_{n=1}^\infty \frac{1}{(n + a)^2} = \int_0^1 \frac{t^a \ln t}{t - 1} dt. \tag{8.27}$$

In particular, when $a = 0$:

$$S(0) = \sum_{n=1}^{\infty} \frac{1}{n^2} = \int_0^1 \frac{\ln t}{t-1} \, dt \, . \tag{8.27a}$$

Note the interesting comparison between (8.24) and (8.27a), that evaluate to $\dfrac{\pi^2}{8}$ and $\dfrac{\pi^2}{6}$, respectively, as shown in formulæ (2.69) and (2.71).

8.10.2   Exercises

1. Using the definite integral $\displaystyle\int_0^1 \frac{1-x}{1-x^4} \, dx$, show that:

$$\sum_{n=0}^{\infty} \frac{1}{(4n+1)(4n+2)} = \frac{\pi}{8} + \frac{1}{4}\ln 2 \, .$$

Hint. Use the partial fraction decomposition:

$$\frac{1}{(1+x)(1+x^2)} = \frac{1-x}{2(1+x^2)} + \frac{1}{2(1+x)} \, .$$

2. Using the definite integral $\displaystyle\int_0^1 \frac{x(1-x)}{1+x} \, dx$, show that:

$$\sum_{n=1}^{\infty} \frac{(-1)^{n-1}}{(n+1)(n+2)} = \frac{3}{2} - \ln 4 \, .$$

3. Show that $\displaystyle\int_0^{\infty} \frac{\sin(ax)}{e^x - 1} \, dx = \sum_{n=1}^{\infty} \frac{a}{n^2 + a^2}$, for any $a \in \mathbb{R}$.

Hint. Use the equality:

$$\int \sin(ax) \, e^{-nx} \, dx = -\frac{e^{-nx}(n\sin(ax) + a\cos(ax))}{a^2 + n^2} \, .$$

4. Show that $\displaystyle\int_0^{\infty} \frac{x^2}{e^x - 1} \, dx = \sum_{n=1}^{\infty} \frac{2}{n^3} \, .$

5. Show that $\displaystyle\sum_{n=1}^{\infty} \frac{1}{n\left(n + \frac{1}{4}\right)} = 2\,(8 - \pi - 6\ln 2) \, .$

Hint: Use the partial fraction decomposition:

$$\frac{x^3}{(x+1)(1+x^2)} = 1 - \frac{1+x}{2(1+x^2)} - \frac{1}{2(1+x)} \, .$$

8.10.3 Dominated convergence

The monotone convergence Theorem 8.45 assumes that the given sequence of functions is increasing. In many circumstances, this hypothesis is not fulfilled, thus, it is important to detect different situations, where the passage to the limit is still possible. The Beppo Levi result remains, however, the key to the proof of Theorem 8.52, on dominated convergence, due to Lebesgue. In order to state and prove such a theorem, we need an important preliminary result, known as *Fatou*[8] *Lemma*, which holds for positive sequences of functions. Note that the notion of *lower limit*, given in equation (8.10), plays here a key role.

Lemma 8.51 (Fatou). Let $(\Omega, \mathcal{A}, \mu)$ be a measure space, and let (f_n) be a sequence of measurable positive functions on Ω. Then:

$$\int_\Omega \liminf_{n \to \infty} f_n \, d\mu \leq \liminf_{n \to \infty} \int_\Omega f_n \, d\mu.$$

Proof. For $n \in \mathbb{N}$, introduce the increasing sequence of measurable functions:

$$g_n := \inf_{p \geq n} f_p.$$

By definition, (g_n) is such that:

$$\lim_{n \to \infty} g_n(x) = \liminf_{n \to \infty} f_n(x).$$

From the monotone convergence Theorem 8.45, since $g_n(x) \leq f_n(x)$, we infer:

$$\int_\Omega \liminf_{n \to \infty} f_n \, d\mu = \lim_{n \to \infty} \int_\Omega g_n \, d\mu \leq \liminf_{n \to \infty} \int_\Omega f_n \, d\mu,$$

which completes the proof. $\qquad\qquad\square$

We are now in the position to state the dominated convergence Theorem 8.52, due to Lebesgue.

Theorem 8.52. Consider a measure space $(\Omega, \mathcal{A}, \mu)$ and let (f_n) be a sequence of measurable functions such that:

$$\lim_{n \to \infty} f_n(x) = f(x). \tag{8.28}$$

Assume that there exists a non–negative summable $g \in \mathcal{L}(\Omega)$ such that, for any $x \in \Omega$ and any $n \in \mathbb{N}$:

$$|f_n(x)| \leq g(x) \tag{8.29}$$

Then, $f \in \mathcal{L}(\Omega)$, and:

$$\lim_{n \to \infty} \int_\Omega f_n \, d\mu = \int_\Omega f \, d\mu. \tag{8.30}$$

---

[8] Pierre Joseph Louis Fatou (1878–1929), French mathematician and astronomer.

Proof. We follow the proof proposed in [52]. Condition (8.29) implies that $f_n + g \geq 0$, thus, Lemma 8.51 can be used to obtain the inequality:

$$\int_A (f + g) \, d\mu \leq \liminf_{n \to \infty} \int_A (f_n + g) \, d\mu .$$

It can then be inferred:

$$\int_A f \, d\mu \leq \liminf_{n \to \infty} \int_A f_n \, d\mu . \tag{8.31}$$

From (8.29) we also get $g - f_n \geq 0$, and, again from Fatou Lemma 8.51, we obtain:

$$\int_A (g - f) \, d\mu \leq \liminf_{n \to \infty} \int_A (g - f_n) \, d\mu .$$

Thus:

$$-\int_A f \, d\mu \leq \liminf_{n \to \infty} \int_A (-f_n) \, d\mu ,$$

that is equivalent to:

$$\int_A f \, d\mu \geq \limsup_{n \to \infty} \int_A f_n \, d\mu . \tag{8.32}$$

Thesis (8.30) follows from (8.31) and (8.32).   □

Remark 8.53. When μ is a complete measure (see Definition 8.6), the dominated convergence Theorem 8.52 holds, also, when conditions (8.28) and (8.29) are true almost everywhere.

8.10.4   Exercises

1. Evaluate $\displaystyle \lim_{n \to +\infty} \int_0^{+\infty} \frac{e^{-(x+1)^n}}{\sqrt{x}} \, dx$.

2. If $f : [0,1] \to \mathbb{R}$ is a continuous function, show that:

$$\lim_{n \to +\infty} \int_0^1 f(t^n) \, dt = f(0) \qquad \text{and} \qquad \lim_{n \to +\infty} \int_0^1 n \, f(t) \, e^{-t} \, dt = f(0) .$$

3. Explain why the passage to the limit $\displaystyle \lim_{n \to \infty} \int_0^1 \frac{e^{-x}}{x^2 + n^2} \, dx$ is possible.

4. Evaluate $\displaystyle \lim_{n \to \infty} \int_0^\infty \frac{\arctan n}{1 + x^2} \, dx$, explaining why it is possible to use the dominated convergence Theorem 8.52.

5. Justify the following passage to the limit $\displaystyle \lim_{n \to \infty} \int_0^1 \frac{x^4}{x^2 + n^2} \, dx$.

6. Consider the sequence of functions $f_n(x) = n^2 \sqrt{x} e^{-nx}$, $x \in [0, +\infty)$. Show that the thesis of Theorem 8.52 does not hold. Explain the reasons, evaluating $\sup\limits_{x \in [0, +\infty)} f_n(x)$.

7. Using any of Theorems 2.15, or 8.45, or 8.52, evaluate the following limits:

(a) $\displaystyle\int_0^\infty n\, x\, e^{-x^2}\, dx$;

(d) $\displaystyle\int_0^\pi \sin(n\, x)\, e^{-n\, x}\, dx$;

(b) $\displaystyle\int_0^1 \sqrt{x^2 + \frac{1}{n}}\, dx$;

(e) $\displaystyle\lim_{n \to \infty} \int_0^1 n\, x\, (1 - x^2)^n\, dx$.

(c) $\displaystyle\int_0^\pi \frac{\sin(n\, x)}{n}\, dx$;

Solution to n. 1 of Exercises 8.10.4

Form the *Bernoulli inequality* [9]:

$$(1 + x)^n \geq 1 + n\, x,$$

it can be inferred, since $x > 0$ and $n \geq 1$:

$$(1 + x)^n \geq 1 + n\, x \geq 1 + x > 0, \qquad \text{so that} \qquad -(1 + x)^n < -x.$$

Then, an uniform estimate for the integrand can be formed:

$$\frac{e^{-(1+x)^n}}{\sqrt{x}} \leq \frac{e^{-x}}{\sqrt{x}} := \varphi(x).$$

Now, observing that $\varphi \in \mathcal{L}(\mathbb{R})$, the dominated convergence Theorem 8.52 can be used, and we can interchange the limit with the integral:

$$\lim_{n \to +\infty} \int_0^{+\infty} \frac{e^{-(x+1)^n}}{\sqrt{x}}\, dx = \int_0^{+\infty} \lim_{n \to +\infty} \frac{e^{-(x+1)^n}}{\sqrt{x}}\, dx = 0.$$

8.10.5 A property of increasing functions

Here, we deal with the Lebesgue measure on the real line. To introduce this topic, we consider again the function f of Example 8.43, which is piecewise constant and almost everywhere differentiable on $[0, 1]$. Its derivative is zero for $x \neq \dfrac{n}{n+1}$ and for any $n \in \mathbb{N}$, meaning that:

$$\int_0^1 f'(x)\, dx = 0.$$

---

[9] See, for example, mathworld.wolfram.com/BernoulliInequality.html

On the other hand, the considered function f verifies:

$$0 = \int_0^1 f'(x)\, dx \; < \; f(1) - f(0) = x_\infty - x_1 \,. \tag{8.33}$$

Note that (8.33) is not against the following celebrated identity[10] which holds true for any function $g \in \mathcal{C}^1([a, b])$:

$$\int_a^b g'(x)\, dx = g(b) - g(a) \,,$$

The following Theorem 8.54 explains the origin of inequality (8.33); to prove it, we follow [45].

Theorem 8.54. Let $f : [a, b] \to \mathbb{R}$ be an increasing function, that is to say, $x, y \in [a, b]$, $x < y \implies f(x) \le f(y)$. Define the set E as the subset of interval (a, b) on which f is differentiable. Then, $f' : E \to \mathbb{R}$ is summable and:

$$\int_E f'(x)\, dx \; \le \; f(b) - f(a) \,. \tag{8.34}$$

Proof. Recall that any increasing function can have, at least, a countable infinity of points of discontinuity, which are indeed jumps. Hence, an increasing function is almost everywhere differentiable, and, then, measurable. Let us, now, extend the domain of $f(x)$ to the interval $[a, b+1]$, defining:

$$\check{f}(x) = \begin{cases} f(x) & \text{if} \quad x \in [a, b], \\ f(b) & \text{if} \quad x \in (b, b+1]. \end{cases}$$

For simplicity, we keep the notation f to indicate $\check{f}$. As in Theorem 8.29, let us introduce the sequence of (measurable) functions $(\varphi_n)_{n \in \mathbb{N}}$, defined for $x \in [a, b]$ as:

$$\varphi_n(x) = \frac{f\left(x + \dfrac{1}{n}\right) - f(x)}{\dfrac{1}{n}} \,.$$

By definition, if $x \in E$, then $\varphi_n(x) \to f'(x)$ as $n \to \infty$, meaning that f' is

---

measurable too. Now, observing that $\ell([a,b] - E) = 0$, we have:

$$\int_E \varphi_n(x)\,dx = \int_a^b \varphi_n(x)\,dx = n\left(\int_a^b f\left(x + \frac{1}{n}\right) dx - \int_a^b f(x)\,dx\right)$$

$$= n\left(\int_{a+\frac{1}{n}}^{b+\frac{1}{n}} f(x)\,dx - \int_a^b f(x)\,dx\right)$$

$$= n\left(\int_b^{b+\frac{1}{n}} f(x)\,dx - \int_a^{a+\frac{1}{n}} f(x)\,dx\right)$$

$$= n\int_b^{b+\frac{1}{n}} f(b)\,dx - n\int_a^{a+\frac{1}{n}} f(x)\,dx$$

$$= f(b) - n\int_a^{a+\frac{1}{n}} f(x)\,dx$$

$$\le f(b) - f(a). \tag{8.35}$$

The last inequality (8.35) follows from the monotonicity assumption on f. In fact, for any $x \in [a, b+1]$, from $f(a) \le f(x)$ and integrating on $[a, a+\frac{1}{n}]$, we obtain:

$$n\int_a^{a+\frac{1}{n}} f(a)\,dx \le n\int_a^{a+\frac{1}{n}} f(x)\,dx \quad \Longleftrightarrow \quad -f(a) \ge -n\int_a^{a+\frac{1}{n}} f(x)\,dx.$$

At this point, using Fatou Lemma 8.51, and passing to the limit, we infer:

$$\int_E f'(x)\,dx = \int_E \liminf_{n\to\infty} \varphi_n(x)\,dx \le \liminf_{n\to\infty} \int_E \varphi_n(x)\,dx \le f(b) - f(a).$$

The proof of Theorem 8.54 is thus completed.

$\square$

Remark 8.55. The inequality in the thesis of Theorem 8.54 can be strict, as shown by the earlier result (8.33).

8.11 Differentiation under the integral sign

Here, the problem is considered, consisting in the differentiation of an integral which depends on one parameter. The solution technique is powerful and allows the evaluation of many definite integrals, which, otherwise, would be impossible to compute. The relevant theorems are stated (without proof), followed by several of their applications, to illustrate their importance.

Theorem 8.56. Consider the real intervals $]a, b[$, $[\alpha, \beta] \subset \mathbb{R}$, and function $f :]a, b[\times [\alpha, \beta] \to \mathbb{R}$, such that:

(i) for any $x \in]a, b[$, function $t \mapsto f(x, t)$ is summable in $[\alpha, \beta]$;

(ii) for almost any $t \in [\alpha, \beta]$, function $x \mapsto f(x, t)$ is differentiable in $]a, b[$;

(iii) for any $x \in]a, b[$ and for almost any $t \in [\alpha, \beta]$, there exists g summable in $[\alpha, \beta]$ such that:

$$\left| \frac{\partial f}{\partial x}(x, t) \right| \leq g(t).$$

Then:

$$F(x) := \int_{\alpha}^{\beta} f(x, t)\, dt$$

is differentiable and:

$$F'(x) = \int_{\alpha}^{\beta} \frac{\partial f}{\partial x}(x, t)\, dt.$$

It is also useful to state an alternative version of Theorem 8.56, which uses the continuity of the partial derivatives of $f(x, t)$.

Theorem 8.57. Let $f : [a, b] \times [\alpha, \beta] \to \mathbb{R}$ be a continuous function, such that the partial derivative $\dfrac{\partial f}{\partial x}$ exists and is continuous on $[a, b] \times [\alpha, \beta]$. Then:

$$F(x) := \int_{\alpha}^{\beta} f(x, t) dt$$

is differentiable and:

$$F'(x) = \int_{\alpha}^{\beta} \frac{\partial f}{\partial x}(x, t)\, dt.$$

Several examples follow, on the use of Theorems 8.56 and 8.57.

8.11.1   The probability integral (1)

The first application of the derivation of parametric integrals is devoted to the evaluation of the probability integral:

$$\int_{0}^{\infty} e^{-x^2}\, dx.$$

Let us introduce the functions:

$$f_1(x) := \left(\int_{0}^{x} e^{-t^2}\, dt \right)^2 \quad \text{and} \quad f_2(x) := \int_{0}^{1} \frac{e^{-(1+t^2)x^2}}{1+t^2}\, dt.$$

The derivative of f_1 is:

$$f_1'(x) := 2\,e^{-x^2} \int_0^x e^{-t^2}\,dt\,,$$

while, for the derivative of f_2, we employ Theorems 8.56 and 8.57, obtaining:

$$f_2'(x) := -2\,x\,e^{-x^2} \int_0^1 e^{-x^2 t^2}\,dt\,.$$

In this last integral, the change of variable $\tau = x\,t$ leads to:

$$f_2'(x) := -2\,e^{-x^2} \int_0^x e^{-\tau^2}\,d\tau\,.$$

Hence, the derivative of function $f(x) = f_1(x) + f_2(x)$ is zero for any x, meaning that $f(x)$ is a constant function, with $f(x) = f(0)$ for all $x \in \mathbb{R}$. And it is clear that:

$$f(0) = \int_0^1 \frac{1}{1+t^2}\,dt = \frac{\pi}{4}\,.$$

In other words, we have shown that, for any $x \in \mathbb{R}$:

$$\left(\int_0^x e^{-t^2}\,dt \right)^2 + \int_0^1 \frac{e^{-(1+t^2)\,x^2}}{1+t^2}\,dt = \frac{\pi}{4}\,.$$

Now, using the dominated convergence Theorem 8.52:

$$\lim_{x \to \infty} \left(\left(\int_0^x e^{-t^2}\,dt \right)^2 + \int_0^1 \frac{e^{-(1+t^2)\,x^2}}{1+t^2}\,dt \right) = \left(\int_0^\infty e^{-t^2}\,dt \right)^2 = \frac{\pi}{4}\,.$$

The value of the probability integral is, therefore:

$$\boxed{\int_0^\infty e^{-x^2}\,dx = \frac{\sqrt{\pi}}{2}}\,. \tag{8.36}$$

Remark 8.58. If the interval of integration is symmetric, $I = [-a, a]$, with $a \in \mathbb{R} \cup \{\pm\infty\}$, then:

$$\int_{-a}^a f(t)\,dt = 0 \qquad \text{for any integrable odd } f\,,$$

while

$$\int_{-a}^a f(t)\,dt = 2\int_0^a f(t)\,dt \qquad \text{for any integrable even } f\,.$$

Remark 8.59. From (8.36) and from the fact that e^{-x^2} is an even function, we infer that:

$$\int_{-\infty}^\infty e^{-x^2}\,dx = \sqrt{\pi}\,. \tag{8.36a}$$

In applications, the following formula is also useful:

$$\int_{-\infty}^{\infty} e^{-\pi x^2}\, dx = 1. \tag{8.36b}$$

Remark 8.60. From (8.36), integrating by parts, it follows:

$$\int_{0}^{\infty} x^2\, e^{-x^2}\, dx = \frac{\sqrt{\pi}}{4}, \tag{8.36c}$$

and, equivalently:

$$\int_{\infty}^{\infty} x^2\, e^{-x^2}\, dx = \frac{\sqrt{\pi}}{2}. \tag{8.36d}$$

Remark 8.61. Let $a > 0$ and $\Delta = b^2 - 4\,a\,c < 0$, and consider the following identity, known as *square completion* [11]:

$$a\,x^2 + b\,x + c = a\left(x + \frac{b}{2\,a}\right)^2 + c - \frac{b^2}{4\,a}.$$

Recalling (8.36a), the Gaussian integral identities follow:

$$\int_{-\infty}^{\infty} e^{-(a x^2 + b x + c)}\, dx \;=\; \sqrt{\frac{\pi}{a}}\; e^{\frac{b^2}{4a} - c}, \tag{8.37}$$

$$\int_{-\infty}^{\infty} x\, e^{-(a x^2 + b x + c)}\, dx \;=\; -\sqrt{\frac{\pi}{a^3}}\, \frac{b}{2}\; e^{\frac{b^2}{4a} - c}, \tag{8.38}$$

$$\int_{-\infty}^{\infty} x^2\, e^{-(a x^2 + b x + c)}\, dx \;=\; \sqrt{\frac{\pi}{a^5}}\, \frac{2a + b^2}{4}\; e^{\frac{b^2}{4a} - c}. \tag{8.39}$$

8.11.2   The probability integral (2)

This second proof of (8.36a) is contained in [59] and, again, it is based on Theorems 8.56 and 8.57 for the derivation of parametric integrals. Assuming that the value of the probability integral is unknown, define:

$$G := \int_{-\infty}^{\infty} e^{-x^2}\, dx = 2\int_{0}^{\infty} e^{-x^2}\, dx, \tag{8.40}$$

and define further, for $t \geq 0$:

$$F(x) := \int_{-\infty}^{\infty} \frac{e^{-x\,(1+t^2)}}{1 + t^2}\, dt. \tag{8.41}$$

Observe that $F(x) \leq \pi\, e^{-x}$. In fact:

$$F(x) = e^{-x} \int_{-\infty}^{\infty} \frac{e^{-x\,t^2}}{1 + t^2}\, dt \leq e^{-x} \int_{-\infty}^{\infty} \frac{1}{1 + t^2}\, dt = \pi\, e^{-x}.$$

---

[11] See, for example, mathworld.wolfram.com/CompletingtheSquare.html

Hence:
$$\lim_{x \to \infty} F(x) = 0.$$

Using Theorems 8.56 and 8.57, for the derivation under the integral sign:
$$F'(x) = -\int_{-\infty}^{\infty} e^{-x(1+t^2)}\, dt = -e^{-x} \int_{-\infty}^{\infty} e^{-xt^2}\, dt.$$

Now, with the change of variable $t = \dfrac{1}{\sqrt{x}}\, z$:
$$F'(x) = -\frac{e^{-x}}{\sqrt{x}} \int_{-\infty}^{\infty} e^{-z^2}\, dz = -\frac{e^{-x}}{\sqrt{x}}\, G.$$

Then, integrating on $[t, \infty)$:
$$\int_{x}^{\infty} F'(s)\, ds = -G \int_{\sqrt{x}}^{\infty} \frac{e^{-s}}{\sqrt{s}}\, ds,$$

that is:
$$F(x) = G \int_{\sqrt{x}}^{\infty} \frac{e^{-s}}{\sqrt{s}}\, ds. \tag{8.42}$$

From (8.41) it follows that $F(0) = \pi$, while (8.42) implies that $F(0) = G^2$. Thus $G = \sqrt{\pi}$, which means that (8.36a) is re-established.

8.11.3 Exercises

1. Consider $F(x) = \displaystyle\int_{0}^{+\infty} e^{-xt}\, \frac{\sin t}{t}\, dt$, where $x > 0$. Show that:

(a) $F'(x) = -\dfrac{1}{1 + x^2}$;

(b) $\displaystyle\lim_{x \to +\infty} F(x) = 0$;

(c) $F(x) = \dfrac{\pi}{2} - \arctan x$;

(d) *Dirichlet integral*:
$$\int_{0}^{+\infty} \frac{\sin t}{t}\, dt = \frac{\pi}{2}. \tag{8.43}$$

2. Using Theorem 8.56, for differentiation under the integral sign, show that, for $x \geq 0$:
$$\int_{0}^{1} \frac{t^x - 1}{\ln t}\, dt = \ln(1 + x).$$

3. For $x > 0$ consider the function:

$$F(x) = \int_0^{+\infty} \frac{\arctan(t) - \arctan(t\,x)}{t}\, dt\,.$$

Explain why it is possible to differentiate under the integral sign; then, show that:

$$F(x) = -\frac{\pi}{2}\,\ln x\,.$$

4. For $x > 0$ consider the function:

$$F(x) = \int_0^{+\infty} \frac{e^{-t} - e^{-t\,x}}{t}\, dt\,.$$

Explain why it is possible to differentiate under the integral sign; then, show that:

$$F(x) = \ln x\,.$$

5. Consider the function, defined for $x > -1$:

$$F(x) = \int_0^\infty \frac{\arctan(t\,x)}{t^3 + t}\, dt\,.$$

Explain why it is possible to differentiate under the integral sign; then, employ the partial fraction decomposition:

$$\frac{1}{(1 + t^2)(1 + t^2\,x^2)} = \frac{1}{1 - x^2}\left(\frac{1}{1 + t^2} - \frac{x^2}{1 + t^2\,x^2}\right),$$

to show that:

$$F(x) = \frac{\pi}{2}\,\ln(1 + x)\,.$$

6. Using the relation:

$$\int_{-\infty}^\infty e^{-x^2}\, dx = \sqrt{\pi}\,,$$

and using the derivation of parametric integrals, show that, for any $x \in \mathbb{R}$:

$$\int_{-\infty}^\infty e^{-t^2}\cos(t\,x)\, dt = \sqrt{\pi}\,e^{-\frac{x^2}{4}}\,.$$

Solutions to n. 1 of Exercises 8.11.3

(a) Observe, first, that, since we assumed x, $t > 0$, then:

$$\left|\frac{\partial f}{\partial x}\right| = \left|-e^{-x\,t}\sin t\right| \le e^{-x\,t} \le 1\,.$$

This means that we can differentiate under the integral sign, and, integrating by parts, we obtain:

$$F'(x) = \int_0^\infty -e^{-xt} \sin t \; dt = -\frac{1}{1+x^2} \; .$$

(b) Make use of the fact that $x > 0$, and that $\dfrac{\sin t}{t} \leq 1$ for any $t \in \mathbb{R}$, to form some estimates:

$$|F(x)| = \left| \int_0^\infty e^{-xt} \frac{\sin t}{t} \; dt \right| \leq \int_0^\infty e^{-xt} \; dt = \frac{1}{x} \; .$$

Now, taking the limit as $x \to \infty$:

$$\lim_{x \to \infty} \left| \int_0^\infty e^{-xt} \frac{\sin t}{t} \; dt \right| \leq \lim_{x \to \infty} \frac{1}{x} = 0 \; .$$

(c) Integrating the formula obtained in (a), we get:

$$F(x) = \int -\frac{1}{1+x^2} \; dx + C = C - \arctan x \; ,$$

where C is some integration constant. Recalling (b) and taking the limit as $x \to \infty$:

$$0 = \lim_{x \to \infty} F(x) = C - \arctan(\infty) = C - \frac{\pi}{2} \; .$$

It follows that $C = \dfrac{\pi}{2}$ and $F(x) = \dfrac{\pi}{2} - \arctan x$.

(d) At this point, it is immediate recognise that:

$$\int_0^\infty \frac{\sin t}{t} \; dt = F(0) = \frac{\pi}{2} \; .$$

The above integral, given in (8.43), is of great importance and is known as *Dirichlet integral.*

Solutions to n. 2 of Exercises 8.11.3

Recall, first, that:

$$\frac{d}{dx} t^x = \frac{d}{dx} e^{\ln(t^x)} = \ln(t) \, t^x \; .$$

Then, set $f(x,t) = \dfrac{t^x - 1}{\ln(t)}$, so that:

$$\frac{\partial f}{\partial x}(x,t) = \frac{\ln(t) \, t^x}{\ln(t)} = t^x \; .$$

The integration is for $t \in [0,1]$, and $x \geq 0$, meaning that $\left|\frac{\partial f}{\partial x}\right| < 1$, and allowing the derivation under the integral sign. Thus:

$$\frac{d}{dx} \int_0^1 \frac{t^x - 1}{\ln t} \, dt = \int_0^1 \frac{d}{dx} \frac{t^x - 1}{\ln t} \, dt = \int_0^1 t^x \, dt = \left[\frac{t^{1+x}}{1+x}\right]_{t=0}^{t=1} = \frac{1}{1+x}.$$

Integrating with respect to t :

$$\int_0^1 \frac{t^x - 1}{\ln t} \, dt = \ln(1 + x) + K.$$

To detect the integration constant K, it suffices to take $x = 0$:

$$0 = \int_0^1 \frac{t^0 - 1}{\ln t} \, dt = \ln(1) + K \qquad \Longrightarrow \qquad K = 0.$$

In conclusion:

$$\int_0^1 \frac{t^x - 1}{\ln t} \, dt = \ln(1 + x).$$

8.12   Basel problem again

We show, here, that differentiation of parametric integrals can be used to provide an alternative solution to the Basel problem, presented in § 2.7 and solved in (2.71), as well as in (8.22) of Example 8.48. As already observed in § 2.7, it is sufficient to demonstrate the equivalent formula (2.69); the proof given here follows [42]. Let us define, for $0 \leq x \leq 1$:

$$F(x) = \int_0^{\frac{\pi}{2}} \arcsin(x \sin t) \, dt.$$

$F(x)$ is continuous for $x \in [0,1]$, and it can be differentiated for $0 \leq x < 1$:

$$F'(x) = \int_0^{\frac{\pi}{2}} \frac{\sin t}{\sqrt{1 - x^2 \sin^2 t}} \, dt.$$

Now, the change of variable $\cos t = u$ yields:

$$F'(x) = \int_0^1 \frac{du}{\sqrt{1 - x^2(1 - u^2)}} = \int_0^1 \frac{du}{x\sqrt{\frac{1 - x^2}{x^2} + u^2}}.$$

Recalling the integration formula:

$$\int \frac{du}{\sqrt{a^2 + u^2}} = \ln\left(u + \sqrt{a^2 + u^2}\right),$$

we find out:

$$F'(x) = \left[\frac{1}{x} \ln \left(u + \sqrt{\frac{1 - x^2}{x^2} + u^2} \right) \right]_{u=0}^{u=1}.$$

Performing the relevant computations leads to:

$$F'(x) = \frac{1}{x} \left(\ln \left(1 + \frac{1}{x} \right) - \ln \left(\frac{\sqrt{1 - x^2}}{x} \right) \right)$$

$$= \frac{1}{x} \ln \left(\frac{1 + x}{\sqrt{1 - x^2}} \right) = \frac{1}{2x} \ln \left(\frac{1 + x}{1 - x} \right) = \sum_{n=0}^{\infty} \frac{x^{2n}}{2n + 1}.$$

The power series (2.36) was used in the very last step. Now, since $F(0) = 0$, then we have that, for $0 \le x < 1$:

$$F(x) = \int_0^x F'(t)\, dt = \int_0^x \sum_{n=0}^{\infty} \frac{t^{2n}}{2n + 1}\, dt.$$

Hence, for $0 \le x < 1$, it is:

$$F(x) = \sum_{n=0}^{\infty} \frac{x^{2n+1}}{(2n + 1)^2}.$$

But $F(x)$ is continuous at $x = 1$ too, therefore:

$$\sum_{n=0}^{\infty} \frac{1}{(2n + 1)^2} = F(1) = \int_0^{\frac{\pi}{2}} \arcsin(\sin t)\, dt = \frac{\pi^2}{8}.$$

In other words, we proved (2.69), obtaining, once more, the solution to the Basel problem.

8.13 Debye integral

We compute a family of definite integrals, introduced by Debye[12], in a Physical Chemistry model: while his focus was on the calculation of heat capacity, depending on a parameter, his model has relevance in Mathematical Finance too. The family of Debye integrals is:

$$D_m(x) = \int_0^x \frac{t^m}{e^t - 1}\, dt, \qquad x > 0; \tag{8.44}$$

---

[12]Peter Joseph William Debye (1884–1966), Dutch–American physicist and physical chemist, and Nobel laureate in Chemistry in 1936.

to ensure the convergence, we have to assume $m > 0$. Let us restrict our attention to the particular case $m = 1$. Taking into account the power series expansion (2.33), we introduce a generalisation of the logarithmic function, $\text{Li}_s(x)$, of order s and argument x, known as *polylogarithm*:

$$\text{Li}_s(x) = \sum_{n=1}^{\infty} \frac{u^n}{n^s} . \qquad (8.45)$$

The case $s = 2$, in particular, was introduced by Euler in 1768 and it is called *dilogarithm*:

$$\text{Li}_2(x) := \sum_{n=1}^{\infty} \frac{x^n}{n^2} = -\int_0^x \frac{\ln(1-t)}{t} \, dt . \qquad (8.45a)$$

We are now in the position to evaluate the Debye integral D_1. When $t > 0$, we can write:

$$\frac{t}{e^t - 1} = t \, e^{-t} \frac{1}{1 - e^{-t}} = t \, e^{-t} \sum_{n=0}^{\infty} e^{-nt} = t \sum_{n=0}^{\infty} e^{-(n+1)t} .$$

Using the monotone convergence Theorem 8.45 and integrating:

$$D_1(x) = \sum_{n=0}^{\infty} \int_0^x t \, e^{-(n+1)t} \, dt = \sum_{n=0}^{\infty} \frac{1 - \left(1 + (n+1)\,x\right) e^{-(n+1)\,x}}{(n+1)^2}$$

$$= \sum_{n=1}^{\infty} \frac{1 - (1 + n\,x)\, e^{-n\,x}}{n^2} = \sum_{n=1}^{\infty} \frac{1}{n^2} - \sum_{n=1}^{\infty} \frac{e^{-n\,x}}{n^2} - x \sum_{n=1}^{\infty} \frac{e^{-n\,x}}{n}$$

$$= \frac{\pi^2}{6} - \text{Li}_2(e^{-x}) + x \, \ln(1 - e^{-x}) .$$

To study the general case of order $m > 0$, it is necessary to use a special function, the *Euler Gamma function*, that is introduced in Chapter 11.

Chapter 9

Radon–Nikodym theorem

This chapter illustrates the Radon–Nikodym[1] Theorem 9.16, which has great implications in Probability theory and in Mathematical Finance, as explained, for example, in the "Applications" section of [18]. The theorem presentation is preceeded by § 9.1, which constitutes a brief account on the topic of signed measures. For further details we refer to [51].

9.1 Signed measures

Definition 9.1. Let $\mathcal{A}$ be a σ–algebra on Ω. A *signed measure*, or *charge*, is any set function ν, defined on $\mathcal{A}$, such that:

(i) μ can take at most one of the values $-\infty$ or $+\infty$;

(ii) $\nu(\emptyset) = 0$;

(iii) for any sequence (E_n) of measurable disjoint sets, it holds:

$$\nu\left(\bigcup_{n=1}^{\infty} E_n \right) = \sum_{n=1}^{\infty} \nu(E_n). \tag{9.1}$$

Remarks 9.2.

- In Definition 9.1, equality (iii) means that, if the measure in the left–hand side of (9.1) is finite, then the infinite series in the right–hand side of (9.1) converges absolutely; otherwise, such a series diverges.

- As a consequence of Definition 9.1, a (positive) measure is a signed measure; vice versa, a signed measure may not be a measure.

Definition 9.3.

(i) A set $P \in \mathcal{A}$ is a *positive* set, with respect to the signed measure ν, if it holds $\nu(E) \geq 0$ for any subset $E \subset P$, $E \in \mathcal{A}$.

---

[1]Johann Karl August Radon (1887–1956), Austrian mathematician.
Otto Marcin Nikodym (1887–1974), Polish mathematician.

(ii) A set $N \in \mathcal{A}$ is a *negative* set, with respect to the signed measure ν, if it holds $\nu(E) \leq 0$ for any subset $E \subset N$, $E \in \mathcal{A}$.

(iii) If a set A is both positive and negative, with respect to the signed measure ν, then A is called *null* set.

Remarks 9.4.

■ Any measurable subset of a positive set is a positive set.

■ The restriction of ν to positive sets is a measure.

Remark 9.5. Note that null sets, for a signed measure, are different from sets of measure zero, for a positive measure. This becomes evident if we observe that, for a null set A, by definition, it holds $\nu(A) = 0$, which means that A may be the union of two non–empty subset of opposite charge. For instance, in $\mathbb{R}$, the set function:

$$\nu(A) = \int_A x \, \mathrm{d}x$$

is a signed measure, defined on the σ–algebra of Lebesgue measurable sets, and:

$$\nu([-2,0]) = \int_{-2}^0 x \, \mathrm{d}x = -2, \qquad \nu([0,2]) = \int_0^2 x \, \mathrm{d}x = 2,$$

$$\nu([-2,2]) = \int_{-2}^2 x \, \mathrm{d}x = 0.$$

The Hahn[2] decomposition Theorem 9.6 explains the behaviour of the signed measures.

Theorem 9.6 (Hahn). Let ν be a signed measure in the measurable space $(\Omega, \mathcal{A})$. Then, there exist a positive set P and a negative set N such that:

$$P \cap N = \emptyset \qquad \text{and} \qquad P \cup N = \Omega.$$

The decomposition of Ω into a positive set P and a negative set N is called *Hahn decomposition* for the signed measure ν. Such a decomposition is not unique.

Remark 9.7. Denote by $\{P, N\}$ a Hahn decomposition of the charge ν. Then, it is possibile to define two positive measures ν^+ and ν^- as follows:

$$\nu^+(E) = \nu(E \cap P), \qquad \nu^-(E) = -\nu(E \cap N).$$

The positive measures ν^+ and ν^- turn out to be *mutually singular*, as they comply with the following Definition 9.8.

---

[2]Hans Hahn (1879–1934), Austrian mathematician.

Definition 9.8. Two measures ν_1 and ν_2, defined on $(\Omega, \mathcal{A})$, are called *mutually singular*, and denoted $\nu_1 \perp \nu_2$, if there exist two measurable disjoint sets A and B such that $\Omega = A \cup B$ and:

$$\nu_1(A) = \nu_2(B) = 0.$$

Various results on the charge measure and Hahn decomposition are contained in the following Theorem 9.9, due to Jordan[3].

Theorem 9.9 (Jordan decomposition). Let ν be a charge, defined on the measurable space $(\Omega, \mathcal{A})$. Then, there are exactly two positive measures ν^+ and ν^- defined on $(\Omega, \mathcal{A})$ and such that:

$$\nu = \nu^+ - \nu^-.$$

ν^+ and ν^- are called, respectively, *positive* and *negative variation* of ν. Since, by definition, μ can take at most one of the values $-\infty$ or $+\infty$, then at least one of the two variations must be finite. If both variations are finite, then ν is a finite signed measure.

Remark 9.10. A consequence of Theorem 9.9 is that a new measure, denoted with the symbol $|\nu|$, can be defined:

$$|\nu|(E) := \nu^+(E) + \nu^-(E).$$

This measure is called *absolute value* or *total variation* of ν. It is possible to show that:

$$|\nu|(E) = \sup \left| \int_E f \, d\nu \right|,$$

where the supremum is taken over all measurable functions f that verify $|f| \leq 1$ everywhere.

Moreover, the following properties hold, for any $E \in \mathcal{A}$:

(i) E is positive if and only if $\nu^-(E) = 0$;

(ii) E is negative if and only if $\nu^+(E) = 0$;

(iii) E is a null set if and only if $|\nu|(E) = 0$.

Definition 9.11. The integral of a function f, with respect to a signed measure ν, is defined by:

$$\int f \, d\nu := \int f \, d\nu^+ - \int f \, d\nu^-, \tag{9.2}$$

where we have to assume that f is measurable with respect to both ν^+ and ν^-, and that the integrals in the right–hand side of (9.2) are not both infinite.

---

[3]Marie Ennemond Camille Jordan (1838–1922), French mathematician.

Example 9.12. Consider $f(x) = (x-1)e^{-|x-1|}$, and define the signed measure on the Lebesgue measurable set in $\mathbb{R}$:

$$\nu(A) = \int_A (x-1)e^{-|x-1|}\,dx\,.$$

We wish to evaluate $\nu^+(\mathbb{R})$, $\nu^-(\mathbb{R})$, $|\nu|(\mathbb{R})$, and:

$$\int_{[0,+\infty]} \frac{1}{x-1}\,d\nu\,.$$

From the Hahn decomposition Theorem 9.6, being ν a signed measure, we know that there exist two sets P and N, positive and negative, respectively, such that $P \cap N = \emptyset$, $P \cup N = \mathbb{R}$, and such that $\nu^+(E) = \nu(E \cap P)$ and $\nu^-(E) = -\nu(E \cap N)$.
Observe that $f(x) \geq 0$ in $[1,+\infty)$, while $f(x) < 0$ on $(-\infty,1)$. Hence, we can choose, for instance, $P = [1,+\infty)$ and $N = (-\infty,1)$. Even though P and N are not uniquely determined, ν^+ and ν^- do not change. We then obtain:

$$\nu^+(\mathbb{R}) = \nu(\mathbb{R} \cap P) = \int_1^\infty (x-1)e^{-|x-1|}\,dx = \int_1^\infty (x-1)e^{-(x-1)}\,dx$$

$$= \int_0^\infty u e^{-u}\,du = 1\,,$$

$$\nu^-(\mathbb{R}) = -\nu(\mathbb{R} \cap N) = -\int_{-\infty}^1 (x-1)e^{-|x-1|}\,dx = -\int_{-\infty}^1 (x-1)e^{(x-1)}\,dx$$

$$= -\int_{-\infty}^0 u e^u\,du = -(-1) = 1\,,$$

$$|\nu|(E) = \nu^+(E) + \nu^-(E) = 2\,.$$

Finally, from equality (9.2), that defines an integral with respect to a signed measure, we get:

$$\int_{[0,+\infty]} \frac{1}{x-1}\,d\nu = \int_{[0,+\infty]\cap P} \frac{1}{x-1}\,d\nu - \int_{[0,+\infty]\cap N} \frac{1}{x-1}\,d\nu$$

$$= \int_1^{+\infty} \frac{1}{x-1}(x-1)e^{-(x-1)}\,dx - \int_0^1 \frac{1}{x-1}(x-1)e^{(x-1)}\,dx$$

$$= \int_1^{+\infty} e^{-(x-1)}\,dx - \int_0^1 e^{(x-1)}\,dx$$

$$= 1 - \frac{e-1}{e} = \frac{1}{e}\,.$$

9.2 Radon–Nikodym theorem

In some sense, the Radon–Nikodym Theorem 9.16 represents the converse of the measure generation Theorem 8.44. From such theorem, in fact, and from the monotone convergence Theorem 8.45, a result of great interest in Mathematical Probability can be inferred. For convenience, we recall the notations of Theorem 8.44.

$(\Omega, \mathcal{A}, \mu)$ denotes a measure space, and $f : \Omega \to [0, +\infty]$ is a non–negative and measurable function. Then, we know that $\phi(A) := \int_A f \, d\mu$ defines a measure on $(\Omega, \mathcal{A}, \mu)$, for any $A \in \mathcal{A}$.

Theorem 9.13. Let $f : \Omega \to [0, +\infty]$ be μ–measurable, and let further $g : \Omega \to [0, +\infty]$ be ϕ–measurable. Then:

$$\int_\Omega g \, d\phi = \int_\Omega g \, f \, d\mu. \tag{9.3}$$

We refer to f as the *density* of the measure ϕ with respect to the measure μ. In particular, if f is such that $\int_\Omega f \, d\mu = 1$, then ϕ is a probability measure.

Proof. Theorem 8.44 implies that (9.3) holds, in particular, for $g = 1_E$, with E measurable, i.e., $E \in \mathcal{A}$. Hence, (9.3) also holds, by linear combination, for any simple function. The general case finally follows, using a passage to the limit, from the monotone convergence Theorem 8.45, in an analogous way to that illustrated in Remark 8.31. □

Remark 9.14. If measure ϕ is built from f and μ, as shown in Theorem 9.13, then, if a set is negligible for μ, it is also negligible for ϕ, that is: $\mu(E) = 0 \implies \phi(E) = 0$.

The situation illustrated in Remark 9.14 is formalised in the following Definition 9.15.

Definition 9.15. Given two measures ϕ and μ, on the same σ–algebra $\mathcal{A}$, we say that ϕ is *absolutely continuous*, with respect to μ, if:

$$\mu(E) = 0 \implies \phi(E) = 0,$$

and we use the notation: $\phi \ll \mu$.

So far, we have shown that, if measure ϕ is obtained from measure μ by integrating a positive measurable function f, then ϕ is absolutely continuous with respect to μ. It is possible to reverse this process: if measure ϕ is absolutely continuous with respect to measure μ, then, under certain essential hypotheses, it is possible to represent μ as an integral of a certain function. Such hypotheses are given in the Radon–Nikodym Theorem 9.16.

Theorem 9.16 (Radon–Nikodym). Let $(\Omega, \mathcal{A}, \mu)$ be a σ–finite measure space, and let ϕ be a measure, on $\mathcal{A}$, absolutely continuous with respect to μ. Then, there exists a non–negative measurable function h such that, for any $E \in \mathcal{A}$:

$$\phi(E) = \int_E h \, d\mu.$$

This function h is almost everywhere unique, it is called *Radon–Nikodym derivative* of ϕ with respect to μ, and it is denoted by:

$$\left[\frac{d\phi}{d\mu} \right]$$

Moreover, for any function $f \geq 0$ and ϕ–measurable:

$$\int_\Omega f \, d\phi = \int_\Omega f \, h \, d\mu.$$

For the proof, see § 6.9 of [53].

The Radon–Nikodym derivative, i.e., the function h which expresses the change of measure, has some interesting properties, stated below.

(a) If $\phi \ll \mu$ and if φ is a non–negative and measurable function, then:

$$\int \varphi \, d\phi = \int \varphi \left[\frac{d\phi}{d\mu} \right] d\mu.$$

(b) For any ϕ_1, ϕ_2, it holds: $\left[\dfrac{d\,(\phi_1 + \phi_2)}{d\mu} \right] = \left[\dfrac{d\phi_1}{d\mu} \right] + \left[\dfrac{d\phi_2}{d\mu} \right].$

(c) If $\phi \ll \mu \ll \lambda$, then: $\left[\dfrac{d\phi}{d\lambda} \right] = \left[\dfrac{d\phi}{d\mu} \right] \left[\dfrac{d\mu}{d\lambda} \right].$

(d) If $\nu \ll \mu$ and $\mu \ll \nu$, then: $\left[\dfrac{d\nu}{d\mu} \right] = \left[\dfrac{d\mu}{d\nu} \right]^{-1}.$

Chapter 10

Multiple integrals

Here, we present the theoretical process that extends the Lebesgue measure from $\mathbb{R}$ to the plane $\mathbb{R}^2$, first, and, then, to the Euclidean space $\mathbb{R}^n$. The Fubini theorem is also described, both for the $\mathbb{R}^2$ case and for the general $\mathbb{R}^n$ case. A brief outline of the theory on products of σ-algebras is given at the end of this chapter.

10.1 Integration in $\mathbb{R}^2$

The process of extending the Lebesgue measure to $\mathbb{R}^2$ provides a method to compute the two-dimensional Lebesgue measure of sets in the plane, using the one-dimensional measure on the line. The basic step is the notion of *section* of a set. Due to our applicative commitment, most of the theorems are stated, but not demonstrated; for their proofs, refer to [51, 6].

Definition 10.1. Let $A \subset \mathbb{R}^2$ and fix $x \in \mathbb{R}$. Then, the *section of foot x* of A is defined as the subset A_x of $\mathbb{R}$:

$$A_x := \{y \in \mathbb{R} \mid (x, y) \in A\} .$$

Vice versa, given $y \in \mathbb{R}$, the *section of foot y* of A is:

$$A_y := \{x \in \mathbb{R} \mid (x, y) \in A\} .$$

Figure 10.1 illustrates the set sections A_x, A_y, for given A, x, y. Sections A_x are also called *vertical* or *y–sections* (as they imply integration along the y–axis). Analogously, sections A_y are also called *horizontal* or *x–sections* The following Theorem 10.2 characterises the measurability of set sections.

Theorem 10.2. Let $A \subset \mathbb{R}^2$ be measurable. Then, the following results hold.

(I) Given $x \in \mathbb{R}$, section A_x is almost everywhere measurable. Moreover, function $x \mapsto \ell(A_x)$ is measurable and:

$$\ell_2(A) = \int_{\mathbb{R}} \ell(A_x) \, dx ,$$

where ℓ_2 denotes the Lebesgue measure in $\mathbb{R}^2$.

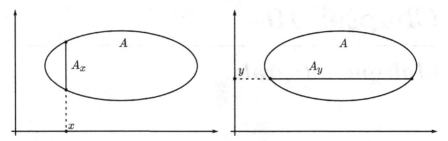

Figure 10.1: Sections of A, respectively of foot x (left) and y (right).

(II) Given $y \in \mathbb{R}$, section A_y is almost everywhere measurable. Moreover, function $y \mapsto \ell(A_y)$ is measurable and:

$$\ell_2(A) = \int_{\mathbb{R}} \ell(A_y) \, dy,$$

where, again, ℓ_2 is the Lebesgue measure in $\mathbb{R}^2$.

Example 10.3. From Theorem 10.2 it is immediate to find the measure of a circle, i.e., its area. Given the unit circle $A = \{(x,y) \mid x^2 + y^2 \leq 1\}$, in fact, we see that its ℓ_2 measure can be found as illustrated in Figure 10.2. Fixed $x \in \mathbb{R}$, the section of foot x is given by $A_x = \left[-\sqrt{1-x^2}, \sqrt{1-x^2}\right]$, if $-1 \leq x \leq 1$, while it is $A_x = \emptyset$ elsewhere. Therefore:

$$\ell_2(A) = \int_{-1}^{1} \ell(A_x) \, dx = 4 \int_0^1 \sqrt{1-x^2} \, dx = 4 \frac{\pi}{4} = \pi.$$

The following theorem is known as Fubini[1] theorem [24, 25]: it rules the evaluation of integrals, with respect to the two–dimensional Lebesgue measure, establishing the method of *nested integration*.

Theorem 10.4 (Fubini). Let $A \subset \mathbb{R}^2$ be a measurable set and let $f \in \mathcal{L}(A)$.

(I) Denote with S_0 the null set where sections A_x are non–measurable, and define S to be the subset of $\mathbb{R}$ where the y–sections of A have positive measure:

$$S = \{x \in \mathbb{R} \setminus S_0 \mid \ell(A_x) > 0\}.$$

Then, function $y \mapsto f(x,y)$ is summable on sections A_x, and it holds true that:

$$\int_A f(x,y) \, dx \, dy = \int_S \left(\int_{A_x} f(x,y) \, dy \right) dx. \tag{10.1}$$

---

[1]Guido Fubini (1879–1943), Italian mathematician.

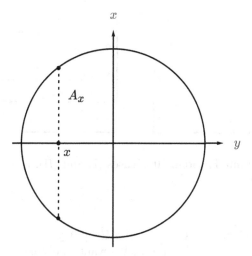

Figure 10.2: Application of Theorem 10.2 to the computation of the unit circle area.

(II) Denote with T_0 the null set where sections A_y are non–measurable, and define $\mathcal{T}$ to be the subset of $\mathbb{R}$ where the x–sections of A have positive measure:

$$\mathcal{T} = \{ y \in \mathbb{R} \setminus T_0 \mid \ell(A_y) > 0 \} .$$

Then, function $x \mapsto f(x,y)$ is summable on sections A_y, and the following holds true:

$$\int_A f(x,y) \, dx \, dy = \int_{\mathcal{T}} \left(\int_{A_y} f(x,y) \, dx \right) dy . \qquad (10.2)$$

Figure 10.3 illustrates cases (I) and (II) of the Fubini Theorem 10.4.

Remark 10.5. In Theorem 10.4, the same integral is evaluated by (10.1) and (10.2); in a practical situation, it usually occurs that one integration path is easier than the other. Moreover, the theoretical equality between integrals (10.1) and (10.2), that is:

$$\boxed{\int_S \left(\int_{A_x} f(x,y) \, dy \right) dx = \int_{\mathcal{T}} \left(\int_{A_y} f(x,y) \, dx \right) dy ,} \qquad (10.3)$$

can be exploited to evaluate integrals that are otherwise difficult to be computed via a direct approach.

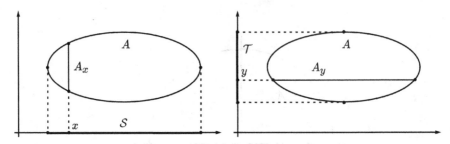

Figure 10.3: Fubini Theorem 10.4: cases (I) and (II), to the left and to the right, respectively.

Example 10.6. Consider $A = \left\{ (x\,,y) \in \mathbb{R}^2 \mid 0 \le x \le 1,\ x^2 \le y \le x + 1 \right\}$; note that $x \in [0\,,1] \implies x^2 \le 1 + x$. We want to evaluate:

$$\iint_A x\, y\, \mathrm{d}x\, \mathrm{d}y\,.$$

The first step consists in the analysis of the integration domain. In this example, the domain is described by a double inequality constraint, of the form $f_1(x) \le y \le f_2(x)$, with $x \in [a\,,b]$. Many Authors describe this type of integration domain as *normal* domain. In our case, $[a\,,b] = [0\,,1]$, and $f_1(x) = x^2$ is a parabola, while $f_2(x) = 1 + x$ is a line, hence, the domain plot is as shown in Figure 10.4.

When working with normal domains, it is natural to adopt the vertical section approach, stated in case (I) of Theorem 10.4. Here $S = [0\,,1]$, $A_x = [x^2\,,1+x]$, and the nested integration formula (10.1) yields:

$$\iint_A x\, y\, \mathrm{d}x\, \mathrm{d}y = \int_0^1 \left(\int_{x^2}^{1+x} x\, y\, \mathrm{d}y \right) \mathrm{d}x = \int_0^1 x \left[\frac{y^2}{2} \right]_{x^2}^{1+x} \mathrm{d}x$$

$$= \frac{1}{2} \int_0^1 \left(-x^5 + x^3 + 2\,x^2 + x \right) \mathrm{d}x = \frac{5}{8}\,.$$

Example 10.7. Given $A = \left\{ (x\,,y) \in \mathbb{R}^2 \mid y \ge 0,\ y \le -x + 3\,,y \le 2x + 3 \right\}$, evaluate:

$$\iint_A y\, \mathrm{d}x\, \mathrm{d}y\,.$$

The integration domain is the triangle with vertices in $\left(-\frac{3}{2}\,,0\right)$, $(3\,,0)$, $(0\,,3)$, as shown in Figure 10.5. We first integrate in x and then in y:

$$\int_0^3 \left(\int_{\frac{y-3}{2}}^{3-y} y\, \mathrm{d}x \right) \mathrm{d}y = \int_0^3 \left(\frac{9}{2} y - \frac{3}{2} y^2 \right) \mathrm{d}y = \frac{27}{4}\,.$$

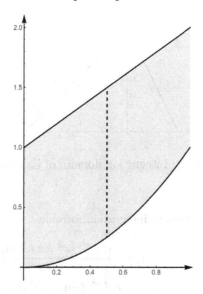

Figure 10.4: Normal integration domain of Example 10.6.

10.1.1 Smart applications of Fubini theorem

10.1.1.1 Fresnel integrals

We show, following [15] (pages 473–474), that:

$$F_1 := \int_{-\infty}^{\infty} \sin x^2 \, dx = \sqrt{\frac{\pi}{8}} = \int_{-\infty}^{\infty} \cos x^2 \, dx := F_2 .$$

The change of variable $x^2 = t$ allows to compute F_1 and F_2 as:

$$F_1 = \int_0^{\infty} \frac{\sin t}{\sqrt{t}} \, dt , \qquad F_2 = \int_0^{\infty} \frac{\cos t}{\sqrt{t}} \, dt .$$

Making use of the probability integral, discussed in § 8.11.1, 8.11.2, we see that:

$$\frac{1}{\sqrt{t}} = \frac{2}{\sqrt{\pi}} \int_0^{\infty} e^{-t x^2} \, dx .$$

Thus, we can write:

$$F_1 = \frac{2}{\sqrt{\pi}} \int_0^{\infty} \int_0^{\infty} e^{-t x^2} \sin t \, dx \, dt , \qquad F_2 = \frac{2}{\sqrt{\pi}} \int_0^{\infty} \int_0^{\infty} e^{-t x^2} \cos t \, dx \, dt .$$

We can employ Fubini Theorem 10.4, to invert the order of integration, and then observe that:

$$\int_0^{\infty} e^{-t x^2} \sin t \, dt = \frac{1}{1 + x^4} , \qquad \int_0^{\infty} e^{-t x^2} \cos t \, dt = \frac{x^2}{1 + x^4} ,$$

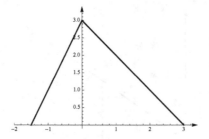

Figure 10.5: Integration domain of Example 10.7.

that follow from the indefinite integration formulæ:

$$\int e^{-t\,x^2} \sin t \, \mathrm{d}t = -\frac{e^{-t\,x^2}\left(x^2 \sin t + \cos t\right)}{1 + x^4},$$

$$\int e^{-t\,x^2} \cos t \, \mathrm{d}t = \frac{e^{-t\,x^2}\left(\sin t - x^2 \cos t\right)}{1 + x^4}.$$

Therefore:

$$F_1 = \frac{2}{\sqrt{\pi}} \int_0^\infty \frac{1}{1 + x^4} \, \mathrm{d}x, \qquad F_2 = \frac{2}{\sqrt{\pi}} \int_0^\infty \frac{x^2}{1 + x^4} \, \mathrm{d}x.$$

The remaining computations are a matter of elementary integration, and they are left to the Reader, or they can be obtained using the integration formulæ (11.25) and (11.26) of the next Chapter 11.

10.1.1.2 Cauchy formula for iterated integration

We study here the problem, solved by Cauchy, of determining a function from the knowledge of its second derivative. The problem can be stated in the following terms; given $f \in \mathcal{L}(\mathbb{R}) \cap \mathcal{C}(\mathbb{R})$, then, the function:

$$x(t) := a_0 + a_1 t + \int_0^t (t - r) \, f(r) \, \mathrm{d}r$$

is the solution to the initial value problem:

$$\begin{cases} x''(t) = f(t), \\ x(0) = a_0, \qquad x'(0) = a_1. \end{cases}$$

The problem above can, indeed, be solved directly by integrating $x'' = f(t)$ twice on $[0, t]$. The first integration provides:

$$\int_0^t x''(\tau) \, \mathrm{d}\tau = \int_0^t f(r) \, \mathrm{d}r \implies x'(t) - x'(0) = \int_0^t f(r) \, \mathrm{d}r,$$

where, since $x'(0) = a_1$, the first integration constant gets determined and we obtain:
$$x'(t) = a_1 + \int_0^t f(r) \, dr \,.$$

Now, integrate for the second time on $[0, t]$:
$$\int_0^t x'(s) \, ds = \int_0^t \left(a_1 + \int_0^t f(r) \, dr \right) ds \,, \tag{10.4}$$

and evaluate the integrals in (10.4):
$$x(t) - x(0) = a_1 t + \int_0^t \left(\int_0^t f(r) \, dr \right) ds \,.$$

Recalling that $x(0) = a_0$, use Fubini Theorem 10.4, to exchange the order of integration:
$$x(t) = a_0 + a_1 t + \int_0^t \left(\int_r^t f(r) \, ds \right) dr = a_0 + a_1 t + \int_0^t (t - r) f(r) \, dr \,.$$

The last passage can be understood by looking at Figure 10.6.

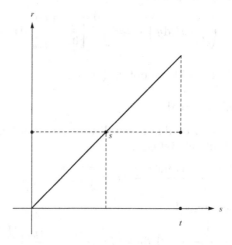

Figure 10.6: Illustration of the solution to Cauchy problem.

It is possibile to extend this argument to any order of integration; the solution of the initial value problem:
$$\begin{cases} x^{(n)}(t) = f(t) \,, \\ x(0) = a_0 \,, \quad x'(0) = a_1 \,, \quad \dots \,, \quad x^{(n-1)}(0) = a_{n-1} \,, \end{cases}$$

is indeed given by:
$$x(t) = a_0 + a_1 t + \dots + a_{n-1} t^{n-1} + \frac{1}{(n-1)!} \int_0^t (t - r)^{n-1} f(r) \, dr \,.$$

10.1.1.3 A challenging definite integral

We use Fubini Theorem 10.4 to compute a definite integral, presented in Chapter 5 of [26], and hard to evaluate otherwise:

$$\int_0^1 \frac{x^b - x^a}{\ln x}\, dx = \ln \frac{1+b}{1+a}, \qquad 0 \le a < b. \tag{10.5}$$

To calculate (10.5), let us define:

$$A = \{(x,y) \in \mathbb{R}^2 \mid 0 \le x \le 1, \quad a \le y \le b\} := [0,1] \times [a,b]$$

and consider the double integral:

$$\iint_A x^y\, dx\, dy.$$

At this point, we integrate first in x and then in y :

$$\iint_A x^y\, dx\, dy = \int_a^b \left(\int_0^1 x^y\, dx \right) dy = \int_a^b \frac{dy}{1+y} = \ln \frac{1+b}{1+a}. \tag{10.6}$$

Reversing the order of integration:

$$\iint_A x^y\, dx\, dy = \int_0^1 \left(\int_a^b x^y\, dy \right) dx = \int_0^1 \left[\frac{x^y}{\ln x} \right]_{y=a}^{y=b} dx = \int_0^1 \frac{x^b - x^a}{\ln x}\, dx. \tag{10.7}$$

Comparing (10.6) and (10.7), we obtain (10.5).

10.1.1.4 Frullani integral

We use again a two–fold double integral, to evaluate a hard single–variable integral, known as Frullani[2] integral:

$$\int_0^\infty \frac{\arctan(b x) - \arctan(a x)}{x}\, dx = \frac{\pi}{2} \ln \frac{b}{a}, \qquad 0 < a < b. \tag{10.8}$$

Consider the double integral:

$$\int_a^b \left(\int_0^\infty \frac{1}{1 + x^2 y^2}\, dx \right) dy = \int_a^b \left[\frac{\arctan(x y)}{y} \right]_{x=0}^{x=\infty} dy$$

$$= \int_a^b \frac{\pi}{2 y}\, dy = \frac{\pi}{2} \ln \frac{b}{a}. \tag{10.9}$$

Now, reverse the order of integration in (10.9):

$$\int_0^\infty \left(\int_a^b \frac{1}{1 + x^2 y^2}\, dy \right) dx = \int_0^\infty \left[\frac{\arctan(x y)}{x} \right]_{y=a}^{y=b} dx$$

$$= \int_0^\infty \frac{\arctan(b x) - \arctan(a x)}{x}\, dx. \tag{10.10}$$

---

[2]Giuliano Frullani (1795–1834), Italian mathematician.

Equation (10.8) follows by comparison of (10.9) with (10.10).

Remark 10.8. An analogous strategy allows to state the identity:

$$\int_0^\infty \frac{e^{-ax} - e^{-bx}}{x} = \ln \frac{b}{a}, \qquad (10.11)$$

which can be obtained starting from the double integral:

$$\int_0^\infty \int_a^b e^{-xy} \, dx \, dy,$$

where it is assumed that $a, b > 0$.

10.1.1.5 Basel problem, once more

Here, we describe two further ways to solve the Basel problem presented in § 2.7; this time, double integration is used and we refer to [28] and [50] for the first and second methods, respectively. In both cases, the aim is to prove the Leibniz integral (8.22) that is linked, through equality (8.24), to identity (2.69). In other words, once more, both methods lead to the Euler summation formula (2.71).

In [28], the double integral is considered:

$$\int_0^{+\infty} \left(\int_0^1 \frac{x}{(1+x^2)(1+x^2y^2)} \, dy \right) dx, \qquad (10.12)$$

which can be re-written as:

$$\int_0^{+\infty} \frac{1}{x(1+x^2)} \left(\int_0^1 \frac{dy}{\frac{1}{x^2} + y^2} \right) dx, \qquad (10.12a)$$

in order to exploit the integration formula:

$$\int \frac{dy}{a^2 + y^2} = \frac{1}{a} \arctan \frac{y}{a},$$

so that:

$$\int_0^{+\infty} \frac{1}{x(1+x^2)} \left(\int_0^1 \frac{dy}{\frac{1}{x^2} + y^2} \right) dx = \int_0^{+\infty} \left[\frac{\arctan xy}{1+x^2} \right]_{y=0}^{y=1} dx.$$

Therefore:

$$\int_0^{+\infty} \left(\int_0^1 \frac{x}{(1+x^2)(1+x^2 y^2)} \, dy \right) dx = \int_0^{+\infty} \frac{\arctan x}{1+x^2} \, dx.$$

Now, since:

$$\int \frac{\arctan x}{1+x^2} \, dx = \frac{\arctan^2 x}{2},$$

it follows:

$$\int_0^{+\infty} \left(\int_0^1 \frac{x}{(1+x^2)(1+x^2\,y^2)}\, dy \right) dx = \frac{\pi^2}{8}. \tag{10.13}$$

At this point, we use the partial fraction:

$$\frac{x}{(1+x^2)\,(1+x^2y^2)} = \frac{1}{2\,(y^2-1)} \left(\frac{2\,y^2 x}{1+x^2\,y^2} - \frac{2\,x}{1+x^2} \right),$$

and we exploit Fubini Theorem 10.4, to change the order of integration in (10.12), and perform the integration:

$$\int_0^1 \left(\int_0^{+\infty} \frac{x}{(1+x^2)\,(1+x^2\,y^2)}\, dx \right) dy = \int_0^1 \frac{1}{2\,(y^2-1)} \left[\ln \frac{1+x^2 y^2}{1+x^2} \right]_0^{+\infty} dy$$

$$= \int_0^1 \frac{\ln y^2}{2\,(y^2-1)} dy = \int_0^1 \frac{\ln y}{y^2-1} dy. \tag{10.14}$$

Comparison of (10.13) with (10.14) yields the Leibniz integral (8.22).

In [50], the double integral is considered:

$$\int_0^\infty \int_0^\infty \frac{dx\, dy}{(1+y)\,(1+x^2\,y)}. \tag{10.15}$$

Let us integrate (10.15) with respect to x, first, and then with respect to y :

$$\int_0^\infty \left(\frac{1}{1+y} \int_0^\infty \frac{dx}{1+x^2\,y} \right) dy = \frac{\pi}{2} \int_0^\infty \frac{dy}{\sqrt{y}\,(1+y)} = \frac{\pi^2}{2}. \tag{10.16}$$

Reversing the order of integration:

$$\int_0^\infty \left(\int_0^\infty \frac{dy}{(1+y)\,(1+x^2\,y)} \right) dx$$

$$= \int_0^\infty \frac{1}{1-x^2} \left(\int_0^\infty \left(\frac{1}{1+y} - \frac{x^2}{1+x^2\,y} \right) dy \right) dx \tag{10.17}$$

$$= \int_0^\infty \frac{1}{1-x^2} \ln \frac{1}{x^2} \, dx = 2 \int_0^\infty \frac{\ln x}{x^2-1} \, dx.$$

Equating (10.16) and (10.17), we get:

$$\int_0^\infty \frac{\ln x}{x^2-1} dx = \frac{\pi^2}{4}. \tag{10.18}$$

Now, split the integration domain into the two subintervals $[0\,,1]$ and $[1,\infty)$, and use the change of variable $x = \dfrac{1}{u}$ in the integral evaluated on $[1\,,\infty)$:

$$\int_0^\infty \frac{\ln x}{x^2-1} dx = \int_0^1 \frac{\ln x}{x^2-1} dx + \int_1^\infty \frac{\ln x}{x^2-1} dx = \int_0^1 \frac{\ln x}{x^2-1} dx + \int_0^1 \frac{\ln u}{u^2-1} du. \tag{10.19}$$

Comparison of (10.18) and (10.19) yields the Leibniz integral (8.22).

10.2 Change of variable

We treat, here, the two–dimensional version of the change of variable for integrals; we begin with presenting the class of well–defined variable transformations.

Definition 10.9. Let A be an open subset of $\mathbb{R}^2$. Map $\varphi : A \to \mathbb{R}^2$ is called *regular* mapping if:

1. φ is injective;

2. φ has continuous partial derivatives;

3. $\det J_\varphi(x) \neq 0$ for any $x \in A$, where $J_\varphi(x)$ is the Jacobian matrix of φ evaluated at x, i.e.:

$$J_\varphi(x) = \begin{pmatrix} \nabla\varphi_1 \\ \nabla\varphi_2 \end{pmatrix} = \begin{pmatrix} \dfrac{\partial\varphi_1(x)}{\partial x_1} & \dfrac{\partial\varphi_1(x)}{\partial x_2} \\ \dfrac{\partial\varphi_2(x)}{\partial x_1} & \dfrac{\partial\varphi_2(x)}{\partial x_2} \end{pmatrix},$$

and

$$\det J_\varphi(x) = \frac{\partial\varphi_1(x)}{\partial x_1} \frac{\partial\varphi_2(x)}{\partial x_2} - \frac{\partial\varphi_1(x)}{\partial x_2} \frac{\partial\varphi_2(x)}{\partial x_1}.$$

The most used regular mapping in the plane is the transformation into the so–called *polar* coordinates.

Example 10.10. Let $A = (0, +\infty) \times [0, 2\pi)$, and let $\varphi : A \to \mathbb{R}^2$ be defined as:

$$\varphi(\rho, \vartheta) := (\rho\cos\vartheta, \rho\sin\vartheta).$$

This is a regular mapping. In fact, the determinant of the Jacobian is $\det J_\varphi(\rho, \vartheta) = \rho > 0$, since:

$$J_\varphi(\rho, \vartheta) = \begin{pmatrix} \dfrac{\partial(\rho\cos\vartheta)}{\partial\rho} & \dfrac{\partial(\rho\sin\vartheta)}{\partial\rho} \\ \dfrac{\partial(\rho\cos\vartheta)}{\partial\vartheta} & \dfrac{\partial(\rho\sin\vartheta)}{\partial\vartheta} \end{pmatrix} = \begin{pmatrix} \cos\vartheta & -\rho\sin\vartheta \\ \sin\vartheta & \rho\cos\vartheta \end{pmatrix}.$$

A second commonly used regular mapping is constituted by *rotations*.

Example 10.11. Fix $\alpha \in (0, 2\pi)$. The map $\varphi : \mathbb{R}^2 \setminus \{(0,0)\} \to \mathbb{R}^2$:

$$\varphi(x, y) := (x\cos\alpha - y\sin\alpha, \ x\sin\alpha + y\cos\alpha)$$

is a regular mapping, and it is called *rotation of angle* α. The Jacobian determinant takes value 1. Note that the composition of two rotations, of angles α and β respectively, is again a rotation of amplitude $\alpha + \beta \mod 2\pi$.

Regular mappings are useful, since they can allow transforming a double integral into a simpler one. We state the Change of variable Theorem 10.12 and, then, study several situations where a good change of variable facilitates the computation of integrals.

Theorem 10.12. Given the open set $A \subset \mathbb{R}^2$, assume that $\varphi : A \to \mathbb{R}^2$ is a regular mapping. Let $f : A \to \overline{\mathbb{R}}$ be a measurable function. Then, $f \in \mathcal{L}(A)$ if and only if $x \mapsto f(\varphi(x)) \, |\det \mathcal{I}_\varphi(x)|$ is summable on the inverse image $\varphi^{-1}(A)$. In such a situation, the equality holds true:

$$\int_A f(u) \, du = \int_{\varphi^{-1}(A)} f(\varphi(x)) \, |\det \mathcal{I}_\varphi(x)| \, dx.$$

Example 10.13. We wish to evaluate, once more, the probability integral G, defined in (8.40). To this aim, a double integral is considered, which equals G^2, thanks to Fubini Theorem 10.4:

$$\iint_{\mathbb{R}^2} e^{-(x^2+y^2)} \, dx \, dy = \int_{\mathbb{R}} e^{-x^2} \, dx \int_{\mathbb{R}} e^{-y^2} \, dy = G^2,$$

while, employing polar coordinates, it also holds:

$$\iint_{\mathbb{R}^2} e^{-(x^2+y^2)} \, dx \, dy = \iint_{(0,+\infty)\times[0,2\pi)} \rho e^{-\rho^2} \, d\rho \, d\vartheta$$

$$= 2\pi \int_0^\infty \rho e^{-\rho^2} \, d\rho = 2\pi \left[-\frac{e^{-\rho^2}}{2} \right]_{\rho=0}^{\rho=\infty} = \pi.$$

Thus $G^2 = \pi$, i.e., $G = \sqrt{\pi}$, as already found in § 8.11.2.

Example 10.14. Let us evaluate the measure of set:

$$E = \left\{ (x,y) \in \mathbb{R}^2 \mid x^2 + y^2 \le 1 \right\} \cap \left\{ (x,y) \in \mathbb{R}^2 \mid 0 \le y \le x \le \sqrt{3}\,y \right\}.$$

As Figure 10.7 shows, E is a circle sector, in which points $A = \left(\dfrac{1}{\sqrt{2}}, \dfrac{1}{\sqrt{2}} \right)$ and $B = \left(\dfrac{1}{2}, \dfrac{\sqrt{3}}{2} \right)$ are obtained solving the systems:

$$\begin{cases} x^2 + y^2 = 1, \\ x = y, \end{cases} \qquad \begin{cases} x^2 + y^2 = 1, \\ x = \sqrt{3}\,y. \end{cases}$$

Passing to polar coordinates, the constraints defining E become:

$$\varphi^{-1}(E) = \left\{ (\rho,\vartheta) \in (0,\infty) \times [0, \tfrac{\pi}{2}] \ \middle| \ \rho < 1, \ 0 < \sin\vartheta < \cos\vartheta < \sqrt{3}\sin\vartheta \right\}.$$

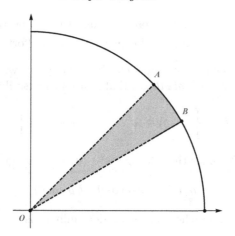

Figure 10.7: Set E of Example 10.14.

From $\vartheta \in [0, \frac{\pi}{2}]$ and $\sin \vartheta < \cos \vartheta$, we infer:

$$0 < \vartheta < \frac{\pi}{4},$$

while, from $\vartheta \in [0, \frac{\pi}{2}]$ and $\cos \vartheta < \sqrt{3} \sin \vartheta$, it follows:

$$\frac{\pi}{6} < \vartheta < \frac{\pi}{2},$$

so that, in conclusion:

$$\varphi^{-1}(E) = (0, 1) \times (\frac{\pi}{6}, \frac{\pi}{4}).$$

Finally, invoking Theorem 10.12, we obtain:

$$\ell_2(E) = \iint_{(0,1) \times (\frac{\pi}{6}, \frac{\pi}{4})} \rho \, d\rho \, d\vartheta = \frac{1}{2} \left(\frac{\pi}{4} - \frac{\pi}{6} \right) = \frac{\pi}{24}.$$

It is obviously possible not to use Theorem 10.12, and compute $\ell_2(E)$ using, instead, Fubini Theorem 10.4. By doing so, though, the relevant computation turns out to be more involved. By looking again at Figure 10.7, in fact, we see that the domain of integration must be split ad follows:

$$\ell_2(E) = \int_0^{\frac{1}{\sqrt{2}}} \left(\int_{\frac{x}{\sqrt{3}}}^x dy \right) dx + \int_{\frac{1}{\sqrt{2}}}^{\frac{\sqrt{3}}{2}} \left(\int_{\frac{x}{\sqrt{3}}}^{\sqrt{1-x^2}} dy \right) dx.$$

To evaluate the second integral, we must employ an appropriate integration formula:

$$\int \sqrt{1 - x^2} \, dx = \frac{1}{2} \left(x \sqrt{1 - x^2} + \arcsin x \right),$$

after which we do arrive at the computation of the same value $\ell_2(E) = \dfrac{\pi}{24}$ obtained using polar coordinates, but with a greater effort.

Remark 10.15. Polar coordinates can be modified, if we wish to compute the measure of the *canonical ellipse*, that is a set E described by:

$$E = \left\{ (x,y) \in \mathbb{R}^2 \;\Big|\; \frac{x^2}{a^2} + \frac{y^2}{b^2} \le 1 \right\},$$

where $a, b > 0$. Consider the map $\varphi : \mathbb{R}^2 \setminus (0,0) \to \mathbb{R}^2$ given by:

$$\varphi(\rho, \vartheta) := (a\rho \cos \vartheta, \; b\rho \sin \vartheta).$$

It is a regular mapping, whose Jacobian determinant is:

$$\det J_\varphi(\rho, \vartheta) = \det \begin{pmatrix} a \cos \vartheta & -b\rho \sin \vartheta \\ a \sin \vartheta & b\rho \cos \vartheta \end{pmatrix} = ab\rho > 0.$$

Observe that $\varphi^{-1}(E) = [0,1] \times [0, 2\pi)$, hence:

$$\ell_2(E) = \iint_E dx\, dy = \iint_{[0,1] \times [0,2\pi)} a\, b\, \rho\, d\rho\, d\vartheta = a\, b\, \pi.$$

In the next Example 10.16, we employ a rotation to compute the measure of a set.

Example 10.16. Let us compute the Lebesgue measure of

$$A := \{ (x,y) \in \mathbb{R}^2 \mid x^2 - xy + y^2 \le 1 \}, \tag{10.20}$$

using an $\alpha = \dfrac{\pi}{4}$ rotation:

$$\begin{cases} x = \dfrac{1}{\sqrt{2}}\, u - \dfrac{1}{\sqrt{2}}\, v, \\[2mm] y = \dfrac{1}{\sqrt{2}}\, u + \dfrac{1}{\sqrt{2}}\, v. \end{cases}$$

In this way, the set of points verifying $x^2 - xy + y^2 \le 1$ is transformed into:

$$\frac{1}{2}(u-v)^2 - \frac{1}{2}(u+v)(u-v) + \frac{1}{2}(u+v)^2 \le 1,$$

that is:

$$\frac{u^2}{2} + \frac{3v^2}{2} \le 1. \tag{10.21}$$

The choice of the rotation angle α is not a lucky guess. In general, we start with an undefined rotation of amplitude α :

$$\begin{cases} x = u \cos \alpha - v \sin \alpha \\ y = u \sin \alpha + v \cos \alpha \end{cases} \tag{10.22}$$

and we insert it into the equation which defines our domain. In the current example, inserting (10.22) into (10.20) yields:

$$(u \sin \alpha + v \cos \alpha)^2 - (u \cos \alpha - v \sin \alpha)(u \sin \alpha + v \cos \alpha) + (u \cos \alpha - v \sin \alpha)^2 \leq 1$$

that is:

$$u^2 + v^2 + (\sin^2 \alpha - \cos^2 \alpha) \, u \, v - (u^2 - v^2) \sin \alpha \cos \alpha \leq 1 . \tag{10.23}$$

We choose the value of α for which the rectangular term $u \, v$ in (10.23) vanishes:

$$\sin^2 \alpha - \cos^2 \alpha = 0 ,$$

namely, $\alpha = \dfrac{\pi}{4}$.

At this point, returning to our transformed set (10.21), and recalling Remark 10.15, we infer that the measure of A is:

$$\sqrt{2} \, \frac{\sqrt{2}}{\sqrt{3}} \pi = \frac{2\pi}{\sqrt{3}} .$$

Remark 10.17. The computational method described in Example 10.16 can be applied to the situation of a general ellipse:

$$E = \{ (x , y) \in \mathbb{R}^2 \mid a \, x^2 + 2 \, b \, x \, y + c \, y^2 \leq d \} , \tag{10.24}$$

where, to make sure we are dealing with an ellipse and a non–empty set, we must assume that $a , c , d > 0$, and that the discriminant $\Delta = b^2 - a \, c$ is negative. That said, we can apply the rotation technique to compute $\ell_2(E)$. The two cases $a = c$ and $a \neq c$ must be distinguished. When $a = c$, the $\dfrac{\pi}{4}$ rotation used in Example 10.16 still works, and transforms E into:

$$\varphi^{-1}(E) = \{ (u , v) \mid u^2 (a + b) + v^2 (a - b) \leq d \} . \tag{10.25}$$

Note that the two conditions $a = c$ and $\Delta < 0$ imply $|b| < a$; hence, $a + b > 0$ and $a - b > 0$. Recalling Remark 10.15, we can thus infer:

$$\ell_2(E) = \frac{d}{\sqrt{a^2 - b^2}} \pi . \tag{10.26}$$

If $a \neq c$, it is always possible to transform the given ellipse into an ellipse of the form (10.25), by means of the variable transformation:

$$\begin{cases} x = x_1 , \\ y = \sqrt{\dfrac{a}{c}} \, y_1 , \end{cases} \tag{10.27}$$

applying which, the general ellipse (10.24) becomes:

$$a\,x_1^2 + 2\,b\,\sqrt{\frac{a}{c}}\,x_1\,y_1 + a\,y_1^2 \le d\,. \tag{10.28}$$

Transformation (10.27) has non–zero Jacobian determinant $\sqrt{\dfrac{a}{c}} > 0$; hence from (10.26), it follows:

$$\ell_2(E) = \frac{d}{\sqrt{a\,c - b^2}}\,\pi\,. \tag{10.29}$$

In other words, (10.29) nicely extends (10.26).

Translations and affine transformations are regular mappings too.

Example 10.18. If $a \in \mathbb{R}^2$, the map $\varphi : \mathbb{R}^2 \to \mathbb{R}^2$, $\varphi(x) = x + a$, is a regular transformation, called *translation of amplitude* a. It is quite immediate to verify that, for such a transformation, it holds $\det J_\varphi = 1$.

Example 10.19. Given $a \in \mathbb{R}^2$, let $M \in \mathbb{R}^{2 \times 2}$ be a non–singular square matrix. Then, the map $\varphi : \mathbb{R}^2 \to \mathbb{R}^2$, $\varphi(x) = M\,x^T + a$ is a regular transformation, called *affine transformation*. Here, the Jacobian determinat is given by $\det J_\varphi = \det M \ne 0$.

Remark 10.20. Any affine transformation is obtained by composing a linear transformation and a translation. As consequence, we can think of a translation as an affine transformation, associated to the identity matrix.

10.3   Integration in $\mathbb{R}^n$

The process of extending the Lebesgue measure, from $\mathbb{R}$ to $\mathbb{R}^2$, can be naturally iterated. First, though, some notations need to be introduced. The Euclidean space $\mathbb{R}^n$ is here decomposed into the Cartesian product of two lower dimensional Euclidean spaces, $\mathbb{R}^p$ and $\mathbb{R}^q$, where $p + q = n$, and we make the identification $\mathbb{R}^n = \mathbb{R}^p \times \mathbb{R}^q$. Moreover, with the writing $(x, y) \in \mathbb{R}^n$, we implicitly mean that $x = (x_1, \ldots, x_p) \in \mathbb{R}^p$ and $y = (y_1, \ldots, y_q) \in \mathbb{R}^q$. This said, following Definition 10.1, we can now provide the idea of *section of a set* $A \subset \mathbb{R}^n$.

Definition 10.21. Consider $A \subset \mathbb{R}^n$ and $x \in \mathbb{R}^p$. Then, the *section of foot* x of A is defined as the following subset A_x of $\mathbb{R}^q$:

$$A_x := \{y \in \mathbb{R}^q \mid (x, y) \in A\}\,.$$

Similarly, if $y \in \mathbb{R}^q$, then the *section of foot* y of A is defined as the following subset A_y of $\mathbb{R}^p$:

$$A_y := \{x \in \mathbb{R}^p \mid (x, y) \in A\} .$$

The Lebesgue measure in $\mathbb{R}^n$ is denoted by ℓ_n ; analogous meaning holds for ℓ_p and ℓ_q . In the following, when the term *measurable* is used, the dimension of the considered Euclidean space will be clear from the context. Symbol d_p indicates that we are integrating with respect to the Lebesgue measure in $\mathbb{R}^p$, and analogously for other dimensions.

The set section Theorem 10.22, in its proof, makes a heavy use of the monotone convergence Theorem 8.45.

Theorem 10.22. Let $A \subset \mathbb{R}^n$ be a measurable set. Then:

(**I**) for almost any $x \in \mathbb{R}^p$, section $A_x \subset \mathbb{R}^q$ is measurable; moreover, function $A_x \mapsto \ell_q(A_x)$ is measurable, and it holds:

$$\ell_n(A) = \int_{\mathbb{R}^p} \ell_q(A_x) \, d_p x ;$$

(**II**) for almost any $y \in \mathbb{R}^q$, section $A_y \subset \mathbb{R}^p$ is measurable; moreover, function $A_y \mapsto \ell_p(A_y)$ is measurable, and it holds:

$$\ell_n(A) = \int_{\mathbb{R}^q} \ell_p(A_y) \, d_q y .$$

We can now formulate the extension of Fubini Theorem 10.4 to the general $\mathbb{R}^n$ case.

Theorem 10.23. (Fubini – general case) Let $A \subset \mathbb{R}^n$ be a measurable set and consider $f \in \mathcal{L}(A)$. Denote with S_0 the null set where sections A_x, for $x \in \mathbb{R}^p$, are non–measurable. Define, further, S to be the subset of $\mathbb{R}^p$ where:

$$S = \{ x \in \mathbb{R}^p \setminus S_0 \mid \ell_q(A_x) > 0 \} . \tag{10.30}$$

Then, $y \mapsto f(x, y)$ is summable on sections A_x, and the equality holds true:

$$\boxed{\int_A f(x, y) \, dx \, dy = \int_S \left(\int_{A_x} f(x, y) \, d_q y \right) d_p x .} \tag{10.31}$$

To apply Fubini Theorem 10.23, it is mandatory that the integrand $f(x, y)$ is a summable function. In many circumstances, summability can be deduced by some a priori considerations. Otherwise, the following Theorem 10.24, due to Tonelli[3], analyses the summability of a given function.

---

[3]Leonida Tonelli (1885–1946), Italian mathematician.

Theorem 10.24. (Tonelli) Let A, S_0 and S be as in Theorem 10.23. Let further $f : A \to \bar{\mathbb{R}}$ be a measurable function. Then, function $y \mapsto f(x, y)$ is measurable on A_x, and the following equality holds true:

$$\int_A |f(x, y)| \, dx \, dy = \int_S \left(\int_{A_x} |f(x, y)| \, d_q y \right) d_p x . \qquad (10.32)$$

We complete our study of integration in the Euclidean space $\mathbb{R}^n$ by describing some applications of the Fubini Theorem 10.23. Some further terminology is also provided.

Definition 10.25. Consider a measurable set $A \subset \mathbb{R}^n$ and a measurable real–valued non–negative function $f : A \to [0, +\infty]$. The following sets are called, respectively, *subgraph* Γ_f and *graph* G_f of f :

$$\Gamma_f = \{(x, t) \in A \times \mathbb{R}^n \mid 0 \leq t < f(x)\} , \quad G_f = \{(x, t) \in A \times \mathbb{R}^n \mid t = f(x)\} .$$

The geometrical meaning of Definition 10.25 is, clearly, that of the n–dimensional generalisation of the concept of *graph* of a function of one variable. The following Theorem 10.26 illustrates the geometric idea of an n–dimensional integral.

Theorem 10.26. Sets Γ_f and G_f, introduced in Definition 10.25, are measurable in $\mathbb{R}^{n+1}$. Moreover:

$$\ell_{n+1}(\Gamma_f) = \int_A f \, dx , \qquad \ell_{n+1}(G_f) = 0 .$$

Combining Theorems 10.26 and 8.13, it is possible to provide an alternative proof to the monotone convergence Theorem 8.45.

Theorem 10.27. Let $A \subset \mathbb{R}^n$ be measurable, and consider a sequence $(f_p)_{p \in \mathbb{N}}$ of real non–negative measurable functions on A, such that, for almost any $x \in A$:

(i) $f_p(x) \leq f_{p+1}(x)$;   (ii) $\lim_{p \to \infty} f_p(x) = f(x)$. $\qquad (10.33)$

Then, the passage to the limit holds true:

$$\int_A f(x) \, d_n x = \lim_{p \to \infty} \int_A f_p(x) \, d_n x .$$

Proof. Following Definition 10.25, consider the subgraphs Γ_f and Γ_{f_p}. From the monotonicity hypothesis given by point (i) of (10.33), it follows that $\Gamma_{f_p} \subset \Gamma_{f_{p+1}}$. Hence, the family of sets $(\Gamma_{f_p})_{p \in \mathbb{N}}$ is a nested increasing family and, due to hypothesis (ii) of (10.33), it is exhaustive, that is:

$$\Gamma_f = \bigcup_{p=1}^{\infty} \Gamma_{f_p} .$$

We can thus invoke Theorem 8.13, to infer:

$$\ell_{n+1}\left(\bigcup_{p=1}^{\infty} \Gamma_{f_p}\right) = \lim_{p\to\infty} \ell_{n+1}\left(\Gamma_{f_p}\right) = \int_A f_p(\boldsymbol{x})\, \mathrm{d}_n\boldsymbol{x} \ .$$

On the other hand, from Theorem 10.26, it follows:

$$\ell_{n+1}\left(\bigcup_{p=1}^{\infty} \Gamma_{f_p}\right) = \ell_{n+1}(\Gamma_f) = \int_A f(\boldsymbol{x})\, \mathrm{d}_n\boldsymbol{x} \ ,$$

and the proof is completed. $\qquad\square$

To conclude § 10.3, we state an alternative formula for integral evaluation, that can be useful in applications.

Theorem 10.28. Consider $A \subset \mathbb{R}^n$ measurable. If $f \in \mathcal{L}(A)$, then the following equality holds true:

$$\int_A f(\boldsymbol{x})\, \mathrm{d}_n\boldsymbol{x} = \int_0^{\infty} \ell_n(\{\boldsymbol{x} \in A \mid f(\boldsymbol{x}) > t\})\, \mathrm{d}t \ .$$

10.3.1 Exercises

1. Consider $A := \left\{ (x,y) \in \mathbb{R}^2 \mid x^2 \le y \le x \right\}$. Show that:

$$\iint_A \frac{1}{y - x^2 - 1}\, \mathrm{d}x\, \mathrm{d}y = \frac{\pi}{\sqrt{3}} - 2 \ .$$

2. Let $A = \left\{ (x,y) \in \mathbb{R}^2 \mid 0 \le x \le 1, \quad x^2 \le y \le x+1 \right\}$. Show that:

$$\iint_A x\, y\, \mathrm{d}x\, \mathrm{d}y = \frac{5}{8} \ .$$

3. Let $A = \left\{ (x,y) \in \mathbb{R}^2 \mid 0 \le x \le y \le 2x \right\}$. Show that:

$$\iint_A x\, e^{-y}\, \mathrm{d}x\, \mathrm{d}y = \frac{3}{4} \ .$$

4. Let $A = \left\{ (x,y) \in \mathbb{R}^2 \mid 0 \le x \le 4, \quad 2 - \frac{x}{2} \le y \le x+2 \right\}$. Show that:

$$\iint_A x\, \mathrm{d}x\, \mathrm{d}y = 32 \ .$$

5. Let $A = \left\{ (x,y) \in \mathbb{R}^2 \mid x^4 \le y \le x^2 \right\}$. Show that:

$$\iint_A x\, y\, \mathrm{d}x\, \mathrm{d}y = \frac{1}{30} \ .$$

6. Let $A = \{ (x, y) \in \mathbb{R}^2 \mid x^2 + 4y^2 \leq 1 \}$. Show that:

$$\iint_A \frac{1}{1 + x^2 + 4y^2} \, dx \, dy = \frac{\pi}{2} \ln 2 .$$

7. Let $A = \{ (x, y) \in \mathbb{R}^2 \mid 5x^2 + 4xy + y^2 \leq 1 \}$. Show that:

$$\iint_A e^{5x^2 + 4xy + y^2} \, dx \, dy = \pi (e - 1) .$$

10.4   Product of σ–algebras

We end this Chapter 10 giving a short outline of the theory on products of σ–algebras. Assume that two measurable spaces $(\Omega_1, \mathcal{A}_1)$ and $(\Omega_2, \mathcal{A}_2)$ are given. We form the Cartesian product $\Omega := \Omega_1 \times \Omega_2$, on which we wish to build a measure μ, that agrees with the given measures on Ω_1 and Ω_2. To do so, we have to provide the domain $\mathcal{A}$ of such a set function μ, that is to say, we must construct a suitable σ–algebra on Ω : the $\mathcal{A}$ that we define is the *product* σ–algebra, built to be the minimum σ–algebra on Ω, containing *rectangles* $A_1 \times A_2$, with $A_1 \in \mathcal{A}_1$ and $A_2 \in \mathcal{A}_2$. In other words, the product σ–algebra is generated by the collection of rectangular sets:

$$\mathcal{R} = \{ A_1 \times A_2 \mid A_1 \in \mathcal{A}_1, \quad A_2 \in \mathcal{A}_2 \} .$$

It is possible to prove the following result.

Theorem 10.29. The product σ–algebra $\mathcal{A}_1 \times \mathcal{A}_2$ verifies what follows:

(i) $\mathcal{A}_1 \times \mathcal{A}_2$ is also generated by cylinders:

$$\mathcal{C} = \{ A_1 \times \Omega_2 \mid A_1 \in \mathcal{A}_1 \} \cap \{ \Omega_1 \times A_2 \mid A_2 \in \mathcal{A}_2 \} ;$$

(ii) $\mathcal{A}_1 \times \mathcal{A}_2$ is the minimum σ–algebra such that the following *projections* are measurable:

$$\begin{aligned}
\mathrm{Pr}_1 &: \Omega \to \Omega_1, & \mathrm{Pr}_1(\omega_1, \omega_2) &= \omega_1, \\
\mathrm{Pr}_2 &: \Omega \to \Omega_2, & \mathrm{Pr}_2(\omega_1, \omega_2) &= \omega_2.
\end{aligned}$$

For applications, the most relevant situation occurs when $\Omega_1 = \Omega_2 = \mathbb{R}$ and $\mathcal{A}_1 = \mathcal{A}_2 = \mathcal{B}$ is the Borel σ–algebra on $\mathbb{R}$. Borel sets, in the plane, can be generated in two equivalent ways.

Theorem 10.30. The σ–algebras generated by:

$$\mathcal{R} = \{ B_1 \times B_2 \mid B_1, B_2 \in \mathcal{B} \} \quad \text{and} \quad \mathcal{I} = \{ I_1 \times I_2 \mid I_1, I_2 \text{ intervals} \}$$

are the same.

Having defined the product σ–algebra, we have to introduce the *product measure*, and we must do so in a consistent manner, that follows the contribution of Fubini to Measure theory, in the case of the Euclidean space $\mathbb{R}^n$.

To build the product measure, we need to work with σ–finite measure spaces $(\Omega_1, \mathcal{A}_1, \mu_1)$ and $(\Omega_2, \mathcal{A}_2, \mu_2)$. Note that this is the case of Lebesgue measure and of probability measures. We denote by μ the product measure defined on the product σ–algebra $\mathcal{A}_1 \times \mathcal{A}_2$. As a first construction step, let us impose the 'natural' condition:

$$\mu(A_1 \times A_2) = \mu(A_1)\,\mu(A_2).$$

The task of assigning a measure to non–rectangular sets requires the notion of *section* of a subset A of $\Omega_1 \times \Omega_2$.

Definition 10.31. Consider $A \subset \Omega_1 \times \Omega_2$ and let $\omega_2 \in \Omega_2$. Then, the *section of foot ω_2* is the subset of Ω_1 defined, and denoted, as:

$$A_{\omega_2} = \{\omega_1 \in \Omega_1 \mid (\omega_1, \omega_2) \in A\}.$$

Analogously, if $\omega_1 \in \Omega_1$, then the *section of foot ω_1* is the subset of Ω_2:

$$A_{\omega_1} = \{\omega_2 \in \Omega_2 \mid (\omega_1, \omega_2) \in A\}.$$

We can now state Theorem 10.32, which concerns the measurability of sections, of a measurable set, for the product measure.

Theorem 10.32. Let $A \in \mathcal{A}_1 \times \mathcal{A}_2$. Then, $A_{\omega_2} \in \mathcal{A}_1$ for any $\omega_2 \in \Omega_2$, and $A_{\omega_1} \in \mathcal{A}_2$ for any $\omega_1 \in \Omega_1$.

Definition 10.33. Given $A \in \mathcal{A}_1 \times \mathcal{A}_2$, we define *measure* of A the number:

$$\mu(A) = \int_{\Omega_2} \mu_1(A_{\omega_2})\,\mathrm{d}\mu_2(\omega_2). \tag{10.34}$$

Theorem 10.34. Consider the product σ–algebra $\mathcal{A}_1 \times \mathcal{A}_2$. Let μ_1, μ_2 be σ–finite measures. Then, the functions:

$$\omega_2 \mapsto \mu_1(A_{\omega_2}) \qquad \text{and} \qquad \omega_1 \mapsto \mu_2(A_{\omega_1})$$

are measurable with respect to $\mathcal{A}_2$ and $\mathcal{A}_1$, respectively. Furthermore, it holds:

$$\int_{\Omega_1} \mu_2(A_{\omega_1})\,\mathrm{d}\mu_1(\omega_1) = \int_{\Omega_2} \mu_1(A_{\omega_2})\,\mathrm{d}\mu_2(\omega_2).$$

Theorem 10.35. The set function μ introduced in (10.34) is a measure. Moreover, μ is unique, since any other measure, that coincides with μ on rectangles, is equal to μ on the product σ–algebra $\mathcal{A}_1 \times \mathcal{A}_2$.

We are, finally, in the position to state the Fubini Theorem 10.36 for nested integrals.

Theorem 10.36. Consider $f \in \mathcal{L}^1(\Omega_1 \times \Omega_2)$. Then, the functions:

$$\omega_1 \mapsto \int_{\Omega_2} f(\omega_1, \omega_2) \, \mathrm{d}\mu_2(\omega_2) \quad \text{and} \quad \omega_2 \mapsto \int_{\Omega_1} f(\omega_1, \omega_2) \, \mathrm{d}\mu_1(\omega_1)$$

belong to $\mathcal{L}^1(\Omega_1)$ and $\mathcal{L}^1(\Omega_2)$, respectively. Moreover, it holds:

$$\int_{\Omega_1 \times \Omega_2} f(\omega_1, \omega_2) \, \mathrm{d}(\mu_1 \times \mu_2)(\omega_1, \omega_1) = \int_{\Omega_1} \left(\int_{\Omega_2} f(\omega_1, \omega_2) \, \mathrm{d}\mu_2(\omega_2) \right) \mathrm{d}\mu_1(\omega_1)$$

$$= \int_{\Omega_2} \left(\int_{\Omega_1} f(\omega_1, \omega_2) \mathrm{d}\mu_1(\omega_1) \right) \mathrm{d}\mu_2(\omega_2).$$

Chapter 11

Gamma and Beta functions

In this chapter, the Gamma function and the Beta function are introduced, together with their properties. The material presented here is based on many excellent textbooks [5, 16, 20, 27, 30, 41, 55, 62, 43, 40], to which the interested Reader is referred.

11.1 Gamma function

The Gamma function was introduced by Euler in relation to the *factorial problem*. From the early beginning of Calculus, in fact, the problem of understanding the behaviour of $n!$, with $n \in \mathbb{N}$, and the related *binomial coefficients*, was under the attention of the mathematical community. In 1730, the formula of Stirling[1] was discovered:

$$\lim_{n \to \infty} \frac{n! \, e^n}{n^n \sqrt{n}} = \sqrt{2\pi}. \tag{11.1}$$

In the same period, Euler introduced a function $e(x)$, given by the explicit formula (11.2) and defined for any $x > 0$, which reduces to $e(n) = n!$ when its argument is $x = n \in \mathbb{N}$. Euler described his results in a letter to Goldbach[2], who had posed, together with Bernoulli, the *interpolation problem* to Euler. The latter problem was inspired by the fact that the additive counterpart of the factorial has a very simple solution:

$$s_n = 1 + 2 + \ldots \cdots + n = \frac{n(n+1)}{2},$$

and by the observation that the above *sum* function admits a continuation to $\mathbb{C}$ given by the following function:

$$f(x) = \frac{x(x+1)}{2}.$$

---

[1] James Stirling (1692–1770), Scottish mathematician.
[2] Christian Goldbach (1690–1764), German mathematician.

Euler's solution for the factorial is the following integral:

$$e(x) = \int_0^1 \left(\ln \frac{1}{t}\right)^x dt . \tag{11.2}$$

Legendre[3] introduced the letter $\Gamma(x)$ to denote (11.2), and he modified its representation as follows:

$$\boxed{\Gamma(x) = \int_0^\infty e^{-t} \, t^{x-1} \, dt , \qquad x > 0 .} \tag{11.3}$$

Observe that (11.2) and (11.3) imply the equality $\Gamma(x + 1) = e(x)$. In fact, the change of variable $t = -\ln u$, in the Legendre integral $\Gamma(x+1)$, yields:

$$\Gamma(x+1) = \int_0^1 e^{\ln u} \left(\ln \frac{1}{u}\right)^x \frac{1}{u} \, du = e(x) .$$

Function Γ is the natural continuation of the discrete factorial, since:

$$\Gamma(n+1) = n! \qquad \text{for} \quad n \in \mathbb{N} , \tag{11.4}$$

and, most importantly, Γ solves the functional equation:

$$\Gamma(x+1) = x \, \Gamma(x) \qquad \text{for any} \quad x > 0 . \tag{11.5}$$

These aspects are treated with the maximal generality in the Bohr–Mollerup[4] Theorems 11.1–11.2. Note that $\Gamma(x)$ appears in many formulæ of Mathematical Analysis, Physics and Mathematical Statistics.

Theorem 11.1. For any $x > 0$, the recursion relation (11.5) is true. In particular, when $x = n \in \mathbb{N}$, then (11.4) holds.

Proof. Consider (11.3) and integrate by parts:

$$\Gamma(x+1) = \int_0^\infty t^x \, e^{-t} \, dt = \left[-t^x \, e^{-t} \right]_0^\infty + x \int_0^\infty t^{x-1} \, e^{-t} \, dt = x \, \Gamma(x) .$$

For the integer argument case, exploit (11.3) to observe that:

$$\Gamma(1) = \int_0^\infty e^{-t} \, dt = 1 ,$$

and use the just proved recursion (11.5) to compute:

$$\Gamma(2) = 1 \cdot \Gamma(1) = 1 , \qquad \Gamma(3) = 2 \cdot \Gamma(2) = 2 , \qquad \Gamma(4) = 3 \cdot \Gamma(3) = 3 \cdot 2 = 6 ,$$

and so on. Hence, $\Gamma(n+1) = n!$ can be inferred by an inductive argument. $\square$

---

[3] Adrien Marie Legendre (1752–1833), French mathematician.
[4] Harald August Bohr (1887–1951), Danish mathematician and soccer player.
Johannes Mollerup (1872–1937), Danish mathematician.

When $x > 0$, function $\Gamma(x)$ is continuous and differentiable at any order. To evaluate its derivatives, we use the differentiation of parametric integrals, obtaining:

$$\Gamma'(x) = \int_0^\infty e^{-t}\, t^{x-1}\, \ln t \, dt\,, \tag{11.6a}$$

$$\Gamma^{(2)}(x) = \int_0^\infty e^{-t}\, t^{x-1}\, (\ln t)^2 \, dt\,, \tag{11.6b}$$

and more generally:

$$\Gamma^{(n)}(x) = \int_0^\infty e^{-t}\, t^{x-1}\, (\ln t)^n \, dt\,. \tag{11.7}$$

From its definition, $\Gamma(x)$ is strictly positive. Moreover, since (11.6b) shows that $\Gamma^{(2)}(x) \geq 0$, it follows that $\Gamma(x)$ is also strictly convex. We thus infer the existence for the following pair of limits:

$$\ell_0 := \lim_{x \to 0^+} \Gamma(x)\,, \qquad \ell_\infty := \lim_{x \to \infty} \Gamma(x)\,.$$

To evaluate ℓ_0, we use the inequality chain:

$$\Gamma(x) > \int_0^1 t^{x-1} e^{-t} \, dt > \frac{1}{e} \int_0^1 t^{x-1} \, dt = \frac{1}{x\, e}\,,$$

which ensures that:

$$\ell_0 = +\infty\,.$$

To evaluate ℓ_∞, since we know, a priori, that such limit exists, we can restrict the focus to natural numbers, so that we have immediately:

$$\ell_\infty = \lim_{n \to \infty} \Gamma(n) = \lim_{n \to \infty} (n-1)! = +\infty\,.$$

Observing that $\Gamma(2) = \Gamma(1)$ and using Rolle theorem[5], we see that there exists $\xi \in\,]1,2[$ such that $\Gamma'(\xi) = 0$. On the other hand, since $\Gamma^{(2)}(x) > 0$, the first derivative $\Gamma'(x)$ is strictly increasing, thus there is a unique ξ such that $\Gamma'(\xi) = 0$. Furthermore, we have that $0 < x < \xi \implies \Gamma'(x) < 0$ and $x > \xi \implies \Gamma'(x) > 0$. This means that ξ represents the absolute minimum for $\Gamma(x)$, when $x \in\,]0,\infty[$. The numerical determination of ξ and $\Gamma(\xi)$ is due to Legendre and Gauss[6]:

$$\xi = 1.4616321449683622\,, \qquad \Gamma(\xi) = 0.8856031944108886\,.$$

An important property of Γ is that its logarithm is a convex function, as stated below.

---

[5] Michel Rolle (1652–1719), French mathematician. For the theorem of Rolle see, for example, mathworld.wolfram.com/RollesTheorem.html

[6] Carl Friedrich Gauss (1777–1855), German mathematician and physicist.

Theorem 11.2. $\Gamma(x)$ is logarithmically convex.

Proof. Recall the Schwarz inequality[7] for functions whose second power is summable:

$$\left(\int_0^\infty f(t)\, g(t)\, dt\right)^2 \leq \int_0^\infty f^2(t)\, dt \cdot \int_0^\infty g^2(t)\, dt.$$

If we take:

$$f(t) = e^{-\frac{t}{2}}\, t^{\frac{x-1}{2}}, \qquad g(x) = f(x)\, \ln(t),$$

recalling (11.3), (11.6a) and (11.6b), we find the inequality:

$$\left(\Gamma'(x)\right)^2 \leq \Gamma(x)\, \Gamma^{(2)}(x).$$

Hence, we conclude that:

$$\frac{d^2}{dx^2} \ln\left(\Gamma(x)\right) = \frac{\Gamma(x)\, \Gamma^{(2)}(x) - \left(\Gamma'(x)\right)^2}{\Gamma^2(x)} \geq 0.$$

$\square$

Property (11.5) can be iterated, therefore, if $n \in \mathbb{N}$ and $x > 0$:

$$\Gamma(x+n) = (x+n-1)\, (x+n-2) \cdots (x+1)\, x\, \Gamma(x). \tag{11.8}$$

Definition 11.3. The quotient:

$$(x)_n := \frac{\Gamma(x+n)}{\Gamma(x)} = (x+n-1)\, (x+n-2) \cdots (x+1)\, x \tag{11.9}$$

is called *Pochhammer symbol* or *increasing factorial*.

Rewriting (11.8) as:

$$\Gamma(x) = \frac{\Gamma(x+n)}{(x+n-1)\, (x+n-2) \cdots (x+1)\, x} \tag{11.8b}$$

allows the evaluation of $\Gamma(x)$ also for negative values of x, except the integers. For instance, when $x \in\,]-1,0[$, Gamma is given by:

$$\Gamma(x) = \frac{\Gamma(x+1)}{x};$$

in particular:

$$\Gamma\left(-\frac{1}{2}\right) = -2\Gamma\left(\frac{1}{2}\right).$$

---

[7]See, for example, mathworld.wolfram.com/SchwarzsInequality.html

When $x \in] -2, -1[$, the evaluation is:

$$\Gamma(x) = \frac{\Gamma(2-x)}{(x+1)\,x}\,;$$

thus, in particular:

$$\Gamma\left(-\frac{3}{2}\right) = \frac{4}{3}\,\Gamma\left(\frac{1}{2}\right)\,.$$

In other words, $\Gamma(x)$ is defined on the real line, except for the singular points $x = 0,\,-1,\,-2,\cdots$, and so on, as shown in Figure 11.1.

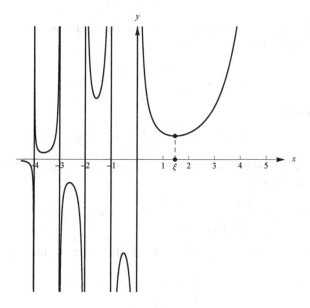

Figure 11.1: Plot of $\Gamma(x)$.

11.2 Beta function

The *Beta* function $B(x,y)$, also called *Eulerian integral of first kind*, is defined as:

$$B(x,y) = \int_0^1 t^{x-1}\,(1-t)^{y-1}\,dt, \qquad x,y > 0. \qquad (11.10)$$

Notice that the change of variable $t = 1 - s$ provides, immediately, the symmetry relation $B(x,y) = B(y,x)$ in (11.10), which yields:

$$B(x,y) = 2 \int_0^{\frac{\pi}{2}} (\cos \vartheta)^{2x-1} (\sin \vartheta)^{2y-1} d\vartheta. \qquad (11.10a)$$

The main property of the Beta function is its relationship with the Gamma function, as expressed by Theorem 11.4 below.

Theorem 11.4. For any real $x,y > 0$, it holds:

$$B(x,y) = \frac{\Gamma(x)\,\Gamma(y)}{\Gamma(x+y)}. \qquad (11.11)$$

Proof. From the usual definition (11.3) for the Gamma function, after the change of variable $t = u^2$, we have:

$$\Gamma(x) = 2 \int_0^{+\infty} u^{2x-1} e^{-u^2} du.$$

In the same way:

$$\Gamma(y) = 2 \int_0^{+\infty} v^{2y-1} e^{-v^2} dv.$$

Now, we form the product of the last two integrals above, and we use Fubini Theorem 10.4, obtaining:

$$\Gamma(x)\Gamma(y) = 4 \iint_{[0,+\infty)\times[0,+\infty)} u^{2x-1} v^{2y-1} e^{-(u^2+v^2)} du\,dv.$$

At this point, we change variable in the double integral, using polar coordinates:

$$\begin{cases} u = \rho \cos \vartheta, \\ v = \rho \sin \vartheta, \end{cases}$$

which leads to the relation:

$$\Gamma(x)\,\Gamma(y) = 4 \left(\int_0^{+\infty} \rho^{2x+2y-1} e^{-\rho^2} d\rho \right) \left(\int_0^{\frac{\pi}{2}} (\cos \vartheta)^{2x-1} (\sin \vartheta)^{2y-1} d\vartheta \right)$$

$$= \Gamma(x+y) \int_0^{\frac{\pi}{2}} 2 (\cos \vartheta)^{2x-1} (\sin \vartheta)^{2y-1} d\vartheta.$$

The thesis follows from (11.10a). □

Remark 11.5. Theorem 11.4 can also be shown using, again, Fubini Theorem 10.4, starting from:

$$\Gamma(x)\,\Gamma(y) = \int_0^\infty \int_0^\infty e^{-(t+s)} t^{x-1} s^{y-1} dt\,ds,$$

and using the changes of variable $t = uv$ and $s = u(1-v)$.

11.2.1 $\Gamma\left(\frac{1}{2}\right)$ and the probability integral

Using (11.11), we can evaluate $\Gamma\left(\frac{1}{2}\right)$ and, then, the probability integral (8.36). Indeed, by taking $x = z$ and $y = 1 - z$, where $0 < z < 1$, we obtain:

$$\Gamma(z)\,\Gamma(1-z) = B(z, 1-z) = \int_0^1 t^{z-1}\,(1-t)^{-z}\,dt = \int_0^1 \left(\frac{t}{1-t}\right)^z \frac{1}{t}\,dt.$$

Now, the change of variable $y = t\,(1-t)^{-1}$, leads to:

$$\Gamma(z)\,\Gamma(1-z) = \int_0^\infty \frac{y^{z-1}}{1+y}\,dy. \tag{11.12}$$

In particular, the choice $z = \dfrac{1}{2}$ yields:

$$\Gamma\left(\frac{1}{2}\right) = \sqrt{\pi}. \tag{11.13}$$

To see it, observe that (11.12) implies:

$$\left(\Gamma\left(\frac{1}{2}\right)\right)^2 = \int_0^\infty \frac{1}{(1+y)\,\sqrt{y}}\,dy, \tag{11.14}$$

where the right–hand side integral is computed by setting $y = x^2$, so that:

$$\left(\Gamma\left(\frac{1}{2}\right)\right)^2 = 2\int_0^\infty \frac{1}{1+x^2}\,dx = 2\lim_{b\to\infty}\,(\arctan b) = \pi.$$

Employing (11.13), it is also possible to evaluate, once more, the probability integral (8.36). Evaluating (11.3) at $x = \dfrac{1}{2}$, and then setting $t = x^2$, we find, in fact:

$$\Gamma\left(\frac{1}{2}\right) = \int_0^\infty \frac{e^{-t}}{\sqrt{t}}\,dt = 2\int_0^\infty e^{-x^2}\,dx,$$

which implies (8.36).

The change of variable $s = t\,(1-t)^{-1}$, used in (11.10), provides an alternative representation for the Beta function:

$$B(x, y) = \int_0^\infty \frac{s^{x-1}}{(1+s)^{x+y}}\,ds. \tag{11.10b}$$

Setting $s = t\,(1-t)^{-1}$, in fact, leads to:

$$B(x, y) = \int_0^1 t^{x-1}\,(1-t)^{y-1}\,dt = \int_0^\infty s^{x-1}\,(1+s)^{-x+1}\,(1+s)^{-2}\,ds$$

$$= \int_0^\infty \frac{s^{x-1}}{(1+s)^{x+y}}\,ds.$$

The following representation Theorem 11.6 describes the family of integrals related to the Beta function.

Theorem 11.6. If $x, y > 0$ and $a < b$, then:

$$\int_a^b (s-a)^{x-1} (b-s)^{y-1} \, ds = (b-a)^{x+y-1} \, B(x,y).$$ (11.15)

Proof. In (11.10), employ the change of variable $t = \dfrac{s-a}{b-a}$, to obtain:

$$B(x,y) = \int_0^1 t^{x-1} (1-t)^{y-1} \, dt$$

$$= \int_a^b (s-a)^{x-1} (b-a)^{-x+1} (b-s)^{y-1} (b-a)^{-y+1} (b-a)^{-1} \, ds$$

$$= (b-a)^{-x-y+1} \int_a^b (s-a)^{x-1} (b-s)^{y-1} \, ds,$$

which ends our argument. □

The particular situation $a = -1$ and $b = 1$ is interesting, since it gives:

$$\int_{-1}^1 (1+s)^{x-1} (1-s)^{y-1} \, ds = 2^{x+y-1} \, B(x,y).$$ (11.15b)

11.2.2   Legendre duplication formula

Legendre formula expresses $\Gamma(2x)$ in terms of $\Gamma(x)$ and $\Gamma(x + \frac{1}{2})$.

Theorem 11.7. It holds:

$$2^{2x-1} \, \Gamma(x) \, \Gamma(x + \tfrac{1}{2}) = \sqrt{\pi} \, \Gamma(2x).$$ (11.16)

Proof. Define the integrals:

$$I = \int_0^{\frac{\pi}{2}} \left(\sin t \right)^{2x} dt \quad \text{and} \quad J = \int_0^{\frac{\pi}{2}} \left(\sin(2t) \right)^{2x} dt.$$

Notice that, with the change of variable $2t = u$ in J, it follows $I = J$. Observe, further, that:

$$I = \frac{1}{2} \, B\left(x + \frac{1}{2}, \frac{1}{2} \right),$$

$$J = \int_0^{\frac{\pi}{2}} (2 \sin t \, \cos t)^{2x} \, dt = 2^{2x-1} \, B\left(x + \frac{1}{2}, x + \frac{1}{2} \right).$$

Hence:

$$\frac{1}{2} \, B\left(x + \frac{1}{2}, \frac{1}{2} \right) = 2^{2x-1} \, B\left(x + \frac{1}{2}, x + \frac{1}{2} \right).$$ (11.17)

Recalling (11.11), equality (11.17) implies (11.16).

□

Formula (11.16) is generalised by the Gauss multiplication Theorem 11.8, which we present without proof.

Theorem 11.8. For each $m \in \mathbb{N}$, the following formula holds true:

$$\Gamma(x) \prod_{k=1}^{m-1} \Gamma\left(x + \frac{k}{m}\right) = m^{\frac{1}{2}-mx} (2\pi)^{\frac{m-1}{2}} \Gamma(mx). \tag{11.18}$$

11.2.3 Euler reflexion formula

This is a famous and beautiful relation, that was found by Euler and that is stated in (11.19). It admits two alternative proofs, which are not presented here, as one uses complex integrals, while the other exploits infinite products.

Theorem 11.9. For any $x \in]0,1[$, it holds:

$$\Gamma(x)\,\Gamma(1-x) = \frac{\pi}{\sin(\pi x)}. \tag{11.19}$$

From the reflexion formula (11.19), it follows immediately the computation of integral (11.20).

Corollary 11.10. For any $x \in]0,1[$, it holds:

$$\int_0^\infty \frac{u^{x-1}}{1+u}\,du = \frac{\pi}{\sin(\pi x)}. \tag{11.20}$$

It is possible to use (11.19) to establish a *cosine* reflexion formula.

$$\Gamma\left(\frac{1}{2}+p\right)\Gamma\left(\frac{1}{2}-p\right) = \frac{\pi}{\sin\left(\left(\frac{1}{2}+p\right)\pi\right)} = \frac{\pi}{\cos(p\pi)} \tag{11.21}$$

in which we must assume $p \neq n + \frac{1}{2}$, for $n = 0,1,\dots$. In terms of the original variable x, we also have:

$$\Gamma(x)\,\Gamma(1-x) = \frac{\pi}{\cos\left(\left(x - \frac{1}{2}\right)\pi\right)}.$$

11.3 Definite integrals

Gamma and Beta functions are extremely useful for the computation of many definite integrals. Here, we present some integrals, that can be found in [40] too: to solve them, reflection formulæ (11.19) and (11.21) are employed. We begin with an integral identity due to Legendre.

Theorem 11.11. If $n \geq 1$ then:

$$\int_0^1 \frac{dx}{\sqrt{1 - x^n}} = \cos\left(\frac{\pi}{n}\right) \int_0^1 \frac{dx}{\sqrt{1 + x^n}} . \tag{11.22}$$

Proof. Legendre established formula (11.22) in equation (z) of his treatise [37]. First observe that, for $n \geq 1$, both integrals converge. Define:

$$I_1 = \int_0^1 \frac{dx}{\sqrt{1 - x^n}} , \qquad I_2 = \int_0^1 \frac{dx}{\sqrt{1 + x^n}} .$$

Employing the change of variable $x^n = t$, in both integrals, yields:

$$I_1 = \frac{1}{n} \int_0^1 \frac{t^{\frac{1}{n} - 1}}{\sqrt{1 - t}} \, dt = \frac{1}{n} B\left(\frac{1}{n}, \frac{1}{2}\right)$$

$$I_2 = \frac{1}{n} \int_0^\infty \frac{t^{\frac{1}{n} - 1}}{\sqrt{1 + t}} \, dt = \frac{1}{n} B\left(\frac{1}{n}, \frac{1}{2} - \frac{1}{n}\right) .$$

In I_2, above, we used the Beta representation (11.10b). Now, form the ratio $\dfrac{I_1}{I_2}$ and exploit Theorem 11.4, to obtain:

$$\frac{I_1}{I_2} = \frac{B\left(\dfrac{1}{n}, \dfrac{1}{2}\right)}{B\left(\dfrac{1}{n}, \dfrac{1}{2} - \dfrac{1}{n}\right)} = \frac{\Gamma\left(\dfrac{1}{n}\right) \Gamma\left(\dfrac{1}{2}\right)}{\Gamma\left(\dfrac{1}{n} + \dfrac{1}{2}\right)} \frac{\Gamma\left(\dfrac{1}{2}\right)}{\Gamma\left(\dfrac{1}{n}\right) \Gamma\left(\dfrac{1}{2} - \dfrac{1}{n}\right)}$$

$$= \frac{\pi}{\Gamma\left(\dfrac{1}{n} + \dfrac{1}{2}\right) \Gamma\left(\dfrac{1}{2} - \dfrac{1}{n}\right)} .$$

Thus, recalling (11.21):

$$\frac{I_1}{I_2} = \frac{\pi}{\dfrac{\pi}{\cos\left(\dfrac{\pi}{n}\right)}} = \cos\left(\frac{\pi}{n}\right) ,$$

which is our statement. □

The same argument followed to demonstrate Theorem 11.11 can be applied to prove the following Theorem 11.12, thus, we leave it as an exercise.

Theorem 11.12. If $2a < n$, then:

$$\int_0^1 \frac{x^{a-1}}{\sqrt{1 - x^n}} \, dx = \cos\left(\frac{a}{n} \pi\right) \int_0^\infty \frac{z^{a-1}}{\sqrt{1 + z^n}} \, dz . \tag{11.23}$$

Theorem 11.13. If $n \in \mathbb{N}$, $n \geq 2$, then:

$$\int_0^1 \frac{dx}{\sqrt[n]{1 - x^n}} = \frac{\pi}{n \sin\left(\dfrac{\pi}{n}\right)} . \tag{11.24}$$

Proof. Using, again, the change of variable $x^n = t$ leads to:

$$\int_0^1 \frac{dx}{\sqrt[n]{1 - x^n}} = \frac{1}{n} \int_0^1 t^{\frac{1}{n} - 1} (1 - t)^{-\frac{1}{n}} \, dt = \frac{1}{n} \int_0^1 t^{\frac{1}{n} - 1} (1 - t)^{\left(1 - \frac{1}{n}\right) - 1} \, dt$$

$$= \frac{1}{n} \, \mathrm{B}\left(\frac{1}{n}, 1 - \frac{1}{n}\right) = \frac{1}{n} \, \Gamma\left(\frac{1}{n}\right) \, \Gamma\left(1 - \frac{1}{n}\right) .$$

Thesis (11.24) follows from reflexion formula (11.19). □

Theorem 11.14. For any $n \geq 2$, it holds:

$$\int_0^\infty \frac{dx}{1 + x^n} = \frac{\pi}{n \sin\left(\dfrac{\pi}{n}\right)} . \tag{11.25}$$

Moreover, if $n - m > 1$, then:

$$\int_0^\infty \frac{x^m}{1 + x^n} \, dx = \frac{\pi}{n \sin\left(\dfrac{\pi}{n}(m + 1)\right)} . \tag{11.26}$$

Proof. The change of variable $1 + x^n = \dfrac{1}{t}$ is employed in the left–hand side integral of formula (11.25), and therefore $dx = -\dfrac{1}{n} \, t^{-\frac{1}{n} - 1} (1 - t)^{\frac{1}{n} - 1} \, dt$:

$$\int_0^\infty \frac{dx}{1 + x^n} = \frac{1}{n} \int_0^1 t^{-\frac{1}{n}} (1 - t)^{\frac{1}{n} - 1} \, dt = \frac{1}{n} \int_0^1 t^{\left(1 - \frac{1}{n}\right) - 1} (1 - t)^{\frac{1}{n} - 1} \, dt$$

$$= \frac{1}{n} \, \mathrm{B}\left(1 - \frac{1}{n}, \frac{1}{n}\right) = \frac{1}{n} \, \Gamma\left(1 - \frac{1}{n}\right) \, \Gamma\left(\frac{1}{n}\right) .$$

Thesis (11.25) follows from reflexion formula (11.19). Formula (11.26) also follows, using an analogous argument. □

Formula (11.27), below, is needed to prove the following Theorem 11.15.

$$\Gamma\left(n - \frac{1}{2}\right) = \frac{\sqrt{\pi}}{2^{n-1}} \prod_{k=1}^{n-1} (2k - 1) = \frac{\sqrt{\pi}}{2^{n-1}} (2n - 3)!! . \tag{11.27}$$

Theorem 11.15. If $n \in \mathbb{N}$, it holds:

$$\int_{-\infty}^\infty \frac{dx}{(1 + x^2)^n} = \frac{\pi}{2^{n-1}} \frac{(2n - 3)!!}{(n - 1)!} . \tag{11.28}$$

Proof. Using the symmetry of the integrand, we have:

$$\int_{-\infty}^{\infty} \frac{dx}{(1+x^2)^n} = 2 \int_0^{\infty} \frac{dx}{(1+x^2)^n} \, .$$

The change of variable $1 + x^2 = \dfrac{1}{t}$, that is, $dx = -\dfrac{1}{2} \, t^{-\frac{3}{2}} \, (1-t)^{-\frac{1}{2}} \, dt$, leads to:

$$\int_{-\infty}^{\infty} \frac{dx}{(1+x^2)^n} = B\left(n - \frac{1}{2}, \frac{1}{2}\right) = \frac{\sqrt{\pi}}{(n-1)!} \, \Gamma\left(n - \frac{1}{2}\right) \, .$$

Exploiting (11.27), we arrive at thesis (11.28). □

By an analogous argument, the following Theorem 11.16 can be demonstrated.

Theorem 11.16. If $np - m > 1$, then:

$$\int_0^{\infty} \frac{x^m}{(1+x^n)^p} \, dx = \frac{\Gamma\left(\dfrac{m+1}{n}\right) \, \Gamma\left(p - \dfrac{m+1}{n}\right)}{n \, \Gamma(p)} \, . \tag{11.29}$$

11.4   Double integration techniques

As it often happens in Analysis, double integration leads to some interesting integral identities, regardless of the order in which the two integrals are evaluated. Here, we obtain a few of such important identities, connecting the Eulerian integrals with double integration reversal. In particular, the *Fresnel integrals* are attained, which are related to the probability integrals, as well as the *Dirichlet integral*, following the presentation in [43].

We start proving the following beautiful identity (11.30), which holds for any $b \in \mathbb{R}$ and for $0 < p < 2$, in order to provide convergence for the integral.

$$\int_0^{+\infty} \frac{\sin(b\,x)}{x^p} \, dx = \frac{\pi \, b^{p-1}}{2 \, \Gamma(p) \, \sin\left(p \, \dfrac{\pi}{2}\right)} \, . \tag{11.30}$$

The starting point, to prove (11.30), is the double integral:

$$I(b, p) = \int_0^{+\infty} \int_0^{+\infty} \sin(b\,x) \, y^{p-1} \, e^{-x\,y} \, dx \, dy \, .$$

The assumption we made for the parameter b ensures summability. Hence, exploiting Fubini Theorem 10.4, the integration can be performed regardless of the order. Let us integrate, first, with respect to x :

$$I(b, p) = \int_0^{+\infty} y^{p-1} \left(\int_0^{+\infty} \sin(b\,x) \, e^{-x\,y} \, dx\right) dy \, .$$

In this way, the integral above turns out to be elementary, since:

$$\int_0^{+\infty} \sin(b\,x)\, e^{-x\,y}\, dx = \frac{b}{b^2 + y^2},$$

and then:

$$I(b,p) = b \int_0^{+\infty} \frac{y^{p-1}}{b^2 + y^2}\, dy,$$

for which, employing the change of variable $t = \dfrac{y}{b}$, we obtain:

$$I(b,p) = b^{p-1} \int_0^{\infty} \frac{t^{p-1}}{1 + t^2}\, dt.$$

The latter formula allows to use identity (11.26) and complete the first computation:

$$I(b,p) = \frac{\pi\, b^{p-1}}{2\, \sin\left(p\, \dfrac{\pi}{2}\right)}. \tag{11.31}$$

Now, reverse the order of integration:

$$I(b,p) = \int_0^{+\infty} \sin(b\,x) \left(\int_0^{+\infty} y^{p-1}\, e^{-x\,y}\, dy \right) dx.$$

The inner integral is immediately evaluated, in terms of the Gamma function, setting $u = x\,y$:

$$\int_0^{+\infty} y^{p-1}\, e^{-x\,y}\, dy = \frac{1}{x^p} \int_0^{\infty} u^{p-1}\, e^{-u}\, du = \frac{1}{x^p}\, \Gamma(p). \tag{11.32}$$

Equating (11.32) and (11.31) leads to:

$$\Gamma(p) \int_0^{\infty} \frac{\sin(b\,x)}{x^p}\, dx = \frac{b^{p-1}\, \pi}{2\, \sin\left(p\, \dfrac{\pi}{2}\right)},$$

which is nothing else but (11.30).

A first consequence of equation (11.30), corresponding to the particular case $b = p = 1$, is the *Dirichlet integral* (8.43), which was obtained in Exercises 8.11.3.

Moreover, from (11.30), it is possible to establish a second integral formula (11.33), which generalises the Dirichlet integral (8.43), and which holds for $q > 1$:

$$\int_0^{\infty} \frac{\sin x^q}{x^q}\, dx = \frac{1}{q-1}\, \Gamma\left(\frac{1}{q}\right) \cos\left(\frac{\pi}{2\,q}\right). \tag{11.33}$$

To prove (11.33), the first step is the quite natural change of variabile $x^q = u$, in the integral:

$$\int_0^{\infty} \frac{\sin x^q}{x^q}\, dx = \frac{1}{q} \int_0^{\infty} \frac{\sin u}{u^{2-\frac{1}{q}}}\, du.$$

The right–hand side integral, above, has the form (11.30), with $b = 1$ and $p = 2 - \dfrac{1}{q}$, therefore:

$$\int_0^\infty \frac{\sin x^q}{x^q}\, dx = \frac{\pi}{2\, q\, \Gamma\left(2 - \dfrac{1}{q}\right)\, \sin\left(\left(2 - \dfrac{1}{q}\right) \dfrac{\pi}{2}\right)}.$$

Now, evaluating the sine:

$$\sin\left(\left(2 - \frac{1}{q}\right) \frac{\pi}{2}\right) = \sin\left(\pi - \frac{\pi}{2q}\right) = \sin\left(\frac{\pi}{2q}\right),$$

and using the reflection formula (11.19):

$$\Gamma\left(2 - \frac{1}{q}\right) = \Gamma\left(1 + 1 - \frac{1}{q}\right) = \left(1 - \frac{1}{q}\right) \Gamma\left(1 - \frac{1}{q}\right) = \frac{q-1}{q} \frac{\pi}{\sin\left(\dfrac{\pi}{q}\right)} \frac{1}{\Gamma\left(\dfrac{1}{q}\right)},$$

we arrive at:

$$\int_0^\infty \frac{\sin x^q}{x^q}\, dx = \frac{1}{2\,(q-1)}\, \Gamma\left(\frac{1}{q}\right) \frac{\sin\left(\dfrac{\pi}{q}\right)}{\sin\left(\dfrac{\pi}{2q}\right)}.$$

Finally, the goniometric identity:

$$\frac{\sin x}{\sin \dfrac{x}{2}} = 2\, \cos \frac{x}{2},$$

implies the equality below, which simplifies to (11.33):

$$\int_0^\infty \frac{\sin x^q}{x^q}\, dx = \frac{1}{2\,(q-1)}\, \Gamma\left(\frac{1}{q}\right)\, 2\, \cos\left(\frac{\pi}{2q}\right).$$

We now show, employing again reversal integration, a cosine relation similar to (11.30), namely:

$$\int_0^\infty \frac{\cos(b\,x)}{x^p}\, dx = \frac{\pi\, b^{p-1}}{2\, \Gamma(p)\, \cos\left(p\, \dfrac{\pi}{2}\right)}, \qquad (11.34)$$

where we must assume that $0 < p < 1$, to ensure convergence of the integral, due to the singularity in the origin. To prove (11.34), we consider the double integral:

$$\int_0^\infty \int_0^\infty \cos(b\,x)\, y^{p-1}\, e^{-x\,y}\, dx\, dy,$$

from which we show that (11.34) can be reached, via the Fubini Theorem 10.4, regardless of the order of integration. The starting point is, then, the equality:

$$\int_0^\infty \cos(b\,x)\left(\int_0^\infty y^{p-1}\,e^{-x\,y}\,dy\right)dx = \int_0^\infty y^{p-1}\left(\int_0^\infty \cos(b\,x)\,e^{-x\,y}\,dx\right)dy.$$

$$(11.35)$$

The inner integral in the right–hand side of (11.35) is elementary:

$$\int_0^\infty \cos(b\,x)\,e^{-x\,y}\,dx = \frac{y}{b^2 + y^2}.$$

Thus, the right–hand side of (11.35) is:

$$\int_0^\infty y^{p-1}\,\frac{y}{b^2+y^2}\,dy = \frac{1}{b^2}\int_0^\infty \frac{y^p}{1+\left(\frac{y}{b}\right)^2}\,dy = b^{p-1}\int_0^\infty \frac{t^p}{1+t^2}\,dt.$$

The latter integral, above, is in the form (11.26). Therefore, the right–hand side integral of (11.35) turns out to be:

$$\int_0^\infty \frac{y^p}{b^2 + y^2}\,dy = \frac{\pi\,b^{p-1}}{2\sin\left(\dfrac{p+1}{2}\,\pi\right)} = \frac{\pi\,b^{p-1}}{2\cos\left(\dfrac{p}{2}\,\pi\right)}. \tag{11.36}$$

The inner integral in the left–hand side of (11.35) is given by (11.32). Hence, the left–hand side integral of (11.35) is:

$$\Gamma(p)\int_0^\infty \frac{\cos(b\,x)}{x^p}\,dx. \tag{11.37}$$

Equating (11.36) and (11.37) leads to (11.34).

There is a further, very interesting consequence of equation (11.30), leading to the evaluation of the *Fresnel* [8] *integrals* (11.38), that holds for $b > 0$ and $k > 1$:

$$\int_0^\infty \sin(b\,x^k)\,dx = \frac{1}{k\,b^{\frac{1}{k}}}\,\Gamma\left(\frac{1}{k}\right)\sin\left(\frac{\pi}{2\,k}\right). \tag{11.38}$$

To prove (11.38), we start from considering its left–hand side integral, inserting into it the change of variable $u = x^k$, i.e., $du = k\,x^{k-1}\,dx = k\,\dfrac{x^k}{x}\,dx$, thus:

$$\int_0^\infty \sin(b\,x^k)\,dx = \int_0^\infty \sin(b\,u)\,\frac{1}{k}\,\frac{u^{\frac{1}{k}}}{u}\,du = \frac{1}{k}\int_0^\infty \frac{\sin(b\,u)}{u^{1-\frac{1}{k}}}\,du.$$

The latter integral, above, is in the form (11.30), with $p = 1 - \dfrac{1}{k}$. Hence:

$$\int_0^\infty \frac{\sin(b\,u)}{u^{1-\frac{1}{k}}}\,du = \frac{\pi\,b^{-\frac{1}{k}}}{2\,\Gamma\left(1-\dfrac{1}{k}\right)\sin\left(\left(1-\dfrac{1}{k}\right)\dfrac{\pi}{2}\right)}, \tag{11.39}$$

---

[8] Augustin–Jean Fresnel (1788–1827), French civil engineer and physicist.

and, thus:

$$\int_0^\infty \sin(b\,x^k)\,dx = \frac{\pi}{2\,k\,b^{\frac{1}{k}}\,\Gamma\left(1 - \frac{1}{k}\right)\,\sin\left(\frac{\pi}{2} - \frac{\pi}{2\,k}\right)}.$$

At this point, we obtain (11.38), by employing the reflection formula (11.19):

$$\Gamma\left(\frac{1}{k}\right)\Gamma\left(1 - \frac{1}{k}\right) = \frac{\pi}{\sin\frac{\pi}{k}},$$

and the goniometric identities:

$$\sin\left(\frac{\pi}{2} - \frac{\pi}{2\,k}\right) = \cos\left(\frac{\pi}{2\,k}\right), \qquad \frac{\sin x}{\cos\frac{x}{2}} = 2\,\sin\frac{x}{2}.$$

In (11.38), the particular choices $b = 1$ and $k = 2$ correspond to the *sine Fresnel integral*:

$$\int_0^\infty \sin(x^2)\,dx = \frac{1}{2}\,\Gamma\left(\frac{1}{2}\right)\,\sin\left(\frac{\pi}{4}\right) = \sqrt{\frac{\pi}{8}}. \tag{11.40}$$

Exploiting the same technique that produced (11.38) from (11.30), it is possible to derive the *cosinus* analogue (11.41) of Fresnel integrals, that holds for $b > 0$ and $k > 1$:

$$\int_0^\infty \cos(b\,x^k)\,dx = \frac{1}{k\,b^{\frac{1}{k}}}\,\Gamma\left(\frac{1}{k}\right)\,\cos\left(\frac{\pi}{2\,k}\right). \tag{11.41}$$

To prove (11.41), the starting point is (11.34). Then, as in the sine case, we introduce the change of variable $u = x^k$, we choose $p = 1 - \frac{1}{k}$, and, via calculations similar to those performed in the sine case, we arrive at:

$$\int_0^\infty \cos(b\,x^k)\,dx = \frac{\pi}{2\,k\,b^{\frac{1}{k}}\,\Gamma\left(1 - \frac{1}{k}\right)\,\cos\left(\frac{\pi}{2} - \frac{\pi}{2\,k}\right)}.$$

Formula (11.41) follows from exploiting the reflection formula (11.19):

$$\Gamma\left(1 - \frac{1}{k}\right) = \frac{\pi}{\Gamma\left(\frac{1}{k}\right)\,\sin\frac{\pi}{k}},$$

and the trigonometric properties:

$$\cos\left(\frac{\pi}{2} - \frac{\pi}{2\,k}\right) = \sin\left(\frac{\pi}{2\,k}\right), \qquad \frac{\sin x}{\sin\frac{x}{2}} = 2\,\cos\frac{x}{2}.$$

Chapter 12

Fourier transform on the real line

A brief presentation of the *Fourier transform* is provided here, limited to those of its aspects that come in handy while integrating a particular partial differential equation, namely, the *heat equation*. The latter constitutes the main tool in solving the *Black-Scholes equation*, which is of great importance in Quantitative Finance. This Chapter 12 is strongly inspired by [44].

12.1 Fourier transform

Definition 12.1. Let f be a real function of a real variable. The Fourier transform of f is the complex–valued function:

$$\boxed{\mathcal{F}f(s) := \int_{-\infty}^{+\infty} e^{-2\pi i s t}\, f(t)\, \mathrm{d}t\,.} \tag{12.1}$$

A sufficient condition for the existence of the Fourier[1] integral is $f \in \mathcal{L}^\infty(\mathbb{R})$. In any Fourier transform, one value is immediate to compute, namely that corresponding to $s = 0$:

$$\mathcal{F}f(0) := \int_{-\infty}^{+\infty} f(t)\, \mathrm{d}t\,.$$

The inverse Fourier transform is realized by a sign change:

$$\mathcal{F}^{-1}g(t) := \int_{-\infty}^{+\infty} e^{2\pi i s t}\, g(s)\, \mathrm{d}s\,. \tag{12.2}$$

The Fourier inversion Theorem 12.2 explaines the inversion process:

Theorem 12.2. Given $f \in \mathcal{L}^1(\mathbb{R})$, then:

$$\mathcal{F}(\mathcal{F}^{-1}g) = g\,, \qquad \mathcal{F}^{-1}(\mathcal{F}f) = f\,.$$

---

[1] Jean Baptiste Joseph Fourier (1768–1830), French mathematician and physicist.

Remark 12.3. A standard definition of the Fourier transform does not exist: the definition presented here is not unique. The reason for non–uniqueness is related to the position of the 2π quantity: it might be part of the exponential, as in (12.1), or it might be an external multiplicative factor, or it might be at all missing. There is also a question of which is the Fourier transform and which is its inverse, that is, where to set the minus sign in the exponential. Various conventions are in common use, according to each particular study branch, and we provide a summary of such conventions, following [35]. Let us consider the general definition:

$$\mathcal{F}f(s) = \frac{1}{A} \int_{-\infty}^{+\infty} e^{iBst} f(t) \, dt \, .$$

The most common choices, found in practice, are the following pairs:

$$A = \sqrt{2\pi}, \qquad\qquad B = \pm 1;$$
$$A = 1, \qquad\qquad B = \pm 2\pi;$$
$$A = 1, \qquad\qquad B = \pm 1.$$

Our choice (12.1) corresponds to $A = 1$, $B = -2\pi$. In computer algebra systems like, for instance, *Mathematica*, the Fourier transform is implemented as:

$$\mathcal{F}_{a,b}f(s) = \sqrt{\frac{|b|}{(2\pi)^{1-a}}} \int_{-\infty}^{\infty} e^{ibst} f(t) \, dt \, .$$

Some results for the Fourier transform are now stated, starting with the *Riemann–Lebesgue* Theorem 12.4, whose proof is omitted for brevity.

Theorem 12.4 (Riemann–Lebesgue). If $f \in \mathcal{L}^1(\mathbb{R})$ then:

$$\lim_{|s|\to\infty} \mathcal{F}(s) = 0 \, .$$

The following *Plancherel*[2] Theorem 12.5 plays a key role in establishing the Fourier transform property in $\mathcal{L}^2(\mathbb{R})$.

Theorem 12.5 (Plancherel). Consider $f \in \mathcal{L}^1(\mathbb{R}) \cap \mathcal{L}^2(\mathbb{R})$. We have that $\mathcal{F}f \in \mathcal{L}^2(\mathbb{R})$, and:

$$\int_{-\infty}^{\infty} |f(t)|^2 \, dt = \int_{-\infty}^{\infty} |\mathcal{F}f(s)|^2 \, ds \, .$$

Finally, Theorem 12.6 illustrates a further interesting property.

Theorem 12.6. Consider $f, g \in \mathcal{L}^1(\mathbb{R})$. Then, it holds that:

$$\int_{-\infty}^{\infty} \mathcal{F}f(s) \, g(s) \, ds = \int_{-\infty}^{\infty} f(x) \, \mathcal{F}g(x) \, dx \, .$$

---

[2]Michel Plancherel (1885–1967), Swiss mathematician.

12.1.1 Examples

Here, some examples are provided, on the computation of the Fourier transform. Before them, let us state some remarks, recalling that *sine* is an odd function, *cosine* is an even function, and $e^{-ix} = \cos(x) - i\sin(x) \quad \forall x \in \mathbb{R}$; recall, also, that the product of two odd (or two even) functions is even.

Remark 12.7. For each even function $f(t) = f(-t)$, the following relation holds:

$$\mathcal{F}f(s) = \int_{-\infty}^{+\infty} \cos(2\pi st)\, f(t)\, dt = 2\int_{0}^{+\infty} \cos(2\pi st)\, f(t)\, dt, \quad (12.3)$$

and, analogously, for each odd function $f(t) = -f(-t)$, it holds:

$$\mathcal{F}f(s) = \int_{-\infty}^{+\infty} \sin(2\pi st)\, f(t)\, dt = 2\int_{0}^{+\infty} \sin(2\pi st)\, f(t)\, dt. \quad (12.4)$$

Example 12.8. The triangle function. Consider the *triangle function*, defined by $\Lambda(x) = \max\{1 - |x|, 0\}$, which is equivalent to the explicit expression:

$$\Lambda(x) = \begin{cases} 1 - |x| & \text{for} \quad |x| \leq 1, \\ 0 & \text{otherwise.} \end{cases}$$

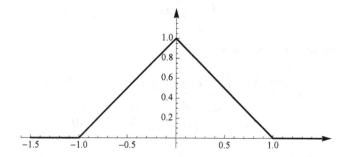

Figure 12.1: Graph of triangle function

To compute the Fourier transform, using the fact that the sine function is odd and Remark 8.58, we evaluate:

$$\mathcal{F}\Lambda(s) = \int_{-\infty}^{+\infty} e^{-2\pi ist}\, \Lambda(t)\, dt$$

$$= \int_{-1}^{1} \left(\cos(2\pi st) - i\sin(2\pi st)\right)(1 - |t|)\, dt$$

$$= \int_{-1}^{1} \cos(2\pi st)\,(1 - |t|)\, dt.$$

Since the cosine function is even, recalling (12.3), it holds:

$$\mathcal{F}\Lambda(s) = 2 \int_0^1 \cos(2\pi s t)\ (1-t)\ dt$$

$$= 2 \left(\frac{\sin(2\pi s)}{2\pi s} - \frac{2\pi s \sin(2\pi s) + \cos(2\pi s) - 1}{4\pi^2 s^2} \right)$$

$$= \frac{1 - \cos(2\pi s)}{2\pi^2 s^2}.$$

Exploiting the trigonometric identity $1 - \cos(2x) = 2\,(\sin x)^2$, we arrive at:

$$\mathcal{F}\Lambda(s) = \left(\frac{\sin(\pi s)}{\pi s} \right)^2.$$

Example 12.9. Exponential even function. Consider $f(t) = e^{-|t|}$; then:

$$\mathcal{F}f(s) = 2 \int_0^{+\infty} \cos(2\pi s t)\ f(t)\ dt = 2 \int_0^{+\infty} \cos(2\pi s t)\ e^{-t}\ dt$$

$$= \left[\frac{2 e^{-t}\,(2\pi s \sin(2\pi s t) - \cos(2\pi s t))}{4\pi^2 s^2 + 1} \right]_{t=0}^{t=+\infty} = \frac{2}{1 + 4\pi^2 s^2}$$

Example 12.10. Gaussian function. Define $f(t) = e^{-\pi t^2}$. Then form:

$$\mathcal{F}f(s) = \int_{-\infty}^{+\infty} e^{-2\pi i s t}\ e^{-\pi t^2}\ dt.$$

Differentiate with respect to s:

$$\frac{d}{ds}\mathcal{F}f(s) = \int_{-\infty}^{+\infty} (-2\pi i t)\ e^{-2\pi i s t}\ e^{-\pi t^2}\ dt$$

$$= \int_{-\infty}^{+\infty} \left(i e^{-\pi t^2} \right)'\ e^{-2\pi i s t}\ dt,$$

and integrate by parts:

$$\frac{d}{ds}\mathcal{F}f(s) = - \int_{-\infty}^{+\infty} i e^{-\pi t^2}\ (-2\pi i s)\ e^{-2\pi i s t}\ dt$$

$$= -2\pi s \int_{-\infty}^{+\infty} e^{-\pi t^2}\ e^{-2\pi i s t}\ dt = -2\pi s\,\mathcal{F}f(s).$$

In other words, $\mathcal{F}f(s)$ satisfies the separable differential equation:

$$\frac{d}{ds}\mathcal{F}f(s) = -2\pi s\,\mathcal{F}f(s),$$

whose unique solution, incorporating the initial condition, is:

$$\mathcal{F}f(s) = \mathcal{F}f(0)\ e^{-\pi s^2}.$$

Since:

$$\mathcal{F}f(0) = \int_{-\infty}^{+\infty} e^{-\pi t^2} dt = 1 \, ,$$

it follows:

$$\mathcal{F}f(s) = e^{-\pi s^2} \, .$$

We have found the remarkable fact that the Gaussian $f(t) = e^{-\pi t^2}$ is equal to its own Fourier transform, that is, the Gaussian function is a fixed point for the Fourier transform as a linear operator.

Remark 12.11. The Fourier transform of the Gaussian function can also be evaluated with a different method, namely, the *square completion of the exponent*, which we now illustrate in detail. Form the Fourier transform for the Gaussian function, according to (12.1):

$$\mathcal{F}f(s) = \int_{-\infty}^{+\infty} e^{-2\pi i s t} \, e^{-\pi t^2} \, dt = \int_{-\infty}^{+\infty} e^{-\pi (2 i s t + t^2)} \, dt \, .$$

Now, rewrite the exponent as:

$$-\pi \left(2 i s t + t^2 \right) = -\pi \left(-s^2 + 2 i s t + t^2 + s^2 \right) = -\pi s^2 - \pi (t + i s)^2 \, .$$

Thus:

$$\mathcal{F}f(s) = e^{-\pi s^2} \int_{-\infty}^{+\infty} e^{-\pi (t + i s)^2} \, dt \, .$$

Employ the change of variable $\sqrt{\pi} \, (t + i s) = \tau$, so that:

$$\mathcal{F}f(s) = \frac{e^{-\pi s^2}}{\sqrt{\pi}} \int_{-\infty}^{+\infty} e^{-\tau^2} \, d\tau \, .$$

Hence:

$$\mathcal{F}f(s) = e^{-\pi s^2} \, .$$

Example 12.12. We follow an ingenious method presented in [43] (on pages 79–82 and leading, there, to integral (3.1.7)), to evaluate the Fourier transform of:

$$f(t) = \frac{1}{b^2 + t^2} \, ,$$

being b a positive parameter.
From Definition (12.1), simplified into (12.3) since the given f is even, we have:

$$\mathcal{F}f(s) = 2 \int_0^\infty \frac{\cos(2 \pi s t)}{b^2 + t^2} \, dt \, .$$

After introducing $x = 2 \pi s$, the parametric integral has to be evaluated:

$$y(x) = \int_0^\infty \frac{\cos(x t)}{b^2 + t^2} \, dt \, . \tag{12.5}$$

Let us integrate (12.5) by parts:

$$x\, y(x) = 2 \int_0^\infty \frac{t\, \sin(x\, t)}{(t^2 + b^2)^2}\, dt\,, \tag{12.5a}$$

and, then, differentiate (12.5a):

$$y(x) + x\, y'(x) = 2 \int_0^\infty \frac{t^2\, \cos(x\, t)}{(t^2 + b^2)^2}\, dt\,. \tag{12.5b}$$

Now, consider the partial decomposition:

$$\frac{t^2}{(t^2 + b^2)^2} = \frac{1}{b^2 + t^2} - \frac{b^2}{(b^2 + t^2)^2}\,,$$

and insert it into (12.5b), to obtain:

$$y(x) + x\, y'(x) = 2\, y(x) - 2\, b^2 \int_0^\infty \frac{\cos(x\, t)}{(t^2 + b^2)^2}\, dt\,,$$

that is:

$$x\, y'(x) - y(x) = 2\, b^2 \int_0^\infty \frac{\cos(x\, t)}{(t^2 + b^2)^2}\, dt\,. \tag{12.5c}$$

We now differentiate (12.5c):

$$x\, y''(x) = 2\, b^2 \int_0^\infty \frac{t\, \sin(x\, t)}{(t^2 + b^2)^2}\, dt\,,$$

which means that, recalling (12.5a), we have arrived at the second–order linear differential equation:

$$y''(x) = b^2\, y(x)\,,$$

whose general solution is:

$$y(x) = c_1\, e^{b\, x} + c_2\, e^{-b\, x}\,.$$

To determine constants c_1, c_2, two conditions are needed; we can obtain the first one from (12.5), that can be evaluated at $x = 0$, yielding:

$$y(0) = \int_0^\infty \frac{dt}{b^2 + t^2} = \frac{1}{b}\left[\arctan \frac{t}{b} \right]_0^\infty = \frac{\pi}{2b}\,, \tag{12.6}$$

which implies $c_1 + c_2 = \dfrac{\pi}{2b}$. Moreover, from (12.5a), we see that:

$$y(x) = \frac{2}{x} \int_0^\infty \frac{t\, \sin(x\, t)}{(t^2 + b^2)^2}\, dt\,,$$

and, thus:

$$\lim_{x \to +\infty}\, y(x) = 0\,. \tag{12.7}$$

Since $b > 0$, condition (12.7) implies $c_1 = 0$, and this means that:

$$y(x) = \int_0^\infty \frac{\cos(x\,t)}{b^2 + t^2}\,dt = \frac{\pi}{2b}\,e^{-bx}.$$ (12.8)

We can finally use (12.8), assuming s real and positive, to obtain the Fourier transform of the initial $f(t)$:

$$\mathcal{F}f(s) = \frac{\pi}{b}\,e^{-2\pi b s}.$$

12.2 Properties of the Fourier transform

The Fourier transform has several useful properties.

12.2.1 Linearity

One of the simplest, and most frequently invoked properties of the Fourier transform is that it is a linear operator. This means:

$$\mathcal{F}(f + g)(s) = \mathcal{F}f(s) + \mathcal{F}g(s)$$

$$\mathcal{F}(\alpha\,f)(s) = \alpha\,\mathcal{F}f(s).$$

where α is any real or complex number.

12.2.2 The Shift theorem

A shift of the variable t, which represents a time delay in many applications, has a simple effect on the Fourier transform. Consider evaluating the Fourier transform of $f(t + b)$, for any constant b. We introduce the special notation $f_b(t) = f(t + b)$, and then perform the following:

$$\mathcal{F}f_b(s) = \int_{-\infty}^{+\infty} f(t + b)\,e^{-2\pi i s t}\,dt = \int_{-\infty}^{+\infty} f(u)\,e^{-2\pi i s (u-b)}\,du$$

$$= e^{2\pi i s b} \int_{-\infty}^{+\infty} f(u)\,e^{-2\pi i s u}\,du = e^{2\pi i s b}\,\mathcal{F}f(s).$$

12.2.3 The Stretch theorem

How does the Fourier transform vary, if we stretch or shrink the variable t in the domain? More precisely, when t is scaled to become $a\,t$, we want to know what happens to the Fourier transform of $_af(t) = f(at)$. First, assume that $a > 0$. Then:

$$\mathcal{F}_a f(s) = \int_{-\infty}^{+\infty} f(at)\ e^{-2\pi i s t}\ dt = \frac{1}{a} \int_{-\infty}^{+\infty} f(u)\ e^{-2\pi i s \frac{u}{a}}\ du$$

$$= \frac{1}{a} \int_{-\infty}^{+\infty} f(u)\ e^{-2\pi i u \frac{s}{a}}\ du = \frac{1}{a} \mathcal{F} f\left(\frac{s}{a}\right).$$

When $a < 0$, the limits of integration are reversed, when we insert the substitution $u = at$, thus, the resulting transform is:

$$\mathcal{F}_a f(s) = -\frac{1}{a} \mathcal{F} f\left(\frac{s}{a}\right).$$

We can combine the two cases and present the Stretch theorem as follows, assuming $a \neq 0$:

$$\mathcal{F}_a f(s) = \frac{1}{|a|} \mathcal{F} f\left(\frac{s}{a}\right).$$

For instance, recalling Example 12.9, the Fourier transform of $g(t) = e^{-a|t|}$, with a positive constant, is:

$$\mathcal{F} g(s) = \frac{2a}{a^2 + 4\pi^2 s^2}.$$

12.2.4   Combining shifts and stretches

We can combine Shift and Stretch theorems, to find the Fourier transform of $f(at + b)$, denoted below as $_a f_b(t)$:

$$\mathcal{F}_a f_b(s) = \frac{1}{|a|}\ e^{2\pi i s \frac{b}{a}}\ \mathcal{F} f\left(\frac{s}{a}\right).$$

Example 12.13. We use the four properties of the Fourier transform to show that, given:

$$f(t) = \frac{1}{\sigma\sqrt{2\pi}}\ e^{-\frac{t^2}{2\sigma^2}},$$

its Fourier transform is:

$$\mathcal{F} f(s) = e^{-2\pi^2 \sigma^2 s^2}$$

We know that, if $g(t) = e^{-\pi t^2}$, then $\mathcal{F} f(s) = e^{-\pi s^2}$. Moreover, using the Stretch theorem:

$$\mathcal{F}_a g(s) = \frac{1}{|a|} \mathcal{F} g\left(\frac{s}{a}\right).$$

Since we want to find a such that $g(at) = e^{-\frac{1}{2\sigma^2} t^2}$, the following relation must hold:

$$-\pi a^2 t^2 = -\frac{1}{2\sigma^2} t^2,$$

that is:

$$a = \frac{1}{\sigma \sqrt{2\pi}}.$$

From relation:

$$\mathcal{F}_a g(s) = \frac{1}{|a|} \mathcal{F} g\left(\frac{s}{a}\right),$$

it finally follows:

$$\mathcal{F}_a g(s) = \sigma \sqrt{2\pi} \, e^{-\pi s^2 \, 2\pi \sigma}.$$

12.3 Convolution

Convolution is an operation which combines two functions, f, g, producing a third function; as shown in (12.9), it is defined as the integral of the pointwise multiplication of f and g, and it has independent variable given by the amount by which either f or g is translated. Convolution finds applications that include Probability, Statistics, Computer vision, Natural language processing, Image and signal processing, Engineering, and differential equations. Our interest, here, is in the last application.

Definition 12.14. The convolution of two functions $g(t)$ and $f(t)$, both defined on the entire real line, is the function defined as:

$$(g \star f)(t) = \int_{-\infty}^{+\infty} g(t - x) \, f(x) \, dx. \qquad (12.9)$$

Remark 12.15. Consider the case in which functions $g(t)$ and $f(t)$ are supported only on $[0, \infty)$, that is, they are zero for negative arguments. Then, the integration limits can be truncated, and the convolution is:

$$(g \star f)(t) = \int_0^t g(t - x) \, f(x) \, dx. \qquad (12.10)$$

Remark 12.16. Convolution is a commutative operation:

$$(g \star f)(t) = (f \star g)(t).$$

This follows from a simple change of variabile, namely $t - x = u$, in which x and u play the role of integration variables, while t acts as a parameter:

$$(g \star f)(t) = \int_{-\infty}^{\infty} g(t - x) \, f(x) \, dx = \int_{\infty}^{-\infty} g(u) \, f(t - u) \, (-du) = (f \star g)(t).$$

Some practical examples of convolution of functions are now provided, in particular to show that this concept is of great importance in the algebra of random variables. Let us begin with the convolution of two power function.

Example 12.17. (Convolution of powers) Consider $g, f : [0, +\infty) \to \mathbb{R}$, respectively defined by $g(x) = x^a$, $f(x) = x^b$, with $a, b > 0$. Then:

$$(g \star f)(t) = t^a \star t^b = \int_0^t g(t - x) \, f(x) \, dx = \int_0^t (t - x)^a \, x^b \, dx.$$

The last integral above can be computed via the Beta function; to see this, let us rewrite it as:

$$\int_0^t (t - x)^a \, x^b \, dx = t^a \int_0^t \left(1 - \frac{x}{t}\right)^a x^b \, dx,$$

so that, after the change of variable $x = tu$, we recognise the Eulerian integral (11.10):

$$t^a \int_0^t \left(1 - \frac{x}{t}\right)^a x^b \, dx = t^a \int_0^1 (1 - u)^a \, t^b \, u^b \, t \, du = t^{a+b+1} \, B(a+1, b+1).$$

In terms of Gamma functions, recalling relation (11.11):

$$t^a \star t^b = t^{a+b+1} \frac{\Gamma(a + 1)\,\Gamma(b + 1)}{\Gamma(a + b + 2)}. \tag{12.11}$$

Formula (12.11) is more expressive, and easy to remember, rewritten as:

$$\frac{t^a}{\Gamma(a + 1)} \star \frac{t^b}{\Gamma(b + 1)} = \frac{t^{a+b+1}}{\Gamma(a + b + 2)},$$

which, when $a = n \in \mathbb{N}$ and $b = m \in \mathbb{N}$, becomes:

$$\frac{t^n}{n!} \star \frac{t^m}{m!} = \frac{t^{n+m+1}}{(n + m + 1)!}.$$

Now, we consider the convolution of two exponential functions.

Example 12.18. (Convolution of exponentials) Let $g, f : [0, +\infty) \to \mathbb{R}$ be respectively defined by $g(x) = e^{ax}$, $f(x) = e^{bx}$. Assume that $a \neq b$. Then:

$$(g \star f)(t) = e^{at} \star e^{bt} = \int_0^t g(t - x) \, f(x) \, dx = \int_0^t e^{a(t-x)} \, e^{bx} \, dx = \frac{e^{at} - e^{bt}}{a - b}.$$

When $a = b$, it is, instead:

$$(g \star f)(t) = e^{at} \star e^{at} = \int_0^t g(t - x) \, f(x) \, dx = \int_0^t e^{a(t-x)} \, e^{ax} \, dx = t e^{at}.$$

Convolution behaves nicely with respect to the Fourier transform, as shown by the following result, known as *Convolution theorem for Fourier transform*.

Theorem 12.19. If $g(t)$ and $f(t)$ are both summable, then:

$$\boxed{\mathcal{F}(g \star f)(s) = \mathcal{F}g(s) \ \mathcal{F}f(s).}$$ (12.12)

Proof. The proof is straightforward and uses Fubini Theorem 10.4, that is why we assume summability of both functions. We can then write:

$$\mathcal{F}(g \star f)(s) = \int_{-\infty}^{\infty} (g \star f)(t) \ e^{-2\pi i s t} \ dt$$

$$= \int_{-\infty}^{\infty} \left(\int_{-\infty}^{\infty} g(t-x) \ f(x) \ e^{-2\pi i s t} \ dx \right) dt$$

$$= \int_{-\infty}^{\infty} \left(\int_{-\infty}^{\infty} g(t-x) \ f(x) \ e^{-2\pi i s t} \ dt \right) dx$$

$$= \int_{-\infty}^{\infty} f(x) \left(\int_{-\infty}^{\infty} g(t-x) \ e^{-2\pi i s t} \ dt \right) dx.$$

With the change of variable $t - x = u$, so that $dt = du$, we obtain:

$$\mathcal{F}(g \star f)(s) = \int_{-\infty}^{\infty} f(x) \left(\int_{-\infty}^{\infty} g(u) \ e^{-2\pi i s (u+x)} \ du \right) dx$$

$$= \int_{-\infty}^{\infty} f(x) \ e^{-2\pi i s x} \left(\int_{-\infty}^{\infty} g(u) \ e^{-2\pi i s u} \ du \right) dx$$

$$= \mathcal{F}g(s) \int_{-\infty}^{\infty} f(x) \ e^{-2\pi i s x} \ dx = \mathcal{F}g(s) \ \mathcal{F}f(s),$$

and this completes the proof.

$\square$

Remark 12.20. The convolution of Fourier transforms has another interesting property, namely:

$$\mathcal{F}(f \ g)(s) = (\mathcal{F}f \star \mathcal{F}g)(s),$$

where we have to assume that functions $f, g \in \mathcal{L}^1(\mathbb{R})$ are such that their product is also $f g \in \mathcal{L}^1(\mathbb{R})$.

12.4 Linear ordinary differential equations

Fourier transform is useful in solving linear ordinary differential equations. To such an aim, a formula is needed, for the Fourier transform of the derivative: it is given in Theorem 12.21.

Theorem 12.21. If $f \in \mathcal{L}^1(\mathbb{R})$ is a differentiable function, then:

$$\mathcal{F}f'(s) = 2\pi i s \, \mathcal{F}f(s). \tag{12.13}$$

Proof. To obtain formula (12.13), it suffices to evaluate the Fourier transform of $f'(t)$:

$$\mathcal{F}f'(s) = \int_{-\infty}^{+\infty} e^{-2\pi i s t} \, f'(t) \, dt,$$

and integrate by parts. $\qquad \square$

Note that differentiation is transformed into multiplication: this represents another remarkable feature of the Fourier transform, providing one more reason for its usefulness.

Formulæ for derivatives of high order also hold, and the relevant result follows by mathematical induction:

$$\mathcal{F}f^{(n)}(s) = (2\pi i s)^n \, \mathcal{F}f(s). \tag{12.14}$$

The derivative Theorem 12.21. is useful for solving linear ordinary, and partial, differential equations. Example 12.22 illustrates its use with an ordinary differential equation.

Example 12.22. Consider the ordinary differential equation:

$$u'' - u = -f,$$

where $f(t)$ is a given function. The problem consists in finding $u(t)$. Form the Fourier transform of both sides of the stated equation:

$$(2\pi i s)^2 \, \mathcal{F}u - \mathcal{F}u = -\mathcal{F}f,$$

and solve with respect to $\mathcal{F}u$:

$$\mathcal{F}u = \frac{1}{1 + 4\pi^2 s^2} \, \mathcal{F}f.$$

Now, observe that quantity $\dfrac{1}{1 + 4\pi^2 s^2}$ is the Fourier transform of $\dfrac{1}{2 \, e^{-|t|}}$, that is,

$$\mathcal{F}u = \mathcal{F}\left(\frac{1}{2} e^{-|t|}\right) \mathcal{F}f$$

The right–hand side, in the above expression, is the product of two Fourier transforms. Therefore, according to the Convolution Theorem 12.19:

$$u(t) = \frac{1}{2} \, e^{-|t|} \star f(t).$$

Written out in full, the solution is:

$$u(t) = \frac{1}{2} \int_{-\infty}^{+\infty} e^{-|t-\tau|} \, f(\tau) \, d\tau.$$

12.5 Exercises

1. Consider the two functions $f, g : [0, +\infty) \to \mathbb{R}$, respectively defined as $f(t) = \sin(t) \, e^{-t}$, $g(t) = \cos(t) \, e^{-t}$. Show that:

$$(g \star f)(t) = \frac{1}{2} t \, e^{-t} \, \sin(t) \, .$$

2. Consider the two functions $f, g : [0, +\infty) \to \mathbb{R}$, respectively defined as $f(t) = \sin(t) \, e^{-2t}$, $g(t) = \cos(t) \, e^{-2t}$. Show that:

$$(g \star f)(t) = \frac{1}{2} t \, e^{-2t} \, \sin(t) \, .$$

3. Show that the Fourier transform of the function $f(t) = e^{-t^2 + t - 1}$ is:

$$\mathcal{F}f(s) = \sqrt{\pi} \, e^{-\pi^2 s^2 - i \pi s - \frac{3}{4}} \, .$$

Chapter 13

Parabolic equations

The Fourier transform method of Chapter 12 is used here, among other methods, to solve partial differential equations of parabolic type, which are of fundamental importance in Mathematical Finance. The exposition presented in this Chapter 13 exploits the material contained in various references; in particular, we mention [14, 54], Chapter 6 of [1], § 2.4 of [8], Chapter 4 of [22], Chapter 4 of [56], and Chapter 6 of [58].

13.1 Partial differential equations

A *partial differential equation* (PDE) is a relation that involves partial derivatives of a function to be determined. Denote by u such an unknown function, and let $t, x, y, \ldots$ and so on, be its independent variables, that is, $u = u(t, x, y, \ldots)$. Among these variables, t often represents *time*, and the expressions we deal with have the following implicit form:

$$F(t, x, y, \ldots, u, u_t, u_x, u_y, \ldots, u_{tt}, u_{tx}, \ldots, u_{ttt}, \ldots) = 0, \qquad (13.1)$$

where the subscript notation indicates partial differentiation:

$$u_t = \frac{\partial u}{\partial t}, \quad u_{tx} = \frac{\partial^2 u}{\partial t\, \partial x}, \quad \cdots\cdots .$$

We assume, always, that the unknown function u is sufficiently well behaved, so that all the necessary partial derivatives exist and the corresponding mixed partial derivatives are equal. As in the case of ordinary differential equations, we define the *order* of (13.1) to be that of the highest–order partial derivative appearing in the equation. Furthermore, we say that (13.1) is *linear* if F is linear as a function of $u, u_t, u_x, u_y, u_{tt}, \cdots$ etcetera, that is to say, F is a linear combination of the unknown u and its derivatives.

Examples of partial differential equations are $u_x + u_y = 3\, u_t - 2\, x^2 - 5\, t$, which is first–order and linear, and $u_{xx} + u_y = x^2$, that is second–order and linear. A *solution* of (13.1) is a continuous function $u = u(t, x, y, \cdots)$ that has continuous partial derivatives and that, when substituted in (13.1), reduces equation (13.1) to an identity. For instance, $u(t, x) = x\, e^{x-t}$ solves equation $F(t, x, u, u_t, u_x, u_{xx}) = u_{xx} - 2\, u_x - u_t = 0$.

Example 13.1. Consider the first–order partial differential equation in the unknown $u = u(x, y)$:

$$u_x + u_y = 0.$$

It is possible to show that a solution is:

$$u = \phi(x - y),$$

where ϕ is any function, of $x - y$, having continuous first–order partial derivatives. Indeed, since:

$$u_x = \phi'(x - y) \quad \text{and} \quad u_y = -\phi'(x - y),$$

it immediately follows:

$$u_x + u_y = \phi'(x - y) - \phi'(x - y) = 0.$$

13.1.1 Classification of second–order linear partial differential equations

Taking into account the nature of the PDE applications, that are of interest here, it is useful to present a classification of second–order linear partial differential equations, where the unknown $u = u(x, y)$ is a function of two independent variables, i.e., PDEs of the form:

$$\begin{aligned} \mathcal{L}[u] := {}& A(x, y)\, u_{xx} + B(x, y)\, u_{xy} + C(x, y)\, u_{yy} \\ & + D(x, y)\, u_x + E(x, y)\, u_y + F(x, y)\, u - G(x, y) = 0. \end{aligned} \tag{13.2}$$

In the linear operator $\mathcal{L}[u]$, above, functions $A(x, y), \ldots, G(x, y)$ are continuous in some open set $\Omega \subset \mathbb{R}^2$. Now, recall that the quadratic equation:

$$a\, x^2 + b\, x\, y + c\, y^2 + d\, x + e\, y + f = 0$$

represents a hyperbola, parabola, or ellipse, according to its discriminant:

$$\Delta = b^2 - 4\, a\, c$$

being positive, zero, or negative. In an analogous way, the differential operator $\mathcal{L}$ and the partial differential equation (13.2) are said to be *hyperbolic*, *parabolic*, or *elliptic*, at a point $(x_0, y_0) \in \Omega$, according to the discriminant:

$$\Delta(x, y) = B^2(x, y) - 4\, A(x, y)\, C(x, y) \tag{13.3}$$

being positive, zero, or negative, when (13.3) is evaluated at $(x, y) = (x_0, y_0)$. Furthermore, $\mathcal{L}$ and the PDE in (13.2) are called hyperbolic, or parabolic, or elliptic, on a *domain* $\Omega \subset \mathbb{R}^2$, if their discriminant (13.3) is positive, or zero, or negative, at each point of Ω.

Example 13.2. The *wave equation* in one dimension:

$$u_{tt} = c^2 u_{xx}$$

is hyperbolic in any domain, since:

$$A = -c^2, \qquad B = 0, \qquad C = 1,$$

so that $\Delta = B^2 - 4AC = 4c^2 > 0$.

Example 13.3. The one–dimensional *heat equation*, which is the main object of our study in § 13.2:

$$u_t = c\, u_{xx}, \qquad c > 0, \tag{13.4}$$

is parabolic in any domain, since:

$$A = -c, \qquad B = 0, \qquad C = 0,$$

so that $\Delta = B^2 - 4AC = 0$.

Example 13.4. The *potential*, or Laplace equation, in two dimensions:

$$u_{xx} + u_{yy} = 0 \tag{13.5}$$

is elliptic in any domain since:

$$A = 1, \qquad B = 0, \qquad C = 1,$$

so that $\Delta = B^2 - 4AC < 0$.

It is useful, here, to introduce the so–called *Laplacian operator* $\nabla^2 u$, for a two–variable function $u \in C^2$, defined as:

$$\nabla^2 u = u_{xx} + u_{yy}. \tag{13.6}$$

Given $u, v \in C^2$, we can apply (13.6), recalling the nabla definition in Theorem 3.8, and the dot (inner) product introduced in Definition 1.1, to show that:

$$\nabla^2(u\,v) = \left(\nabla^2 u\right) v + 2\left(\nabla u\right) \bullet \left(\nabla v\right) + u\left(\nabla^2 v\right).$$

A C^2 function u is called *harmonic* if:

$$\nabla^2 u = 0.$$

Functions $u(x,y) = y^3 - 3x^2 y$ and $u(x,y) = e^{kx} \sin(ky)$ constitute examples of harmonic functions.

13.2   The heat equation

In this section, we treat the one–dimensional *heat equation*, that is a partial differential equation of the form:

$$u_t(x,t) = c\ u_{xx}(x,t) + P(x,t)\,, \qquad x \in \mathbb{R}, \quad t > 0\,, \qquad (13.7)$$

where $P(x,t)$ is a given real function of the two independent variables (x,t). The n–dimensional heat equation is:

$$u_t(\boldsymbol{x},t) = c\ \nabla_{\boldsymbol{x}}^2\, u(\boldsymbol{x},t) + P(\boldsymbol{x},t)\,, \qquad \boldsymbol{x} \in \mathbb{R}^n, \quad t > 0\,, \qquad (13.8)$$

where the Laplace differential operator $\nabla_{\boldsymbol{x}}^2$ acts on the so–called *state variables* $\boldsymbol{x} = (x_1,\dots,x_n)$.

The heat equation is of great importance in Mathematical Physics, and it is of high interest, also, in Physics and Probability. Here, only the one–dimensional case is treated, as it is useful in Mathematical Finance. We begin with considering the homogeneous case, as well as some equations that can be transformed into homogeneous ones. We then examine non–homogeneous equations, introducing the *Duhamel principle* in § 13.5.

13.2.1   Uniqueness of solution: homogeneous case

For Mathematical Finance applications, we have to associate a given parabolic partial differential equation with an appropriate initial condition. We treat, here, the homogeneous version of equation (13.7), which means that we make the assumption that $P(x,t) = 0$, for any (x,t). We further assume that $c = 1$ in (13.7), since this does not affect the generality of our argument, as shown in Remark 13.5. Hence, we deal with the Cauchy problem:

$$\begin{cases} u_t(x,t) = u_{xx}(x,t)\,, & \text{for} \quad t > 0\,, \\[2mm] u(x,0) = f(x)\,, \end{cases} \qquad (13.9)$$

where both $f(x)$ and $u(x,t)$ are defined for $x \in \mathbb{R}$, that is, $-\infty < x < +\infty$. The parabolic partial differential equation in (13.9):

$$u_t = u_{xx}\,, \qquad (13.10)$$

is sometimes called *diffusion equation*, given its fundamental connections with *Brownian motion*, i.e., the limit of a *random walk*, which is a mathematical formalisation of a path that contains random steps, and which turns out to be linked to the heat equation [36].

Remark 13.5. Generality is not affected by considering problem (13.9) instead of the following problem (13.11), in which $c \neq 1$ and $x \in \mathbb{R}$:

$$\begin{cases} u_t(x,t) = c\, u_{xx}(x,t) \,, & \text{for } t > 0, \\ u(x,0) = f(x) \,. \end{cases} \tag{13.11}$$

To see it, let us assume that $u(x,t)$ solves $u_t = c\, u_{xx}$, and introduce $w(x,t) = u\left(x,\dfrac{t}{c}\right)$. Then, w solves $w_t = w_{xx}$, since it verifies:

$$w_t(x,t) = \frac{1}{c}\, u_t\left(x,\frac{t}{c}\right) = \frac{1}{c}\, c\, u_{xx}\left(x,\frac{t}{c}\right) = u_{xx}\left(x,\frac{t}{c}\right) = w_{xx}(x,t).$$

This means that, without any loss of generality, we can assume $c = 1$, and deal only with (13.9).

The first step towards solving the Cauchy problem (13.9) is to establish a uniqueness result for its solution. To such an aim, some assumptions are needed, as stated in the following *Energy* Theorem 13.6.

Theorem 13.6. Problem (13.9) admits solution $u \in C^2(\mathbb{R},[0,\infty))$, which is unique and satisfies the asymptotic condition:

$$\lim_{|x| \to \infty} u_x(x,t) = 0. \tag{13.12}$$

Proof. Assume that function $u(x,t)$ satisfies the differential equation in (13.9). Define the function:

$$W(t) = \frac{1}{2} \int_{-\infty}^{\infty} u^2(x,t)\, \mathrm{d}x.$$

$W(t)$ is called the *energy* of the solution to the heat equation in (13.9). Now, assume that there exist two solutions to problem (13.9), say, $r(x,t)$ and $s(x,t)$, and assume that both r and s satisfy condition (13.12). At this point, if we define u to be the function $u(x,t) = r(x,t) - s(x,t)$, since the heat equation in (13.9) is linear, we see that u solves, for $x \in \mathbb{R}$:

$$\begin{cases} u_t(x,t) = u_{xx}(x,t) \,, & \text{for } t > 0, \\ u(x,0) = 0 \,, \\ \lim_{|x| \to \infty} u_x(x,t) = 0 \,. \end{cases} \tag{13.13}$$

We aim to prove that the unique solution to (13.13) is the zero function, as such a result implies the thesis of Theorem 13.6. Now, condition (13.12) allows to differentiate the energy function $W(t)$:

$$W'(t) = \int_{-\infty}^{\infty} u(x,t)\, u_t(x,t)\, \mathrm{d}x = \int_{-\infty}^{\infty} u(x,t)\, u_{xx}(x,t)\, \mathrm{d}x.$$

In the last equality step, above, we used the assumption that $u(x,t)$ solves the heat equation in (13.9). Integrating by parts, and using condition (13.12), leads to:

$$W'(t) = -\int_{-\infty}^{\infty} u_x^2(x,t) \, dx \leq 0,$$

which means that $W(t)$ is a decreasing function. On the other hand, we know, a priori, that $W(t) \geq 0$. Evaluating $W(0) = 0$, we see that it must be $W(t) = 0$ for any $t > 0$. Hence, it must also be $u(x,t) = 0$ for any $t > 0$, which shows that solutions $r(x,t)$ and $s(x,t)$ are equal.   □

It is possible to state a deeper result then Theorem 13.6, but its proof require tools that go beyond the undergraduate curriculum in Economics and Management; hence, we just state such a result in Theorem 13.7, and refer to Theorem 4.4 of [22] for its proof. We remark that hypothesis (13.12) is replaced by conditions (13.14)–(13.15).

Theorem 13.7. Assume that u is a real continuous function on $\mathbb{R} \times (0,\infty)$, that u is C^2 on $\mathbb{R} \times [0,\infty)$, and that $u(x,t)$ solves the following Cauchy problem, where $x \in \mathbb{R}$:

$$\begin{cases} u_t = u_{xx}, & \text{for } t > 0, \\ u(x,0) = 0. \end{cases}$$

If, for any $\varepsilon > 0$, there exists $C > 0$ such that, for any $(x,t) \in \mathbb{R} \times [0,\infty)$:

$$|u(x,t)| \leq C \, e^{\varepsilon \, x^2}, \tag{13.14}$$

$$|u_x(x,t)| \leq C \, e^{\varepsilon \, x^2}, \tag{13.15}$$

then $u(x,t)$ is identically zero on $\mathbb{R} \times [0,\infty)$.

13.2.2   Fundamental solutions: heat kernel

In force of Theorems 13.6 and 13.7, the Cauchy problem (13.9) can now be solved via analytical methods, since, if a solution is found, then it has to be unique. To understand the type of solutions that are obtained, it is interesting to take Remarks 13.8 into account: among them, in particular, the last one is called *scale invariance* and it represents the most important property of solutions to (13.10).

Remarks 13.8.

- If $u(x,t)$ solves (13.10), then function $(x,t) \mapsto u(x-y,t)$ also solves (13.10), for any y.

- If $u(x,t)$ solves (13.10), then any of its derivatives, u_t, u_x, u_{tt}, $\cdots$ etcetera, is also a solution to (13.10).

- Any linear combination of solutions of (13.10) is also a solution to (13.10).

- *Scale invariance.* If $u(x,t)$ solves (13.10), then $(x,t) \mapsto u(\alpha x, \alpha^2 t)$ is a function that also solves (13.10), for any $\alpha > 0$.

The scale invariance property provides an educated guess on the structure of the solutions to (13.10); namely, we look for solutions of the form:

$$u(x,t) = w\left(\frac{x}{\sqrt{t}}\right), \tag{13.16}$$

being $w = w(r)$ a differentiable real function of a real variable. Differentiating (13.16), we obtain:

$$u_t(x,t) = -\frac{x}{2t\sqrt{t}}\, w'\left(\frac{x}{\sqrt{t}}\right), \qquad u_{xx}(x,t) = \frac{1}{t}\, w''\left(\frac{x}{\sqrt{t}}\right).$$

Equating $u_t(x,t)$ and $u_{xx}(x,t)$ yields a second–order linear differential equation, with variable coefficients, in the unknown $w = w(r)$, where $r := \dfrac{x}{\sqrt{t}}$:

$$w''(r) = -\frac{r}{2}\, w'(r),$$

which can be integrated via separation of variables, in the unknown $w'(r)$:

$$w'(r) = w'(0)\, e^{-\frac{r^2}{4}}. \tag{13.17}$$

Integration of (13.17) provides:

$$w(r) = w'(0) \int_0^r e^{-\frac{s^2}{4}}\, ds + w(0),$$

that, recovering the original coordinates, can be rewritten as:

$$u(x,t) = k_1 \int_0^{\frac{x}{\sqrt{t}}} e^{-\frac{s^2}{4}}\, ds + k_2, \tag{13.18}$$

being k_1 and k_2 arbitrary constants of integration. Due to its construction, function (13.18) solves (13.10) for $t > 0$, and so does, by Remarks 13.8, its partial derivative, with respect to x :

$$h(x,t) := \frac{\partial u}{\partial x}(x,t) = k_1\, \frac{e^{-\frac{x^2}{4t}}}{\sqrt{t}}. \tag{13.19}$$

To fix one solution, we choose the integration constant k_1 such that:

$$\int_{-\infty}^{\infty} h(x,t)\, dx = 1.$$

Recalling the probability integral computation, illustrated in § 8.11.1 and § 11.2.1 and given by formula (8.36), we take $k_1 = \dfrac{1}{\sqrt{4\pi}}$ and obtain the particular solution to (13.10) known as *heat kernel* (or *Green function* or fundamental solution for the heat equation) and given by the Gaussian function:

$$\mathcal{H}(x,t) = \frac{1}{\sqrt{4\pi t}}\, e^{-\frac{x^2}{4t}}. \tag{13.20}$$

This preliminary discussion is essential to understand the following integral transform approach to solving the Cauchy problem (13.9). Starting from the Fourier transform of the Gaussian function (see Examples 12.10 and 12.13), the idea is to compute the Fourier transform of both sides of the heat equation in (13.9), *with respect to* x and thinking of t as a *parameter*. Following [22], we thus state Theorem 13.9, namely, the analytical treatment of the heat equation based on Fourier transforms; we provide an euristic proof for it, and refer to Theorem 4.3 of [22] for a more rigorous demonstration.

Theorem 13.9. Assume that $f \in \mathcal{L}^p$, with $1 \le p \le \infty$. The solution to problem (13.9) is then given by the following formula, which holds true on $\mathbb{R} \times (0, +\infty)$:

$$u(x,t) = \frac{1}{\sqrt{4\pi t}} \int_{-\infty}^{+\infty} e^{\frac{-(x-y)^2}{4t}}\, f(y)\, \mathrm{d}y, \tag{13.21}$$

or, equivalently, recalling the heat kernel (13.20), by the *convolution* formula:

$$u(x,t) = \int_{-\infty}^{\infty} \mathcal{H}(x-y,t)\, f(y)\, \mathrm{d}y. \tag{13.22}$$

Moreover, if f is bounded and continuous, then the solution (13.22) is continuous on $\mathbb{R} \times (0, +\infty)$.

Proof. The Fourier transform of the right–hand side $u_{xx}(x,t)$ in (13.10) is:

$$\mathcal{F}u_{xx}(s,t) = (2\pi i s)^2\, \mathcal{F}u(s,t) = -4\pi^2 s^2\, \mathcal{F}u(s,t),$$

while, for the left–hand side $u_t(x,t)$, we note that it holds:

$$\mathcal{F}u_t(s,t) = \int_{-\infty}^{+\infty} u_t(x,t)\, e^{-2\pi i s x}\, \mathrm{d}x$$

$$= \frac{\partial}{\partial t} \int_{-\infty}^{+\infty} u(x,t)\, e^{-2\pi i s x}\, \mathrm{d}x = \frac{\partial}{\partial t}\, \mathcal{F}u(s,t).$$

In other words, forming the Fourier transform (with respect to x) of both sides of equation $u_t(x,t) = u_{xx}(x,t)$ leads to:

$$\frac{\partial}{\partial t}\, \mathcal{F}u(s,t) = -4\pi^2 s^2\, \mathcal{F}u(s,t).$$

This is an ordinary differential equation in t, despite the partial derivative symbol; solving it yields:

$$\mathcal{F}u(s,t) = \mathcal{F}u(s,0)\ e^{-4\pi^2 s^2 t},$$

where the initial condition $\mathcal{F}u(s,0)$ can be computed as follows:

$$\mathcal{F}u(s,0) = \int_{-\infty}^{+\infty} u(x,0)\ e^{-2\pi i s x}\ \mathrm{d}x = \int_{-\infty}^{+\infty} f(x)\ e^{-2\pi i s x}\ \mathrm{d}x = \mathcal{F}f(s).$$

Putting it all together:

$$\mathcal{F}u(s,t) = \mathcal{F}f(s)\ e^{-4\pi^2 s^2 t}.$$

Recalling Example 12.13, we recognise that the exponential factor, in the right–hand side, above, is the Fourier transform of the heat kernel $\mathcal{H}(s,t)$ in (13.20). In other words, the Fourier transform of u is given by the product of two Fourier transforms:

$$\mathcal{F}u(s,t) = \mathcal{F}f(s)\ \mathcal{F}\mathcal{H}(s,t). \tag{13.23}$$

Inverting (13.23) yields a convolution in the x–domain:

$$u(x,t) = \mathcal{H}(x,t) \star f(x),$$

that can be written out as (13.22). □

Remark 13.10. In the case of the (apparently) more general Cauchy problem (13.11), the solution is:

$$u(x,t) = \frac{1}{\sqrt{4\pi c t}} \int_{-\infty}^{+\infty} e^{\frac{-(x-y)^2}{4 c t}}\ f(y)\ \mathrm{d}y. \tag{13.24}$$

Remark 13.11. Observe that $\mathcal{H}(x,t)$ vanishes very rapidly as $|x| \to \infty$. Then, the convolution integral (13.22) is well defined, for $t < T$, if the following *growth condition* is fulfilled:

$$|f(x)| \le c\, e^{\frac{x^2}{4T}}. \tag{13.25}$$

Moreover, under condition (13.25), and if f is assumed to be continuous, then, as $t \to 0^+$, $u(x,t)$ approaches $f(x)$ uniformly on bounded sets.

Remark 13.12. The heat kernel $\mathcal{H}(x,t)$ is defined, by (13.20), for $t > 0$ only. Moreover, $\mathcal{H}(x,t)$ is an even function of x.

Remark 13.13. Via the change of variable $y = x + 2\,s\,\sqrt{t}$, i.e., $\mathrm{d}y = 2\sqrt{t}\,\mathrm{d}s$, solution (13.21), to the Cauchy problem (13.9), can be written as:

$$u(x,t) = \frac{1}{\sqrt{\pi}} \int_{-\infty}^{\infty} e^{-s^2} f(x + 2\,s\,\sqrt{t})\ \mathrm{d}s. \tag{13.26}$$

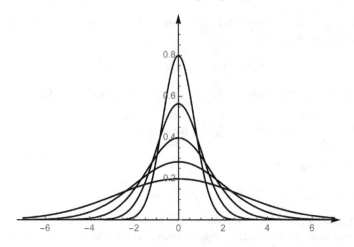

Figure 13.1: Graph of the heat kernel (13.20) for $t = 1, 2, 4, \frac{1}{2}, \frac{1}{4}$.

Remark 13.14. If the initial value $f(x)$, appearing in problem (13.9), is an odd function, i.e., $f(-x) = -f(x)$ for any $x \in \mathbb{R}$, then (13.21) implies that solution $u(x,t)$ is such that

$$u(-x,t) = -u(x,t), \quad \text{for any } x \in \mathbb{R} \text{ and for any } t > 0.$$

Hence, $u(0,t) = 0$, by a continuity argument.

If, instead, $f(x)$ is an even function, i.e., $f(-x) = f(x)$ for any $x \in \mathbb{R}$, then:

$$u(-x,t) = u(x,t), \quad \text{for any } x \in \mathbb{R} \text{ and for any } t > 0.$$

The heat kernel has some interesting properties, as Proposition 13.15 outlines.

Proposition 13.15. The heat kernel introduced in (13.20) verifies:

(i) $\mathcal{H}_t(x,t) = \mathcal{H}_{xx}(x,t)$,   for $t > 0$, $x \in \mathbb{R}$;

(ii) $\displaystyle\int_{-\infty}^{\infty} \mathcal{H}(x,t)\, dx = 1$,   for $t > 0$;

(iii) $\displaystyle\lim_{t \to 0^+} \mathcal{H}(x,t) = \begin{cases} 0 & \text{for } x \neq 0, \\ +\infty & \text{for } x = 0. \end{cases}$

Proof. Equality (i) stems from a direct calculation. The integral value (ii) follows from the change of variable $z = \dfrac{x}{\sqrt{4t}}$, i.e., $dx = \sqrt{4t}\, dz$, which implies:

$$\int_{-\infty}^{+\infty} \mathcal{H}(x,t)\, dx = \frac{1}{\sqrt{\pi}} \int_{-\infty}^{+\infty} e^{-z^2}\, dz = 1.$$

Limit (iii) is trivial if $x = 0$; when $x \neq 0$, by the change of variable $s = \dfrac{1}{t}$, and using *L'Hospital Rule* [1], we obtain:

$$\lim_{t \to 0^+} \frac{1}{\sqrt{4\pi t}} e^{\frac{-x^2}{4t}} = \lim_{s \to +\infty} \frac{\sqrt{s}}{\sqrt{4\pi}\, e^{\frac{s\,x^2}{4}}} = \lim_{s \to +\infty} \frac{1}{x^2 \sqrt{s\pi}\, e^{\frac{s\,x^2}{4}}} = 0.$$

$\square$

We now provide some examples on how to solve some heat equations for given initial value functions $f(x)$. The first example, due to the nature of the f considered, has applications in Finance.

Example 13.16. Recall the notion of positive part of a function, introduced in Proposition 8.25, that is $f^+(x) := \max\{f(x), 0\}$, and consider the initial value problem:

$$\begin{cases} u_t = u_{xx}, & \text{for} \quad t > 0, \\ u(x,0) = x^+, \end{cases} \tag{13.27}$$

where $x \in \mathbb{R}$. Using (13.26), we can write:

$$\begin{aligned}
u(x,t) &= \frac{x}{\sqrt{\pi}} \int_{-\frac{x}{2\sqrt{t}}}^{\infty} e^{-s^2}\, ds + \frac{2\sqrt{t}}{\sqrt{\pi}} \int_{-\frac{x}{2\sqrt{t}}}^{\infty} s\, e^{-s^2}\, ds \\
&= \frac{x}{\sqrt{\pi}} \left(\int_{0}^{\infty} e^{-s^2}\, ds - \int_{0}^{-\frac{x}{2\sqrt{t}}} e^{-s^2}\, ds \right) + \frac{2\sqrt{t}}{\sqrt{\pi}} \left(\frac{1}{2} e^{-\frac{x^2}{4t}} \right) \\
&= \frac{x}{\sqrt{\pi}} \left(\frac{\sqrt{\pi}}{2} - \frac{\sqrt{\pi}}{2} \operatorname{erf}\left(-\frac{x}{2\sqrt{t}} \right) \right) + \frac{\sqrt{t}}{\sqrt{\pi}} e^{-\frac{x^2}{4t}} \\
&= \frac{x}{2} \left(1 - \operatorname{erf}\left(-\frac{x}{2\sqrt{t}} \right) \right) + \frac{\sqrt{t}}{\sqrt{\pi}} e^{-\frac{x^2}{4t}}.
\end{aligned}$$

Example 13.17. Using (13.26), we show that $u(x,t) = x^2 + 2t$ solves:

$$\begin{cases} u_t = u_{xx}, & \text{for} \quad t > 0, \\ u(x,0) = x^2, \end{cases} \tag{13.28}$$

where $x \in \mathbb{R}$. Observe that the given initial value function $f(x) = x^2$ is even, thus the solution obeys Remark 13.14. Furthermore, from (13.26), we can write:

$$u(x,t) = \frac{1}{\sqrt{\pi}} \int_{-\infty}^{\infty} e^{-s^2} \left(x + 2s\sqrt{t} \right)^2 ds.$$

---

[1] Guillaume Franois Antoine, Marquis de L'Hospital (1661–1704), French mathematician. For his Rule, see, for example, mathworld.wolfram.com/LHospitalsRule.html

Now:

$$\left(x + 2s\sqrt{t}\right)^2 = x^2 + 4xs\sqrt{t} + 4s^2 t.$$

Observe that $s \mapsto e^{-s^2} 4xs\sqrt{t}$ is an odd function of s; thus, by Remark 8.58:

$$u(x,t) = \frac{1}{\sqrt{\pi}} \int_{-\infty}^{\infty} e^{-s^2} \left(x^2 + 4s^2 t\right) \, ds$$

$$= \frac{x^2}{\sqrt{\pi}} \int_{-\infty}^{\infty} e^{-s^2} \, ds + \frac{4t}{\sqrt{\pi}} \int_{-\infty}^{\infty} s^2 e^{-s^2} \, ds = x^2 + \frac{4t}{\sqrt{\pi}} c.$$

To find the value of the constant c, impose that the found family of functions $u(x,t)$ solves (13.28):

$$u_t = \frac{4}{\sqrt{\pi}} c, \qquad u_{xx} = 2, \qquad \Longrightarrow \qquad c = \frac{\sqrt{\pi}}{2}.$$

Finally, the solution to (13.28) is:

$$u(x,t) = x^2 + \frac{4t}{\sqrt{\pi}} \frac{\sqrt{\pi}}{2} = x^2 + 2t.$$

Note that, in the previous calculation, we also re–established result (8.36d).

Example 13.18. This example is taken from [54] and consists in solving the Cauchy problem:

$$\begin{cases} u_t = u_{xx}, & x \in \mathbb{R}, \quad t > 0, \\ u(x,0) = \sin x, & x \in \mathbb{R}, \end{cases}$$

and, then, inferring the integration formula:

$$\int_{-\infty}^{+\infty} e^{-s^2} \cos(a\,s) \, ds = \sqrt{\pi}\, e^{\frac{-a^2}{4}}, \qquad \forall a \in \mathbb{R}. \tag{13.29}$$

In the case of the given Cauchy problem, formula (13.26) yields:

$$u(x,t) = \frac{1}{\sqrt{\pi}} \int_{-\infty}^{\infty} e^{-s^2} \sin(x + 2s\sqrt{t}) \, ds.$$

The trigonometric identity $\sin(\alpha + \beta) = \sin\alpha\,\cos\beta + \cos\alpha\,\sin\beta$ implies:

$$\sin(x + 2s\sqrt{t}) = \sin(x)\,\cos(2s\sqrt{t}) + \cos(x)\,\sin(2s\sqrt{t}).$$

Since function $s \mapsto e^{-s^2} \cos(x) \sin(2s\sqrt{t})$ is odd, by Remark 8.58, we obtain:

$$u(x,t) = \frac{\sin x}{\sqrt{\pi}} \int_{-\infty}^{\infty} e^{-s^2} \cos(2s\sqrt{t}) \, ds.$$

In other words, the solution to the given Cauchy problem has the form:

$$u(x,t) = \frac{\sin x}{\sqrt{\pi}} \, y(t) \,,$$

where we set:

$$y(t) = \int_{-\infty}^{\infty} e^{-s^2} \, \cos(2 s \sqrt{t}) \, ds \,.$$

Since equality $u_t = u_{xx}$ must hold, by computing:

$$u_t = \frac{\sin x}{\sqrt{\pi}} \, y'(t) \,, \qquad u_{xx} = -\frac{\sin x}{\sqrt{\pi}} \, y(t) \,,$$

we see that $y(t)$ solves the initial value problem, for ordinary differential equations:

$$\begin{cases} y'(t) = -y(t) \,, \\ y(0) = \sqrt{\pi} \,. \end{cases}$$

Thus:

$$y(t) = \sqrt{\pi} \; e^{-t} \,,$$

and then:

$$u(x,t) = e^{-t} \, \sin x \,.$$

Finally, statement (13.29) follows from the equality:

$$\frac{\sin x}{\sqrt{\pi}} \int_{-\infty}^{\infty} e^{-s^2} \, \cos(2 s \sqrt{t}) \, ds = e^{-t} \, \sin x \,.$$

Example 13.19. Solve the Cauchy problem for the heat equation:

$$\begin{cases} u_t = u_{xx} \,, \\ u(x,0) = x^2 + x \,. \end{cases}$$

Its solution is given by formula (13.26), with $f(x) = x^2 + x$, so that the integrand is:

$$e^{-s^2} \left(4 s^2 t + 4 s \sqrt{t} \, x + 2 s \sqrt{t} + x^2 + x \right) \,.$$

Discarding the integrand odd components, we arrive at:

$$u(x,t) = \frac{1}{\sqrt{\pi}} \int_{-\infty}^{\infty} e^{-s^2} \left(4 s^2 t + x^2 + x \right) \, ds = x^2 + x + \frac{4t}{\sqrt{\pi}} \int_{-\infty}^{\infty} s^2 \, e^{-s^2} \, ds \,.$$

Now, imposing equality $u_t = u_{xx}$, we see that:

$$\frac{4}{\sqrt{\pi}} \int_{-\infty}^{\infty} s^2 \, e^{-s^2} \, ds = 2 \,,$$

which leads, again, to formula (8.36d). In conclusion, the required solution is:

$$u(x,t) = x^2 + x + 2t \,.$$

Example 13.20. Here, we use some formulæ that follow from the probability integral (8.36), and that are extremely useful in many various applications. Consider the initial value problem:

$$\begin{cases} u_t = u_{xx}, & x \in \mathbb{R}, \quad t > 0, \\ u(x,0) = x \, e^x, & x \in \mathbb{R}. \end{cases}$$

To obtain its solution $u(x,t)$, let us use formula (13.26), with $f(x) = x \, e^x$. Since:

$$f(x + 2s\sqrt{t}) = 2s\sqrt{t} \, e^x \, e^{2s\sqrt{t}} + x \, e^x e^{2s\sqrt{t}},$$

then:

$$u(x,t) = \frac{1}{\sqrt{\pi}} \left(2\sqrt{t} \, e^x \int_{-\infty}^{\infty} s \, e^{-s^2 + 2\sqrt{t}\,s} \, ds + x \, e^x \int_{-\infty}^{\infty} e^{-s^2 + 2\sqrt{t}\,s} \, ds \right).$$

Now, formulæ (8.37)–(8.38), that are linked to the probability integral, yield, respectively:

$$\int_{-\infty}^{\infty} e^{-s^2 + 2\sqrt{t}\,s} \, ds = \sqrt{\pi} \, e^t, \qquad \int_{-\infty}^{\infty} s \, e^{-s^2 + 2\sqrt{t}\,s} \, ds = \sqrt{\pi} \, \sqrt{t} \, e^t.$$

In conclusion:

$$u(x,t) = \frac{1}{\sqrt{\pi}} \left(2\sqrt{t} \, e^x \, \sqrt{\pi} \, \sqrt{t} \, e^t + x \, e^x \, \sqrt{\pi} \, e^t \right) = (x + 2t) \, e^{t+x}.$$

13.2.3   Initial data on $(0, \infty)$

It is interesting to associate the diffusion equation (13.10) with an initial data, which is defined only on $(0, \infty)$, and not on the whole real line $\mathbb{R}$. To do so, a technique is employed, here, that is similar to *d'Alembert*[2] *method* for the *wave equation*, for which we refer to § 1 of Chapter 5 in [1].

Given a continuous function $f : (0, \infty) \to \mathbb{R}$, we deal, here, with two types of initial value problems, respectively:

$$\begin{cases} u_t = u_{xx}, & x > 0, \quad t > 0, \\ u(0,t) = 0, & t > 0, \\ u(x,0) = f(x), \end{cases} \tag{13.30}$$

and

$$\begin{cases} u_t = u_{xx}, & x > 0, \quad t > 0, \\ u_x(0,t) = 0, & t > 0, \\ u(x,0) = f(x). \end{cases} \tag{13.31}$$

To solve (13.30), we extend the initial data $f(x)$ to an odd function $f^o(x)$, called *odd continuation* of f, that is defined on the whole real axis as follows:

---

[2] Jean–Baptiste le Rond d'Alembert (1717–1783), French mathematician, physicist, encyclopaedist, philosopher, and music theorist.

$$f^o(x) = \begin{cases} f(x) & \text{if } x > 0, \\ -f(-x) & \text{if } x < 0, \end{cases}$$

and, then, we write the solution to the Cauchy problem with initial data f^o, using the convolution formula (13.22):

$$\begin{aligned} u(x,t) &= \int_{-\infty}^{\infty} \mathcal{H}(x-y,t) \; f^o(y) \, dy \\ &= -\int_{-\infty}^{0} \mathcal{H}(x-y,t) \; f(-y) \, dy + \int_{0}^{\infty} \mathcal{H}(x-y,t) \; f(y) \, dy \\ &= \int_{0}^{\infty} \left(\mathcal{H}(x-y,t) - \mathcal{H}(x+y,t) \right) \; f(y) \, dy \\ &= \int_{0}^{\infty} \mathcal{H}_1(x,y,t) \; f(y) \, dy, \end{aligned}$$

where $\mathcal{H}_1(x,y,t)$ is the so-called *Green function of first kind*:

$$\mathcal{H}_1(x,y,t) = \mathcal{H}(x-y,t) - \mathcal{H}(x+y,t) = \frac{1}{\sqrt{4\pi t}} \left(e^{-\frac{(x-y)^2}{4t}} - e^{-\frac{(x+y)^2}{4t}} \right).$$

Note that condition $u(0,t) = 0$ comes from the argument illustrated in § 13.14.

To solve (13.31), we form the *even continuation* of f :

$$f^e(x) = \begin{cases} f(x) & \text{if } x > 0, \\ f(-x) & \text{if } x < 0, \end{cases}$$

and, then, we solve the Cauchy problem with initial data f^e, using the convolution formula (13.22):

$$\begin{aligned} u(x,t) &= \int_{-\infty}^{\infty} \mathcal{H}(x-y,t) \; f^e(y) \, dy \\ &= \int_{-\infty}^{0} \mathcal{H}(x-y,t) \; f(-y) \, dy + \int_{0}^{\infty} \mathcal{H}(x-y,t) \; f(y) \, dy \\ &= \int_{0}^{\infty} \left(\mathcal{H}(x-y,t) + \mathcal{H}(x+y,t) \right) \; f(y) \, dy \\ &= \int_{0}^{\infty} \mathcal{H}_2(x,y,t) \; f(y) \, dy, \end{aligned}$$

where $\mathcal{H}_2(x,y,t)$ is the so-called *Green function of second kind*:

$$\mathcal{H}_2(x,y,t) = \mathcal{H}(x-y,t) + \mathcal{H}(x+y,t) = \frac{1}{\sqrt{4\pi t}} \left(e^{-\frac{(x-y)^2}{4t}} + e^{-\frac{(x+y)^2}{4t}} \right)$$

Note that condition $u_x(0,t) = 0$ comes from the argument described in § 13.14.

13.3   Parabolic equations with constant coefficients

A partial differential equation, which is linear, parabolic, and with constant coefficients, has the form:

$$v_t = v_{xx} + a\,v_x + b\,v\,, \tag{13.32}$$

where $a, b \in \mathbb{R}$. The following Theorem 13.21 shows that (13.32) can always be reduced to the heat equation, by a suitable change of variable.

Theorem 13.21. The solution to (13.32) is given by:

$$v(x,t) = e^{(b-\frac{a^2}{4})t}\; e^{-\frac{a\,x}{2}}\; h(x,t)\,, \tag{13.33}$$

where $h(x,t)$ is a solution to the heat equation (13.9).

Proof. We seek for a function that solves equation (13.32) and has the form:

$$v(x,t) = e^{\alpha t}\; e^{\beta x}\; h(x,t)\,. \tag{13.34}$$

To do so, we impose that $v(x,t)$, above, is a solution to (13.32), finding α and β accordingly. Let us compute:

$$v_t = e^{\alpha t}\; e^{\beta x}\; (\alpha\,h + h_t)\,,$$
$$v_x = e^{\alpha t}\; e^{\beta x}\; (\beta\,h + h_x)\,,$$
$$v_{xx} = e^{\alpha t}\; e^{\beta x}\; (\beta^2\,h + 2\,\beta\,h_x + h_{xx})\,,$$

and substitute them into (13.32), rewritten as $v_t - v_{xx} - a\,v_x - b\,v = 0$, obtaining:

$$e^{\alpha t}\; e^{\beta x}\; \big(h_t - (b - \alpha + a\,\beta + \beta^2)\,h - (a + 2\,\beta)\,h_x - h_{xx}\big) = 0\,.$$

The hypothesis that h solves the heat equation allows to consider the system:

$$\begin{cases} a + 2\,\beta = 0\,, \\ b - \alpha + a\,\beta + \beta^2 = 0\,, \end{cases}$$

which can be solved with respect to α and β :

$$\begin{cases} \alpha = b - \dfrac{a^2}{4}\,, \\ \beta = -\dfrac{a}{2}\,, \end{cases}$$

showing that, when $h_t = h_{xx}$, equation (13.32) is solved by (13.33). $\qquad\square$

The proof to Theorem 13.21 can be adapted to the case of a Cauchy problem.

Corollary 13.22. The Cauchy problem, defined for $x \in \mathbb{R}$ and $t \geq 0$:

$$\begin{cases} u_t(x,t) = u_{xx}(x,t) + a\, u_x(x,t) + b\, u(x,t)\,, & t > 0\,, \\ u(x,0) = f(x)\,, \end{cases}$$

(13.35)

is solved by:

$$u(x,t) = e^{(b-\frac{a^2}{4})t}\ e^{-\frac{a\,x}{2}}\ h(x,t)\,,$$

where $h(x,t)$ solves the Cauchy problem for the heat equation, given below:

$$\begin{cases} h_t(x,t) = h_{xx}(x,t)\,, \\ h(x,0) = e^{\frac{a\,x}{2}}\ f(x)\,. \end{cases}$$

Proof. The proof follows from combining Theorems 13.9 and 13.21. □

Corollary 13.22 can be further generalised to the case in which the second derivative term is multiplied by $c \neq 1$.

Corollary 13.23. The Cauchy problem, defined for $x \in \mathbb{R}$ and $t \geq 0$:

$$\begin{cases} u_t(x,t) = c\, u_{xx}(x,t) + a\, u_x(x,t) + b\, u(x,t)\,, & t > 0\,, \\ u(x,0) = f(x)\,, \end{cases}$$

(13.36)

is solved by:

$$u(x,t) = e^{(b-\frac{a^2}{4c})t}\ e^{-\frac{a\,x}{2c}}\ h(x,t)\,,$$

where $h(x,t)$ solves the following Cauchy problem for the heat equation:

$$\begin{cases} h_t(x,t) = c\ h_{xx}(x,t)\,, \\ h(x,0) = e^{\frac{a\,x}{2c}}\ f(x)\,. \end{cases}$$

Example 13.24. We want to solve the parabolic Cauchy problem:

$$\begin{cases} u_t = u_{xx} - 2\,u_x\,, \\ u(x,0) = x\,e^x\,. \end{cases}$$

Observe that it is of the form (13.35), with $f(x) = x\,e^x$, $a = -2$, and $b = 0$. In order to use Corollary 13.22, we have to solve, first, the following Cauchy problem for the heat equation:

$$\begin{cases} h_t(x,t) = h_{xx}(x,t)\,, \\ h(x,0) = e^{\frac{a\,x}{2}}\ x\,e^x\,, \end{cases} \quad \text{i.e.,} \quad \begin{cases} h_t(x,t) = h_{xx}(x,t)\,, \\ h(x,0) = x\,, \end{cases}$$

whose solution, recalling Remark 13.13, is:

$$h(x,t) = \frac{1}{\sqrt{\pi}} \int_{-\infty}^{\infty} e^{-s^2} e^{\frac{a(x+2s\sqrt{t})}{2}} (x+2s\sqrt{t}) e^{x+2s\sqrt{t}} ds$$

$$= \frac{1}{\sqrt{\pi}} \int_{-\infty}^{\infty} e^{-s^2} (x+2s\sqrt{t}) ds$$

$$= \frac{x}{\sqrt{\pi}} \int_{-\infty}^{\infty} e^{-s^2} ds + \frac{2\sqrt{t}}{\sqrt{\pi}} \int_{-\infty}^{\infty} s\, e^{-s^2} ds$$

$$= \frac{x}{\sqrt{\pi}} \sqrt{\pi} + \frac{2\sqrt{t}}{\sqrt{\pi}} \, 0$$

$$= x\,,$$

where the chain of equalities rely on (8.36a) and on the fact that $s \mapsto s\, e^{-s^2}$ is an odd function, for which Remark 8.58 holds. In conclusion, the given problem is solved by:

$$u(x,t) = e^{(b-\frac{a^2}{4})t} \, e^{-\frac{ax}{2}} \, h(x,t) = e^{-t} \, e^x \, x = x\, e^{x-t}.$$

Example 13.25. We wish to solve the parabolic Cauchy problem:

$$\begin{cases} u_t = u_{xx} + 4\, u_x\,, \\ u(x,0) = x^2\, e^{-2x}\,. \end{cases}$$

This problem is of the form (13.35), with $f(x) = x^2\, e^{-2x}$, $a = 4$, and $b = 0$. In order to use Corollary 13.22, we have to solve, first, the following Cauchy problem for the heat equation:

$$\begin{cases} h_t(x,t) = h_{xx}(x,t)\,, \\ h(x,0) = x^2\,, \end{cases}$$

whose solution, recalling Remark 13.13, is:

$$h(x,t) = \frac{1}{\sqrt{\pi}} \int_{-\infty}^{\infty} e^{-s^2} (x+2s\sqrt{t})^2 \, ds$$

$$= \frac{1}{\sqrt{\pi}} \int_{-\infty}^{\infty} e^{-s^2} (x^2 + 4xs\sqrt{t} + 4s^2 t) \, ds$$

$$= \frac{1}{\sqrt{\pi}} \int_{-\infty}^{\infty} e^{-s^2} (x^2 + 4s^2 t) \, ds$$

$$= \frac{x^2}{\sqrt{\pi}} \int_{-\infty}^{\infty} e^{-s^2} \, ds + \frac{4t}{\sqrt{\pi}} \int_{-\infty}^{\infty} s^2\, e^{-s^2} \, ds$$

$$= \frac{x^2}{\sqrt{\pi}} \sqrt{\pi} + \frac{4t}{\sqrt{\pi}} \frac{\sqrt{\pi}}{2}$$

$$= x^2 + 2t\,,$$

where the chain of equalities relies on (8.36a) and (8.36d), and on the fact that $s \mapsto s e^{-s^2}$ is an odd function, for which Remark 8.58 holds. In conclusion, the given problem is solved by:

$$u(x,t) = e^{-4t-2x} (x^2 + 2t).$$

Example 13.26. Solve the parabolic initial value problem:

$$\begin{cases} u_t = u_{xx} - 4u_x, \\ u(x,0) = x^3 e^{2x}. \end{cases}$$

This problem is of the form (13.35), with $f(x) = x^3 e^{2x}$, $a = -4$, and $b = 0$. In order to use Corollary 13.22, we have to solve, first, the following Cauchy problem for the heat equation:

$$\begin{cases} h_t(x,t) = h_{xx}(x,t), \\ h(x,0) = x^3, \end{cases}$$

whose solution, recalling Remark 13.13, is:

$$\begin{aligned} h(x,t) &= \frac{1}{\sqrt{\pi}} \int_{-\infty}^{\infty} e^{-s^2} (x + 2s\sqrt{t})^3 \, ds \\ &= \frac{1}{\sqrt{\pi}} \int_{-\infty}^{\infty} e^{-s^2} (x^3 + 6x^2 s\sqrt{t} + 12x\, s^2 t + 8 s^3 t\sqrt{t}) \, ds \\ &= \frac{1}{\sqrt{\pi}} \int_{-\infty}^{\infty} e^{-s^2} (x^3 + 12x\, s^2 t) \, ds \\ &= \frac{x^3}{\sqrt{\pi}} \sqrt{\pi} + \frac{12xt}{\sqrt{\pi}} \frac{\sqrt{\pi}}{2} \\ &= x(x^2 + 6t), \end{aligned}$$

where the chain of equalities relies on (8.36a) and (8.36d), and on the fact that functions $s \mapsto s e^{-s^2}$ and $s \mapsto s^3 e^{-s^2}$ are both odd, thus they verify Remark 8.58. In conclusion, the given problem is solved by:

$$u(x,t) = e^{2x-4t} \, x(x^2 + 6t).$$

13.3.1   Exercises

1. Show that function $u(x,t) = (2t + x) e^{t+2x}$ solves the parabolic Cauchy problem:

$$\begin{cases} u_t = u_{xx} - 2u_x + u, \qquad t > 0, \\ u(x,0) = x e^{2x}. \end{cases}$$

2. Show that function $u(x,t) = (2t + x) e^{2t+2x}$ solves the parabolic Cauchy problem:

$$\begin{cases} u_t = u_{xx} - 2u_x + 2u, & t > 0, \\ u(x,0) = x \, e^{2x}. \end{cases}$$

3. Consider a positive $\mathcal{C}^2$ function $u(x,t)$, which solves (13.10) for $t > 0$. Then, the following function:

$$\theta(x,t) = -2\frac{u_x}{u}$$

satisfies, for $t > 0$, the differential equation:

$$\theta_t + \theta \, \theta_x = \theta_{xx}.$$

4. Consider the initial value problem (13.9), in which we set:

$$f(x) = \begin{cases} 1 & \text{if } x > 0, \\ 0 & \text{if } x < 0. \end{cases}$$

Then, the solution of the so-formed initial value problem is given by:

$$u(x,t) = \frac{1}{2} \left(1 + \phi\left(\frac{x}{\sqrt{4t}} \right) \right),$$

being $\phi(s)$ the error function, defined by:

$$\phi(s) = \frac{2}{\sqrt{\pi}} \int_0^s e^{-t^2} dt.$$

13.4   Black–Scholes equation

We are finally in the position to solve analytically the *Black–Scholes* equation, namely, a parabolic partial differential equation, with variable coefficients, which is at the base of Quantitative Finance, as it governs the price evolution of a *European call option* or *put option,* under the Black–Scholes[3] model. This is not the place to explain the origin and the economic foundation of such a model, for which we refer the Reader to [7]. The Black-Scholes equation is:

$$\frac{\partial V}{\partial t} + \frac{1}{2}\sigma^2 S^2 \frac{\partial^2 V}{\partial S^2} + rS\frac{\partial V}{\partial S} - rV = 0, \qquad S \geq 0, \quad t \in [0,T],$$
$$(13.37)$$

where:

---

[3]Fischer Sheffey Black (1938–1995), American economist.
Myron Samuel Scholes (1941–living), Canadian–American financial economist.

- t is the time;

- $S = S(t)$ is the price of the underlying asset, at time t;

- $V = V(S,t)$ is the value of the option;

- T is the expiration date;

- σ is the volatility of the underlying asset;

- r is the risk–free interest rate.

We assume that r and σ are constant. The treatment of (13.37) is based on the interesting on–line material provided by [14]. We reduce the Black-Scholes equation to a general parabolic equations, with constant coefficients. For clarity, and for consistency with the previous chapters of this book, we employ the (x,t) notation to model the Black-Scholes equation, namely:

$$u_t + \left(\frac{1}{2}\sigma^2\right) x^2 \, u_{xx} + r\,x\,u_x - r\,u = 0, \qquad x \geq 0, \quad 0 \leq t \leq T.$$

In other words, the Black-Scholes equation can be described as:

$$\begin{cases} u_t = a\,x^2\,u_{xx} + b\,x\,u_x + c\,u, \\ u(x,0) = f(x), \end{cases} \tag{13.38}$$

where a, b, c are given real numbers. Equation (13.38) can be turned into a parabolic equation, with constant coefficients, using the change of variable:

$$\begin{cases} x = e^y, \\ t = \frac{1}{a}\,\tau, \end{cases} \qquad \text{that is to say,} \qquad \begin{cases} y = \ln x, \\ \tau = a\,t. \end{cases} \tag{13.39}$$

Equating $u(x,t) = v(y,\tau)$, the transformed differential equation is obtained:

$$\begin{cases} v_\tau = v_{yy} + \left(\frac{b}{a} - 1\right) v_y + \frac{c}{a}\,v, \\ v(y,0) = f(e^y). \end{cases} \tag{13.40}$$

To see it, let us compute the partial derivatives of $u(x,t)$ in terms of the transformed function $v(y,\tau)$:

$$
\begin{aligned}
u_t &= \frac{\partial v}{\partial \tau}\frac{\partial \tau}{\partial t} = \frac{\partial v}{\partial \tau}\frac{\partial (a\,t)}{\partial t} = a\,\frac{\partial v}{\partial \tau} = a\,v_\tau\;; \\[2mm]
u_x &= \frac{\partial v}{\partial y}\frac{\partial y}{\partial x} = \frac{\partial v}{\partial y}\frac{\partial (\ln x)}{\partial x} = \frac{1}{x}\frac{\partial v}{\partial y} = \frac{1}{x}\,v_y\;; \\[2mm]
u_{xx} &= \frac{\partial u_x}{\partial x} = \frac{\partial}{\partial x}\left(\frac{1}{x}\,v_y\right) = \frac{\partial}{\partial x}\left(\frac{1}{x}\right)v_y + \frac{1}{x}\frac{\partial}{\partial x}\left(\frac{\partial v}{\partial y}\right) \\[2mm]
&= -\frac{1}{x^2}\,v_y + \frac{1}{x}\left(\frac{\partial}{\partial y}\frac{\partial y}{\partial x}\right)\left(\frac{\partial v}{\partial y}\right) \\[2mm]
&= -\frac{1}{x^2}\,v_y + \frac{1}{x}\frac{\partial}{\partial y}\left(\frac{\partial y}{\partial x}\right)\left(\frac{\partial v}{\partial y}\right) \\[2mm]
&= -\frac{1}{x^2}\,v_y + \frac{1}{x^2}\frac{\partial}{\partial y}\left(\frac{\partial v}{\partial y}\right) = \frac{1}{x^2}\,(v_{yy} - v_y)\,.
\end{aligned}
\tag{13.41}
$$

Note that, in computing u_{xx}, above, we wrote the differential operator as:

$$
\frac{\partial}{\partial x} = \frac{\partial}{\partial y}\frac{\partial y}{\partial x}\,.
$$

Inserting (13.41) into (13.38) yields:

$$
\begin{cases}
a\,v_\tau = a\,x^2\,\dfrac{1}{x^2}\,(v_{yy} - v_y) + b\,x\,\dfrac{1}{x}\,v_y + c\,v\,, \\[3mm]
v(y,0) = f(e^y)\,,
\end{cases}
$$

which is indeed (13.40). At this point, observe that (13.40) is a constant–coefficients problem of the form (13.35), which we rewrite here, for convenience, as:

$$
\begin{cases}
v_\tau = v_{yy} + A\,v_y + B\,v\,, \\[2mm]
v(y,0) = g(y)\,.
\end{cases}
\tag{13.42}
$$

Recalling Corollary 13.22, we know that, in order to solve (13.42), we have to consider, first, the following heat equation:

$$
\begin{cases}
h_\tau = h_{yy}\,, \\[2mm]
h(y,0) = e^{\frac{A\,y}{2}}\,g(y)\,,
\end{cases}
$$

whose solution $h(y,\tau)$ is a component of the following function, that solves (13.42):

$$
v(y,\tau) = e^{\left(B - \frac{A^2}{4}\right)\tau}\;e^{-\frac{A\,y}{2}}\;h(y,\tau)\,.
$$

In other words, since in our case it is:

$$
A = \frac{b}{a} - 1 = \frac{b-a}{a}\,, \qquad B = \frac{c}{a}\,,
$$

we have to consider the heat equation:

$$\begin{cases} h_\tau = h_{yy}\,, \\ h(y,0) = e^{\frac{(b-a)y}{2a}}\ f(e^y)\,, \end{cases}$$

and solve it, using (13.26), thus obtaining:

$$h(y,\tau) = \frac{1}{\sqrt{\pi}} \int_{-\infty}^{+\infty} e^{-s^2}\ e^{\frac{(b-a)(y+2s\sqrt{\tau})}{2a}}\ f(e^{y+2s\sqrt{\tau}})\ ds\,. \tag{13.43}$$

Then, we can insert (13.43) into the solution of the transformed problem (13.40), which is:

$$v(y,\tau) = e^{\left(\frac{c}{a} - \frac{(b-a)^2}{4a^2}\right)\tau}\ e^{-\frac{(b-a)y}{2a}}\ h(y,\tau)\,. \tag{13.44}$$

Given the variable–coefficient problem (13.38), equations (13.43)–(13.44) lead automatically to the solution of the transformed problem (13.40), via the evaluation of the integral involved. Once such a solution is computed, the solution to the given problem (13.38) can be obtained recovering the original variables through (13.39).

Summarising, we can express the solution of (13.38) as:

$$\boxed{u(x,t) = \frac{1}{\sqrt{\pi}} \int_{-\infty}^{\infty} e^{-s^2 + (b-a)s\sqrt{\frac{t}{a}} + \left(c - \frac{(b-a)^2}{4a}\right)t}\ f\left(x\,e^{2s\sqrt{at}}\right)\ ds\,.}$$

The following Examples 13.27, 13.28 and 13.29 illustrate the solution procedure.

Example 13.27. Compute the solution to the parabolic problem:

$$\begin{cases} u_t = x^2\,u_{xx} + x\,u_x + u\,, \\ u(x,0) = x\,. \end{cases} \tag{13.45}$$

The given problem has the form (13.38), with $a = b = c = 1$ and $f(x) = x$. Let us form (13.43):

$$h(y,\tau) = \frac{1}{\sqrt{\pi}} \int_{-\infty}^{+\infty} e^{-s^2}\ e^0\ e^{y+2\sqrt{\tau}s}\ ds = \frac{1}{\sqrt{\pi}} \int_{-\infty}^{+\infty} e^{-s^2 + 2\sqrt{\tau}s + y}\ ds = e^{y+\tau}\,,$$

where we used the integration formula (8.37). From (13.44), we arrive at the solution of the transformed problem (13.40):

$$v(y,\tau) = e^{y+2\tau}\,.$$

Finally, we use (13.39), to recover the original variables:

$$\tau = t\,, \qquad\qquad y = \ln x\,,$$

so that solution to (13.45) is:

$$u(x,t) = x\,e^{2t}\,.$$

Example 13.28. Compute the solution to the parabolic problem:

$$\begin{cases} u_t = 2\,x^2\,u_{xx} + 4\,x\,u_x + u\,, \\ u(x,0) = x\,. \end{cases} \tag{13.46}$$

This problem has the form (13.38), with $a = 2$, $b = 4$, $c = 1$ and $f(x) = x$. As in the previous Example 13.28, problem (13.46) can be solved working in exact arithmetic. Here, (13.43) becomes:

$$\begin{aligned} h(y,\tau) &= \frac{1}{\sqrt{\pi}} \int_{\infty}^{\infty} e^{-s^2}\, e^{\frac{y+2\,s\,\sqrt{\tau}}{2}}\; e^{y}+2\,s\,\sqrt{\tau}\;\, ds \\ &= \frac{1}{\sqrt{\pi}} \int_{\infty}^{\infty} e^{-s^2 + 3\sqrt{\tau}\,s + \frac{3\,y}{2}}\; ds \;=\; e^{\frac{3}{2}y+\frac{9}{4}\tau}\,, \end{aligned}$$

where the last equality is obtained via the integration formula (8.37). Applying (13.44), we arrive at the solution of the transformed problem (13.40), that is:

$$v(y,\tau) = e^{y+\frac{5}{2}\tau}\,.$$

Finally, recovering the original variables by means of the change of variable (13.39), that here is:

$$\tau = 2t\,, \qquad\qquad y = \ln x\,,$$

we arrive at the solution of the differential equation (13.46), namely:

$$u(x,t) = x\,e^{5t}\,.$$

Example 13.29. Solve the Cauchy problem:

$$\begin{cases} u_t = 2\,x^2\,u_{xx} + x\,u_x + u\,, & x > 0\,, & t > 0\,, \\ u(x,0) = \ln x\,. \end{cases}$$

This problem has the form (13.38), with $a = 2$, $b = c = 1$ and $f(x) = \ln x$. The associated heat equation, modelled in (13.43), is:

$$\begin{aligned} h(y,\tau) &= \frac{1}{\sqrt{\pi}} \int_{-\infty}^{+\infty} e^{-s^2}\, e^{-\frac{y+2\,s\,\sqrt{\tau}}{4}}\; (y + 2\,s\,\sqrt{\tau})\; ds \\ &= \frac{y}{\sqrt{\pi}} \int_{-\infty}^{+\infty} e^{-(s^2+\frac{s\sqrt{\tau}}{2}+\frac{y}{4})}\; ds + \frac{2\sqrt{\tau}}{\sqrt{\pi}} \int_{-\infty}^{+\infty} s\; e^{-(s^2+\frac{s\sqrt{\tau}}{2}+\frac{y}{4})}\; ds \\ &= \frac{y}{\sqrt{\pi}}\, \sqrt{\pi}\, e^{\frac{\tau}{16}-\frac{y}{4}} + \frac{2\sqrt{\tau}}{\sqrt{\pi}}\, (-\sqrt{\pi})\, \frac{\sqrt{\tau}}{4}\, e^{\frac{\tau}{16}-\frac{y}{4}} \\ &= \left(y - \frac{\tau}{2}\right)\, e^{\frac{\tau-4\,y}{16}}\,, \end{aligned}$$

where the Gaussian integration formulæ (8.37)–(8.38) were employed. Using (13.44), the solution of the transformed problem (13.40) is obtained:

$$v(y,\tau) = e^{\frac{7\,\tau}{16}}\; e^{\frac{y}{4}}\; \left(y - \frac{\tau}{2}\right)\, e^{\frac{\tau-4\,y}{16}} \;=\; e^{\frac{\tau}{2}}\, \left(y - \frac{\tau}{2}\right)\,.$$

Finally, we use (13.39), to recover the original variables:

$$\tau = 2t, \qquad y = \ln x,$$

so that solution to (13.45) is:

$$u(x,t) = e^t \left(\ln(x) - t \right).$$

Remark 13.30. In some (very) particular situations, we can seek for solutions of (13.38) using the *separation of variables* approach. With boundary conditions of monomial form $f(x) = x^n$, we can look for a solution of (13.38) of the form $u(x,t) = g(t)\, x^n$, with $g \in C^1$ and $g(0) = 1$. By construction, $u(x,0) = x^n$ meets the boundary condition in (13.38) and, imposing that $u(x,t)$ solves the parabolic equation, we obtain the ordinary differential equation for $g(t)$:

$$g'(t) = (a\, n\, (n-1) + b\, n + c)\, g(t),$$

which provides:

$$g(t) = e^{(a\, n\, (n-1) + b\, n + c)\, t}.$$

Hence, in the special case of monomial boundary conditions, we can avoid the Green kernel procedure and go directly to the solution of (13.38), which is:

$$u(x,t) = x^n\, e^{(a\, n\, (n-1) + b\, n + c)\, t}.$$

Furthermore, from the linearity of the parabolic operator, we can use the same technique with boundary conditions of polynomial form. For instance, the boundary value problem:

$$\begin{cases} u_t = a\, x^2\, u_{xx} + b\, x\, u_x + c\, u \\ u(x,0) = x + x^2 \end{cases}$$

is solved by the sum of the solutions $u_1(x,t)$ and $u_2(x,t)$ of the boundary value problems:

$$\begin{cases} u_t = a\, x^2\, u_{xx} + b\, x\, u_x + c\, u, \\ u(x,0) = x, \end{cases} \qquad \begin{cases} u_t = a\, x^2\, u_{xx} + b\, x\, u_x + c\, u, \\ u(x,0) = x^2, \end{cases}$$

that is:

$$u(x,t) = u_1(x,t) + u_2(x,t) = x\, e^{(b+c)\, t} + x^2\, e^{(2\, a + 2\, b + c)\, t}.$$

13.4.1 Exercises

1. Show that function $u(x,t) = x^2\, e^{5t}$ solves the parabolic Cauchy problem:

$$\begin{cases} u_t = x^2\, u_{xx} + x\, u_x + u, \qquad t > 0, \\ u(x,0) = x^2. \end{cases}$$

2. Show that function $u(x,t) = e^{2t}(\ln x - 2t)$ solves the parabolic Cauchy problem:

$$\begin{cases} u_t = x^2 u_{xx} - x u_x + 2u, & t > 0, \\ u(x,0) = \ln x. \end{cases}$$

13.5   Non–homogeneous equation: Duhamel integral

We study here the non–homogeneous initial value problem:

$$\begin{cases} u_t(x,t) = c\ u_{xx}(x,t) + P(x,t), & x \in \mathbb{R}, \quad t > 0, \\ u(x,0) = f(x), \end{cases} \tag{13.47}$$

which differs from (13.11) in the *source* term $P(x,t)$. For problem (13.47), a general solution formula is obtained, which is analogous to (13.21). To do so, we follow an approach, that generalises the variation of parameters method, illustrated in Theorem 5.26 and in § 6.2.1. Such a generalisation apply to linear partial differential equations, and it is known as *Duhamel integral* or *principle*. To understand how it works, we first present Example 13.31, which refers to an ordinary differential equation, revisited having in mind the Duhamel[4] approach.

Example 13.31. Let $a \in \mathbb{R}$ be a given real number, and let $f(t)$ be a continuous function, defined on $[0,+\infty)$. Then, the linear initial value problem:

$$\begin{cases} y'(t) = a\ y(t) + f(t), & t > 0, \\ y(0) = 0, \end{cases} \tag{13.48}$$

has the solution illustrated in Theorem 5.26, namely:

$$y(t) = e^{at} \int_0^t e^{-as} f(s)\ \mathrm{d}s = \int_0^t f(s)\ e^{a(t-s)}\ \mathrm{d}s. \tag{13.49}$$

Consider, now, the following set of linear homogeneous equations, depending on the one parameter a, and in which s plays the role of a dummy variable:

$$\begin{cases} u'(t) = a\ u(t), & t > 0, \\ u(0) = f(s). \end{cases} \tag{13.50}$$

At this point, observe that the solution (which is a function of t) of the parametric problem (13.50) is:

$$u(t,s) = f(s)\ e^{at}. \tag{13.51}$$

---

[4] Jean–Marie Constant Duhamel (1797–1872), French mathematician and physicist.

Hence, if we look back at solution (13.49) of problem (13.48), we can write it as:

$$y(t) = \int_0^t u(t - s, s) \ ds \ .$$ (13.52)

We can interpret the found solution in the following way. The solution of the non–homogeneous equation $y' = a\,y + f(t)$, corresponding to the zero–value initial condition $y(0) = 0$, is obtained from the solution of the homogeneous equation $u' = a\,u$, when it is parametrized by the non–homogeneous initial condition $u(0) = f(s)$.

The argument set out in Example 13.31 constitutes the foundation of the Duhamel method for partial differential equations. Though the method works not only for parabolic equations, we use it, here, for solving non–homogeneous parabolic equations. Let us consider, in fact, the initial value problem:

$$\begin{cases} u_t(x,t) = c\ u_{xx}(x,t) + P(x,t), & x \in \mathbb{R}, \quad t > 0, \\ u(x,0) = 0. \end{cases}$$ (13.53)

Applying a procedure similar to that of Example 13.31, we build the homogeneous parametric initial value problem:

$$\begin{cases} h_t = c\ h_{xx}, & x \in \mathbb{R}, \quad t > 0, \\ h(x,0) = P(x,s). \end{cases}$$ (13.54)

Observe that (13.54) has the form of the homogeneous initial value problem (13.11), whose solution is given by (13.24). Thus, the solution to the parametric problem (13.54) can be expressed in terms of the heat kernel $\mathcal{H}(x,t)$, introduced in (13.20), modified as follows:

$$\mathcal{H}(x,t) = \frac{1}{\sqrt{4\pi ct}}\ e^{-\frac{x^2}{4ct}},$$

and employed to define:

$$h(x,t;s) = \int_{-\infty}^{\infty} \mathcal{H}(x - y, t)\ P(y,s)\ dy.$$ (13.55)

Finally, motivated by considerations similar to those described in Example 13.31, we conclude that the solution to (13.53) should be:

$$\begin{aligned} u(x,t) &= \int_0^t h(x,t-s;s)\ ds \\ &= \int_0^t \left(\int_{-\infty}^{\infty} \mathcal{H}(x-y,t-s)\ P(y,s)\ dy \right) ds \\ &= \frac{1}{2\sqrt{\pi c}} \int_0^t \left(\int_{-\infty}^{\infty} \frac{1}{\sqrt{t-s}}\ e^{-\frac{(x-y)^2}{4c(t-s)}}\ P(y,s)\ dy \right) ds . \end{aligned}$$ (13.56)

As a matter of fact, the function $u(x,t)$, defined in (13.56), is indeed solution to the initial value problem (13.53). The technical details, concerning differentiation under the integral sign, are omitted here, but we point out that formula (13.56) can be used to compute solutions to (13.53) in explicit form.

Example 13.32. We solve, here, the non–homogeneous initial value problem:

$$\begin{cases} u_t(x,t) = u_{xx}(x,t) + x\,t\,, & x \in \mathbb{R}, \qquad t > 0, \\ u(x,0) = 0\,. \end{cases} \tag{13.57}$$

We build (13.56), with $P(x,t) = x\,t$, i.e.:

$$u(x,t) = \frac{1}{2\sqrt{\pi}} \int_0^t s \left(\int_{-\infty}^{\infty} \frac{y}{\sqrt{t-s}}\, e^{-\frac{(x-y)^2}{4(t-s)}}\, dy \right)\, ds\,. \tag{13.58}$$

Now, perform the change of variable, from y to z :

$$z = \frac{x-y}{2\sqrt{t-s}}\,,$$

so that the evaluation of the inner integral in (13.58) simplifies as follows:

$$\frac{1}{2\sqrt{\pi}} \int_{-\infty}^{\infty} \frac{y}{\sqrt{t-s}}\, e^{-\frac{(x-y)^2}{4(t-s)}}\, dy = \frac{1}{\sqrt{\pi}} \int_{-\infty}^{\infty} (x - 2z\sqrt{t-s})\, e^{-z^2}\, dz$$

$$= \frac{x}{\sqrt{\pi}} \int_{-\infty}^{\infty} e^{-z^2}\, dz = x\,,$$

where the chain of equalities relies on (8.36a) and on the fact that function $z \mapsto z\,e^{-z^2}$ is odd, thus it verifies Remark 8.58. In conclusion, the solution to the initial value problem (13.57) is:

$$u(x,t) = \int_0^t s\,x\,ds = \frac{1}{2}x\,t^2\,.$$

Note that formula (13.56) provides the solution to the particular initial value problem (13.53), with zero–value initial condition. To arrive at the solution in the general case (13.47), we exploit the linearity of the differential equation, using a *superposition* technique, that is based on solving two initial value problems, namely:

$$\begin{cases} v_t(x,t) = c\,v_{xx}(x,t)\,, & x \in \mathbb{R}, \qquad t > 0, \\ v(x,0) = f(x)\,, \end{cases} \tag{13.59}$$

and

$$\begin{cases} w_t(x,t) = c\,w_{xx}(x,t) + P(x,t)\,, & x \in \mathbb{R}, \qquad t > 0, \\ w(x,0) = 0\,. \end{cases} \tag{13.60}$$

It turns out that function $u(x,t) = v(x,t) + w(x,t)$ solves the initial value problem (13.47). In other words, using formulæ (13.24) and (13.56) jointly, we can state that the solution $u(x,t)$ to the initial value problem (13.47) is given by:

$$u(x,t) = \frac{1}{2\sqrt{\pi c t}} \int_{-\infty}^{+\infty} e^{-\frac{(x-y)^2}{4ct}} \, f(y) \, dy$$

$$+ \frac{1}{2\sqrt{\pi c}} \int_0^t \int_{-\infty}^{\infty} \frac{1}{\sqrt{t-s}} \, e^{-\frac{(x-y)^2}{4c(t-s)}} \, P(y,s) \, dy \, ds.$$

(13.61)

Bibliography

[1] Kuzman Adzievski and Abdul Hasan Siddiqi. *Introduction to partial differential equations for scientists and engineers using Mathematica.* CRC Press, Boca Raton, Florida, U.S.A., 2014.

[2] James L. Allen and F. Max Stein. On the solution of certain Riccati differential equations. *American Mathematical Monthly*, 71:1113–1115, 1964.

[3] Tom Mike Apostol. *Calculus: Multi variable calculus and linear algebra, with applications to differential equations and probability.* John Wiley & Sons, New York, U.S.A., 1969.

[4] Daniel J. Arrigo. *Symmetry analysis of differential equations: an introduction.* John Wiley & Sons, New York, U.S.A., 2015.

[5] Emil Artin and Michael Butler. *The Gamma function.* Holt, Rinehart and Winston, New York, U.S.A., 1964.

[6] Robert G. Bartle. *The elements of integration and Lebesgue measure.* John Wiley & Sons, New York, U.S.A., 2014.

[7] Fischer Black and Myron Scholes. The pricing of options and corporate liabilities. *The Journal of Political Economy*, JSTOR:637–654, 1973.

[8] David Borthwick. *Partial differential equations.* 2nd ed. Springer, Cham, Switzerland, 2016.

[9] David Brannan. *A first course in mathematical analysis.* Cambridge University Press, Cambridge, U.K., 2006.

[10] Douglas S. Bridges. *Foundations of real and abstract analysis.* Springer–Verlag, New York, U.S.A.

[11] Boo Rim Choe. An elementary proof of $\sum_{n=1}^{\infty} (1/n^2) = (\pi^2/6)$. *American Mathematical Monthly*, 94(7):662–663, 1987.

[12] Donald L. Cohn. *Measure theory.* 2nd ed. Birkhäuser, Basel, Switzerland, 2013.

[13] Lothar Collatz. *Differential equations: an introduction with applications.* John Wiley & Sons, New York, U.S.A., 1986.

[14] François Coppex. Solving the Black-Scholes equation: a demystification. *private communication*, 2009.

[15] Richard Courant and Fritz John. *Introduction to calculus and analysis*, volume 2. John Wiley & Sons, New York, U.S.A., 1974.

[16] Peter L. Duren. *Invitation to Classical Analysis*, volume 17. American Mathematical Society, Providence, Rhode Island, U.S.A., 2012.

[17] Costas J. Efthimiou. Finding exact values for infinite sums. *Mathematics magazine*, 72(1):45–51, 1999.

[18] *Radon-Nikodym theorem*. Applications. Wikipedia, https://en.wikipedia.org/wiki/Radon–Nikodym_theorem, 2019.

[19] Leonhard Euler. De summis serierum reciprocarum. *Commentarii Academiæ Scientiarum Petropolitanae*, 7:123–134, 1740.

[20] Orin J. Farrell and Betram Ross. *Solved problems: gamma and beta functions, Legendre polynomials, Bessel functions*. Macmillan, New York, U.S.A., 1963.

[21] Angelo Favini, Ermanno Lanconelli, Enrico Obrecht, and Cesare Parenti. *Esercizi di analisi matematica: equazioni differenziali*, volume 2. CLUEB, Bologna, Italy, 1978.

[22] Gerald B. Folland. *Introduction to partial differential equations*. 2nd ed. Princeton University Press, Princeton, New Jersey, U.S.A., 1995.

[23] Herbert I. Freedman. *Deterministic mathematical models in population ecology*. M. Dekker, Inc., 1980.

[24] Guido Fubini. Sugli integrali multipli. *Rendiconti Accademia Nazionale Lincei*, 16:608–614, 1907.

[25] Guido Fubini. Il teorema di riduzione per gli integrali multipli. *Rendiconti Seminario Matematico Univ. e Politecnico Torino*, 9:125–133, 1949.

[26] Claude George. *Exercises in integration*. Springer, New York, U.S.A., 1984.

[27] Aldo Ghizzetti, Alessandro Ossicini, and Luigi Marchetti. *Lezioni di complementi di matematica*. 2nd ed. Libreria eredi Virgilo Veschi, Rome, Italy, 1972.

[28] James Harper. Another simple proof of $1+(1/2^2)+(1/3^2)+\cdots = (\pi^2/6)$. *American Mathematical Monthly*, 110(6):540–541, 2003.

[29] Phillip Hartman. *Ordinary differential equations* 2nd ed. SIAM, Philadelphia, Pennsylvania, U.S.A., 2002.

[30] Omar Hijab. *Introduction to calculus and classical analysis*. Springer, New York, U.S.A., 2011.

[31] Peter E. Hydon. *Symmetry methods for differential equations: a beginner's guide*. Cambridge University Press, Cambridge, U.K., 2000.

[32] Edward L. Ince. *Ordinary differential equations*. Dover, New York, U.S.A., 1956.

[33] Wilfred Kaplan. *Advanced calculus*. Addison–Wesley, Boston, Massachusetts, U.S.A., 1952.

[34] John L. Kelley. *General topology*. D. Van Nostrand Company Inc., Princeton, New Jersey, U.S.A., https://archive.org/details/GeneralTopology, 1955.

[35] Thomas William Körner. *Fourier analysis*. Cambridge University Press, Cambridge, U.K., 1989.

[36] Gregory F. Lawler. *Random walk and the heat equation*. American Mathematical Society, Providence, Rhode Island, U.S.A., 2010.

[37] Adrien-Marie Legendre. *Traité des fonctions elliptiques et des intégrales Euleriennes*, volume 2. Huzard-Courcier, Paris, France, 1826.

[38] Derrick Henry Lehmer. Interesting series involving the central binomial coefficient. *The American Mathematical Monthly*, 92(7):449–457, 1985.

[39] Andy Roy Magid. Lectures on differential galois theory. *Notices of the American Mathematical Society*, 7:1041–1049, 1994.

[40] C.C. Maican. *Integral evaluations using the Gamma and Beta functions and elliptic integrals in engineering: A Self-study Approach*. International Press of Boston Inc., Boston, Massachusetts, U.S.A., 2005.

[41] Arakaparampil Mathai Mathai and Hans Joachim Haubold. *Special functions for applied scientists*. Springer, New York, U.S.A., 2008.

[42] Habib Bin Muzaffar. A new proof of a classical formula. *American Mathematical Monthly*, 120(4):355–358, 2013.

[43] Paul J. Nahin. *Inside interesting integrals*. Springer, Berlin, Germany, 2014.

[44] Brad Osgood. *Lectures on the Fourier transform and its applications*. American Mathematical Society, Providence, Rhode Island, U.S.A., 2019.

[45] Bruno Pini. *Terzo corso di analisi matematica*, volume 1. CLUEB, Bologna, Italy, 1977.

[46] Earl David Rainville. *Intermediate differential equations.* Macmillan, New York, U.S.A., 1964.

[47] Earl David Rainville and Phillip E. Bedient. *Elementary differential equations.* 6th ed. Macmillan, New York, U.S.A., 1981.

[48] P.R.P. Rao. The Riccati differential equation. *American Mathematical Monthly*, 69:995–996, 1962.

[49] P.R.P. Rao and V.H. Hukidave. Some separable forms of the Riccati equation. *American Mathematical Monthly*, 75:38–39, 1968.

[50] Daniele Ritelli. Another proof of $\zeta(2) = \dfrac{\pi^2}{6}$ using double integrals. *American Mathematical Monthly*, 120:642–645, 2013.

[51] Halsey Lawrence Royden and Patrick Fitzpatrick. *Real analysis.* 4th ed. Macmillan, New York, U.S.A., 2010.

[52] Walter Rudin. *Principles of mathematical analysis.* 3rd ed. McGraw–Hill, New York, U.S.A., 1976.

[53] Walter Rudin. *Real and Complex Analysis.* Tata McGraw–Hill, New York, U.S.A., 2006.

[54] Fabio Scarabotti. *Equazioni alle derivate parziali.* Esculapio, Bologna, Italy, 2010.

[55] Laurent Schwartz. *Mathematics for the physical sciences.* Addison–Wesley, New York, U.S.A., 1966.

[56] William Shaw. *Modelling financial derivatives using Mathematica.* Cambridge University Press., Cambridge, U.K., 1988.

[57] Harry Siller. On the separability of the Riccati differential equation. *Mathematics Magazine*, 43:197–202, 1970.

[58] Walter A. Strauss. *Introduction to partial differential equations.* John Wiley & Sons, New York, U.S.A., 2007.

[59] J. Van Yzeren. Moivre's and Fresnel's integrals by simple integration. *American Mathematical Monthly*, JSTOR:690–693, 1979.

[60] Wolfgang Walter. *Ordinary differential equations.* Springer, Berlin, Germany, 1998.

[61] James S.W. Wong. On solution of certain Riccati differential equations. *Mathematics Magazine*, 39:141–143, 1966.

[62] Robert C. Wrede and Murray R. Spiegel. *Schaum's Outline of advanced calculus.* McGraw–Hill, New York, U.S.A., 2010.

Analytic Index

Chapters summary

1. Euclidean space

Chapter 1 introduces basic, though fundamental notions, on vector spaces and $\mathbb{R}^n$ topology, that are necessary throughout the book; this is done both to keep to a minumum the requirement of familiarity with the concepts presented, here and in the following chapters, and for reasons of completeness.

2. Sequences and series of functions

The notions of sequences and series, of numbers and of functions, are presented in this chapter, with the related concepts of pointwise and uniform convergence. The aim is mainly to minimise assumptions on the mathematical background possessed by the Reader; a purpose of notational introduction is also involved. The Basel problem is described, which will be met again in Chapters 8 and 10.

3. Multidimensional differential calculus

Chapter 3 presents the concept on differentiability of vector–valued functions, which is crucial for studying the theory of ordinary and partial differential equations, that will be presented in the following chapters of this book, and in particular in Chapters 4, 5 and 13. The notion of critical point, and the related definitions of gradient vector, Jacobian and Hessian matrices, and Lagrange multipliers, are recalled. The important Implicit Function Theorem is also stated and proved.

4. Ordinary differential equations of first order: general theory

Our goal, in introducing ordinary differential equations, is to provide a brief account on methods of explicit integration, for the most common types of ordinary differential equations. However, it is not taken for granted the main theoretical problem, concerning existence and uniqueness of the solution of the *Initial Value Problem*, modelled by (4.3). Indeed, the proof of the Picard–Lindelöhf Theorem 4.17 is presented in detail: to do this, we will use some notions from the theory of uniform convergence of sequences of functions, already discussed in Theorem 2.15. An abstract approach followed, for instance, in Chapter 2 of [60], is avoided here.

5. Ordinary differential equations of first order: methods for explicit solutions

We present some classes of ordinary differential equations for which, using suitable techniques, the solution can be described in terms of known functions: in this case, we say that we are able to find an *exact solution* of the given ordinary differential equation.

In Chapter 4 we exposed the general theory, concerning conditions for existence and uniqueness of an initial value problem. Here, we consider some important particular situations, in which, due to the structure of certain kind of scalar ordinary differential equations, it is possible to establish methods to determine their explicit solution

6. Linear differential equations of second order

The general form of a differential equation of order $n \in \mathbb{N}$ was briefly introduced in equation (4.10) of Chapter 4. The current Chapter 6 is devoted to the particular situation of linear equations of second order:

$$a(x)\, y'' + b(x)\, y' + c(x)\, y = d(x),$$

where a, b, c and d are continuous real functions of the real variable $x \in I$, being I an interval in $\mathbb{R}$ and $a(x) \neq 0$. Equation (6.1) may be represented, at times, in operational notation:

$$M\, y = d(x),$$

where $M : \mathcal{C}^2(I) \to \mathcal{C}(I)$ is a differential operator that acts on the function $y \in \mathcal{C}^2(I)$:

$$M\, y = a(x)\, y'' + b(x)\, y' + c(x)\, y.$$

In this situation, existence and uniqueness of solutions are verified, for any initial value problem associated to (6.1).

Before dealing with the simplest case, in which the coefficient functions $a(x), b(x), c(x)$ are constant, we examine general properties, that hold in any situation. We will study some variable–coefficient equations, that are meaningful in applications. Our treatment can be easily extended to equations of any order; for details, refer to Chapter 5 of [47] or Chapter 6 of [3].

7. Prologue to Measure theory

Some basic notions are presented in this chapter, that are needed as introduction to *Measure theory*. Some familiarity with the concepts presented, here and in the following Chapters 8 to 10, is also assumed and recalled, briefly, for the sake of completeness.

8. Lebesgue integral

Here and in Chapters 9 and 10, we deal with function μ, called *measure*, which returns area, or volume, or probability, of a given set. We assume that μ is already defined, adopting an axiomatic approach which turns out advantageous, as the same theoretical results apply to other situations, besides area in $\mathbb{R}^2$ or volume in $\mathbb{R}^3$, and which is particularly fruitful in Probability theory. A general domain that can be assumed for μ is a σ-algebra, defined in § 8.1. For completeness, a few basic concepts are recalled, for which some familiarity is assumed.

9. Radon–Nikodym theorem

This chapter illustrates the Radon–Nikodym Theorem 9.16, which has great implications in Probability theory and in Mathematical Finance, as explained, for example, in the 'Applications' section of [18]. The theorem presentation is preceeded by § 9.1, which constitutes a brief account on the topic of signed measures. For further details we refer to [51].

10. Multiple integrals

In this chapter, we expose the theoretical process that extends the Lebesgue measure from $\mathbb{R}$ to the plane $\mathbb{R}^2$, first, and, then, to the Euclidean space $\mathbb{R}^n$. The Fubini theorem is also described, both for the $\mathbb{R}^2$ case and for the general $\mathbb{R}^n$ case. A brief outline of the theory on products of σ-algebras is given at the end of this chapter.

11. Gamma and Beta functions

In this chapter, the Gamma function and the Beta function are introduced, together with their properties. The material presented here is based on many excellent textbooks [5, 16, 20, 27, 30, 41, 55, 62, 43, 40], to which the interested Reader is referred.

12. Fourier transform on the real line

A brief presentation of the *Fourier transform* is provided here, limited to those of its aspects that come handy while integrating a particular partial differential equation, namely, the *heat equation*. The latter constitutes the main tool in solving the *Black-Scholes equation*, which is of great importance in Quantitative Finance. This Chapter 12 is strongly inspired by [44].

13. Parabolic equations

The Fourier transform method of Chapter 12 is used here, among other methods, to solve partial differential equations of parabolic type, which are of fundamental importance in Mathematical Finance. The exposition presented in this Chapter 13 exploits the material contained in various references; in particular, we refer to [14], Chapter 6 of [1], Chapter 4 of [56], § 2.4 of [8], Chapter 6 of [58], Chapter 4 of [22], and [54].

Authors Bios

Daniele Ritelli is associate professor of mathematical analysis at the Department of Statistical Sciences of the University of Bologna, in Italy. He teaches ordinary and partial differential equations and special functions in various three–year and magister degree courses in statistics and finance. His research interests concern special functions, ordinary differential equations and pure and applied mathematics.

Giulia Spaletta is associate professor of numerical analysis at the Department of Statistical Sciences of the University of Bologna, in Italy. She lectures in various three–year and master degree courses in computer science, statistics and economics. Her research interests include numerical error analysis in linear algebra and differential equations and numerical modeling in biomedical engineering.

Back Cover

'Introductory Mathematical Analysis for Quantitative Finance' is a textbook designed to enable students with little knowledge of mathematical analysis to fully engage with modern quantitative finance. A basic understanding of Calculus and Linear Algebra is assumed.

The exposition of the topics is as concise as possible, since the chapters are intended to represent a preliminary contact with the mathematical concepts used in Quantitative Finance. The aim is that this book can be used as a basis for an intensive one–semester course.

Features

- Written with applications in mind, and maintaining mathematical rigor.

- Suitable for undergraduate or master's students with an Economic or Management background.

- Complemented with various solved examples and exercises, to support the understanding of the subject.

Printed in the United States
by Baker & Taylor Publisher Services